FUNDAMENTALS OF LASER DYNAMICS

FUNDAMENTALS OF LASER DYNAMICS

Ya I Khanin

Viva Books

New Delhi | Mumbai | Chennai | Kolkata | Bangalore | Hyderabad | Kochi

First Indian Edition 2008

VIVA BOOKS PRIVATE LIMITED

- 4737/23 Ansari Road, Daryaganj, New Delhi 110 002
  E-mail: vivadelhi@vivagroupindia.net, Tel. 42242200
- Plot No. 76, Service Industries, Shirvane, Sector 1, Nerul, Navi Mumbai 400 706
  E-mail: vivamumbai@vivagroupindia.net, Tel. 27721273, 27721274
- Jamals Fazal Chambers, 26 Greams Road, Chennai 600 006
  E-mail: vivachennai@vivagroupindia.net, Tel. 28290304, 28294241
- B-103, Jindal Towers, 21/1A/3 Darga Road, Kolkata 700 017
  E-mail: vivakolkata@vivagroupindia.net, Tel. 22836381, 22816713
- 7, Sovereign Park Apartments, 56-58, K. R. Road, Basavanagudi, Bangalore 560 004
  E-mail: vivabangalore@vivagroupindia.net, Tel. 26607409, 26607410
- 101-102 Mughal Marc Apartments, 3-4-637 to 641, Narayanguda, Hyderabad 500 029
  E-mail: vivahyderabad@vivagroupindia.net, Tel. 27564481, 27564482
- First Floor, Beevi Towers, SRM Road, Kaloor, Kochi 682 018, Kerala
  E-mail: vivakochi@vivagroupindia.net, Tel: 0484-2403055, 2403056

www.vivagroupindia.com

Published by arrangement with

Cambridge International Science Publishing
7, Meadow Walk, Great Abington
Cambridge CB1 6AZ, UK

ISBN-10: 81-309-0936-7
ISBN-13: 978-81-309-0936-3

Published by Vinod Vasishtha for Viva Books Private Limited, 4737/23, Ansari Road, Daryaganj, New Delhi-110 002. Printed and bound by Anand Sons, Delhi - 110 092.

About the Author

Prof Ya I Khanin was a Deputy Director of the Institute of Applied Physics of the Russian Academy of Sciences, Head of the Division of Nonlinear Dynamics and Optics, an expert in quantum electronics, nonlinear and quantum optics, laser dynamics, and the author of three monographs in these fields. He was a Fellow of the Optical Society of America and a Humboldt Research Award Winner.

CONTENTS

Preface

The resonant interaction of the atomic system with the radiation field is the main member of the functional circuit of the laser. The redistribution of the population of the energy levels of the medium, accompanying this interaction, indicates the principle of the unremovable nonlinearity of the system. The lasing process is also affected by other nonlinearities, if they are typical of the media located inside the laser cavity. The nature of the processes may change as a result of the external effects, accompanying the variation of the laser parameters with time.

Examination of the role of interactions and effects of this type and also their influence on the time, spatial and spectral characteristics of laser radiation and the justification of practical methods of controlling the lasing process are the subjects of tasks in laser dynamics.

In the history of laser dynamics, which is now more than 30 years old, there are two periods of the most rapid advances. The considerable interest in the problems of dynamics arose with the construction of the first solid-state lasers. This interest was stimulated by the fact that the experimentally detected spike nature of lasing could not be satisfactorily explained. The universal nature of the spike regime and of the absence of visible reasons for the occurrence of this regime may have indicated that the non-stationarity is the only property of the process of induced radiation of condensed active media in the cavity. Therefore, in the initial investigations, special attention was given to determining the possible mechanisms of the instability of this process.

The first period, which included the 60s of the previous century, was extremely fruitful for laser dynamics. The basic models were formulated, their properties were investigated in general features, investigations were carried out into the conditions of free lasing, active and passive modulation of the *Q*-factor of the laser, and the fundamentals of the theory of formation of giant pulses and mode synchronisation were laid. Using the well-known definition of radiophysics, proposed by S.M. Rytov (in the preface to his book, Introduction into statistical radiophysics, Nauka, Moscow, 1966, S.M. Rytov subdivided radiophysics to 'physics for radio' and 'radio for physics'), this period may be regarded as 'physics for lasers'. This period ended at the start of the 70s when it was established that the spike regimes of solid-state lasers are formed mainly by technical fluctuations of the parameters, and other problems of this type were solved.

The 70s were not marked by any significant achievements in laser dynamics. The individual 'grey areas' remained but it appeared that their number was smaller. For example, the regime of non-attenuating coherent pulsations of the single-mode laser, which was mentioned many times by the theoreticians, could not be realised. Doubts remained regarding the true reasons for the irregular nature of spike lasing. Only a small amount of work was carried out in the low-frequency dynamics of multi-mode lasers. Nevertheless, it appeared that these are only small fragments and no significant achievements were expected in laser dynamics.

However, these years were marked by the rapid progress in the theory of nonlinear oscillations (nonlinear dynamics, as it is called now). The scientific practice included concepts such as determined chaos and strange attractor. The new concepts of nonlinear dynamics had a significant effect on developments in greatly differing areas of science.

In this period, it was found that the laser belongs to the group of systems which are capable not only of demonstrating complicated behaviour but are also greatly suitable for the examination of general relationships of nonlinear dynamics. This led to a new and rapid progress in the dynamics of lasers, which started in the 80s and took place, in contrast to the first period, under the name 'Lasers for physics!'. The renewed attention paid to the possibilities of examination of the determined chaos in lasers resulted in the first experimental successes in 1982. Now, we have a large amount of original literature on the subject, including both experimental and theoretical investigations of complicated lasing regimes. This aspect of laser dynamics was described in the book Dynamics of Lasers, by C. Weiss and R. Vilaseca, Weinheim, New York, 1991.

However, the number of monographs on laser dynamics was smaller. They all were published in the period to the 80s and were written in most cases by Soviet authors. The latter cannot be regarded as surprising because the contribution of Soviet authors to laser dynamics was significant. In this connection, it is important to mention the monograph Molecular Generators by A.N. Opaevskii (Nauka, Moscow, 1964), the book by K.G. Folin and A.V. Gainer Dynamics of free lasing of solid-state lasers (Nauka, Novosibirsk, 1979), the book Dynamics of radiation of semiconductor quantum generators by L.A. Rivlin, and the book Dynamics and emission spectra of semiconductor lasers, by L.A. Rivlin, et al (Radio i svyaz', Moscow, 1983). This list should be supplemented by a later monograph Self-oscillations in lasers, by A.M. Samson et al (Nauka i tekhnika, Minsk, 1990). In each book special attention is given to the specific type of laser or specific operating regime.

A significant contribution to the dynamic theory of lasers by the well-known German physicist H. Haken has been reflected in his monograph

Laser light dynamics published in the West in 1985 (Amsterdam, North Holland). A list of foreign publications on the subject was supplemented in 1997 by the book 'Theoretical problems in cavity nonlinear optics (Cambridge: Cambridge University press).

The attempts for systematization of the material on the subject and explanation of the general situation were explained in the monograph by the author of the present book, Dynamics of quantum generators. However, events requiring that attention be paid to different subjects occurred with time. The true position of a number of studies, which were previously disregarded, was found. The positions of other studies, on the other hand, proved to be less important than previously thought. Therefore, retaining this material from the book Dynamics of quantum generators which withstood the test of time, the author of the present book published a new book under the name Principles of laser dynamics. The book was published in 1995 (Amsterdam: North Holland). Russian readers could not obtain this book because of the extremely high price and inefficient advertising.

In this monograph, the free lasing of lasers in different conditions is examined in detail. In the chapters, concerned with the subject (chapters 3 to 5), special attention is given to the stationary states and their stability, the behaviour of lasers in the unstable region, the characteristics of regular and random self-modulation processes and the nature of mechanisms responsible for them.

The processes in the lasers, accompanying changes of the parameters with time, are the subject of chapter 6. Special attention is given to the response of the laser to low-intensity low-frequency modulation of the parameters. Problems of the resonant amplification of modulation, transition to the nonlinear regime, random response to the periodic effect, spike lasing under the effect of changing geometry of the cavity and the drift of temperature of the active element are studied.

The behaviour of the laser may change quantitatively if its optical elements show nonlinear properties. More detailed investigations have been carried out into the effect of a saturated absorber, leading to the instability of stationary lasing and ensuring, in specific conditions, the passive modulation of the Q-factor of the resonator. Less attention has been paid to the effect, on the processes in the laser, of other nonlinear effects, in particular self-focusing, which is also capable of having a strong effect on lasing dynamics. All this is discussed in chapter 7 of the book.

The five main chapters are preceded by two chapters of introductory nature. In one of the chapters, essential information is given on quantum generators, requirements are formulated on the main elements of the laser, and the extent to which these requirements are fulfilled in the generators of different types is shown. The information on the types of dynamic behaviour of lasers in relation to the ratio between the parameters is also provided. In the second introductory chapter, the most general mathematical models, used in the semiclassical laser theory are presented. Discussion of idealisations and simplifications, used in different specific situations, and also the ranges of the applicability are transferred to the chapters in which the specific models are analysed. It is assumed that the reader is acquainted with the main considerations of the modern theory of nonlinear fluctuations. It is necessary, this information may be found in, from various books and other sources, presented in the literature list. In this connection, it is again necessary to mention the monograph by C. Weiss and R. Vilaseca.

It is also important to mention the sections of laser dynamics which for some reasons have not been reflected in this monograph. For example, no mention is made of the concepts relating to the time evolution of the spatial structures in the laser emission field. The section of nonlinear dynamics of the optical systems is being developed. The author decided not to include in the monograph the material on different methods of lasing of the giant pulses because no significant changes have been made in this area of quantum electronics from the date of publication of his previous book. This may also be said of the theory of sweep lasers.

Finally, it is important to also mention inverse problems of laser dynamics. This larger area of activity must be mentioned, but only small parts have been developed. At the same time, the problem of extracting information on the laser parameters and individual intra-cavity elements are of considerable practical importance and, in this case, special use is made of new concepts based on the advanced concepts of nonlinear dynamics.

Evidently, it is clear to the reader that the book is a physical rather than mathematical version of the subject. This is justified by the fact that in this case we are not concerned with nonlinear dynamics in general but with laser dynamics, and this does not correspond to the scientific activity of the author.

Laser dynamics is a rapidly developing science in which events take place very frequently. In these conditions it is obvious that the material in the book may not be complete and the situation will become worse with time. Nevertheless, the already existing system of 'eternal values' makes it possible to hope that the usefulness of the book will not decrease very rapidly.

Chapter 1

Quantum Oscillators: General Considerations

This chapter should not be treated as detailed analysis of the principles of the quantum electronics. The reader can find them in [1–9]. Below there is given only some information related to the dynamical laser behaviour.

1.1. Principle of Operation and Practical Implementation

There are three problems, which led to the advent of quantum electronics: lowering the noise threshold of amplifiers, raising the stability of oscillators, and generating millimetre and shorter waves. In solving these problems the strategies of classical electronics encountered fundamental difficulties; many of them were overcome by using stimulated emission in systems of bound particles.

1.1.1. Induced and Spontaneous Emission

The possibility of enhancing electromagnetic fields by quantum systems is based on the induced (stimulated) emission of radiation. Under the influence of incident radiation a quantum system such as an atom, molecule or a crystal is capable to pass to a lower energy state by emitting a photon. The induced emission is fully identical to the incident radiation. The inverse of induced emission is the absorption of a photon when the quantum system makes a transition to an upper energy level.

In quantum electronics we deal with a medium composed of a large number of molecules rather with a single molecule. In thermal equilibrium the higher the energy of molecular levels the lower their population. Since downward and upward induced transitions have equal probabilities, media in thermal equilibrium are net absorbers. An excess of transitions with emission over those with absorption can be achieved only in nonequilibrium systems, in which the upper level population exceeds the population of a lower level. Such a population inversion of a couple of levels is sufficient provided the energy level spacings are unequal. The last condition is required to ensure that the monochromatic field resonant with the inverted transition is not resonant with other transitions, which have an excess of absorption.

The unequal level spacing of the energy spectrum permits a simplified description of a quantum system with a minimum number of levels taken into account. A two-level approximation is often successfully used in quantum electronics.

The simplest version of a quantum amplifier is a layer of material, in which a population inversion is achieved on the chosen transition in one way or another. An electromagnetic wave of the proper frequency is amplified when it propagates in such a medium. More effective use of the amplifying medium can be achieved by slowing down the wave or by arranging for multiple passes through the active element, by placing in a cavity. Thus, in quantum electronics a resonant cavity is used to achieve not only frequency selection but also feedback.

Induced emission leads to an increase of the field energy density inside the active medium. Various processes of the field dissipation and the field emission into the ambient space act in opposite manner. When these processes equilibrate each other or the induced emission dominates the dissipation, such a device operates as a generator of radiation.

Unlike the induced photons, the spontaneously emitted photons are not correlated with the field in the laser medium volume. Hence the spontaneous emission processes play the role of a natural source of noise, which limits sensitivity of quantum amplifiers and the stability of oscillators, as well as the role of triggering mechanism of the sustained operation as a coherent generator of radiation. The spontaneous emission process determines the finite lifetime of an excited isolated molecule and the related natural width of the spectral lines.

The active medium, or the working substance, of a quantum amplifier or an oscillator should ensure the needed gain with reasonable consumption of energy for the population inversion. The gain is given by

$$k_{\text{gain}} = \sigma_{\text{tr}}(N_a - N_b)\,, \tag{1.1}$$

where σ_{tr} is a quantity termed the quantum transition cross-section, which describes the active molecule, $N_a - N_b$ is the population difference on the working levels, defined both by the properties of the substance and the potential of the pumping system. Further we need a relation [3]

$$\sigma_{tr} = \frac{8\pi\omega_0 |\mathbf{d}|^2}{\hbar c \delta\omega_0}, \tag{1.2}$$

which express the transition cross-section in terms of other constants of the medium such as the dipole moment **d**, the frequency of radiation ω_0 and spectral linewidth $\delta\omega_0$.

A large transition cross-section means a small lifetime of the upper level, which impedes the accumulation of population there. In practice, both alternative ways are used to reach the necessary gain: the creation of large overpopulation on a metastable upper level when transition has a small cross-section, or the use of a large cross-section with relatively weak population on the upper level. Typical media of the first kind are the weakly doped crystals such as ruby or yttrium aluminium garnet and glasses with the Nd^{3+} ion as impurity. The active media of thc second type are exemplified by organic dye solutions. In any case, the pumping energy must be used effectively, characterized by the fraction of excited molecules brought finally to the upper energy level. This gives a general requirement for a useful working substance – high quantum yield.

Since the gain is determined by the population difference rather than by the upper level population, active media with fast depletion of the lower level are advantageous. Rapid depletion is possible if this level is higher than the ground state by k_BT or more, as in crystals and glasses doped by neodymium. An opposite example is a medium with self-limited transitions, which terminate on a metastable lower level. Examples of such media are metal vapours and erbium-doped crystals. Generally speaking, ruby is also of the second type since its lower laser level corresponds to the ground state, but its depletion is combined with the pumping process. Self-limitation of the transition hinders the use of such material for CW operation.

1.1.2. Methods of Producing an Inverted Population

Quantum electronics began with centimetre wavelength molecular beam masers [10, 11]. The term *maser*, proposed in [11], came to denote microwave quantum devices using the principle of amplification by stimulated emission. An attempt to introduce the term 'optical maser', which was made in the seminal theoretical paper [12], was not a long-time success. After the practical breakthrough into the optical range [13, 14] a new acronym *laser* was adopted. In turned out to be extremely stable, such that even nowadays we read and hear about 'X-ray lasers', 'γ-lasers' and even 'submillimetre lasers'! Habit began to dominate sensibility, but this is not a rare case with the terminology. Nevertheless, wavelength range is one basis for classification in quantum electronics. Another basis for classification is the material phase of the medium (gas, liquid, solid) and a third basis is the concrete process, which creates the population inversion in the laser medium.

Brief comments should be made on lasers with condensed and gaseous active media. The difference is due to the greater interaction between material particles with increasing particle density. This leads to greater *homogeneous broadening* of the spectral lines in condensed media. Conversely, the higher mobility of gas molecules is exhibited in the *Doppler broadening* of transitions.

The third basis of classification needs a somewhat longer discussion. There are lasers with optical, electrical, thermal and chemical pumping. For a complete description of the system we should also specify the type of the process in the medium, which leads to the inversion. This can be photon absorption, collision of a molecule with an electron or a molecule of a different species, a chemical reaction or a dynamical process like molecules separation by their motion between domains of different physical conditions.

Optical pumping is undoubtedly the most common way to achieve a population inversion. When a photon is absorbed directly by an active molecule, bringing it finally to the upper working level, we are dealing with an optical pumped laser – that of solid-state, liquid or gaseous type. The laser provides a direct transformation of the pumping light to the laser emission. In particular, another laser can serve as a source of pumping.

Direct transformation of electric energy to laser emission occurs in an injection semiconductor laser: this is an example of a *device with electric pumping*.

Very often, the pumping energy is transformed to laser emission in a more complicated way than in the cases cited above. Suppose that the energy required for the laser operation is provided by the chemical reaction. It is important that one of the reaction products be formed in the excited state but it is not necessary that exactly this product be the active component of the laser medium. The excited product can serve as an auxiliary gas to transfer energy to the working gas molecules through collisions. Such a laser should be called a *collisional laser with chemical pumping*.

Following the same principle, gas discharge lasers should be categorized as *collisional lasers with electrical pumping*. The discharge electrons, accelerated in a static or high-frequency (rf) electric field, impact energy to the molecules they collide with. This can be molecules of the working gas but often it is more advantageous to convey the energy through an intermediate stage of excitation of an auxiliary gas, thus, for example, neon and carbon dioxide are excited in discharges with higher efficiency through intermediary gases, helium and nitrogen, respectively.

Differences in relaxation times for different molecular states offer the opportunity to obtain an inversion during the time of establishing the thermal equilibrium after a fast variation of temperature of the medium. Fast cooling of a gas can be obtained by the methods of gas dynamics using adiabatic expansion. In carbon dioxide, antisymmetrical vibrations are 'cooled' more slowly so that under the favourable conditions the population of these vibrational levels can exceed the population of other vibrational levels, at least for a while.

To obtain the necessary inversion in a gas flow, a sufficient number of molecules must initially be brought to the excited states. In a *gas-dynamic laser with thermal pumping* this can be achieved by preheating the gas. A similar result can be obtained in other ways using, for example, optical pumping or by excitation in a gas discharge. Then we are dealing with a *gas-dynamic laser with optical* or *electrical* pumping, respectively.

Within the framework of this classification we should also mention beam masers. In such devices, inversion is achieved by means of the spatial separation of molecules in different energy states, when the molecular beam is transmitted through a region with a nonuniform electric or magnetic field. The degree of inversion is determined by the parameter of the sorting system and by the temperature to which the gas is heated in the beam source.

Making other combinations of the form of consumed energy and the mode of the active molecule excitation, it is possible to obtain a complete list of ways to create the population inversion.

One significant feature of all processes discussed so far is that they are noncoherent. An exception is optical pumping directly to the upper level by means of a laser source. The specific point is that the action of coherent pumping is not merely the variation of population distribution but it leads to more complicated nonlinear processes in the active media, contributed both by pumping and the laser field.

1.1.3. Amplification in Quantum Systems without Population Inversion

We have mentioned above that amplification could be achieved only in nonequilibrium media. The simplest example of such a medium is a two-level atomic system with population inversion. But the set of variables characterizing the state of the medium is not limited by populations of the energy levels, and inversion does not present the unique type of nonequilibrium state that gives amplification.

We can illustrate this using the example of a paramagnetic, which is a medium consisting of atoms possessing magnetic properties. The energy levels of the elementary magnetic dipole in a static external magnetic field correspond to two orientations of the dipole: parallel to the magnetic field (lower level) and antiparallel to the magnetic field (upper level). In the state of thermal equilibrium the majority of the dipoles oriented parallel to the field, *i.e.* they occupy the lower energy level. In other words, the resultant magnetization vector is oriented parallel to the magnetic field. The population inversion is equivalent to reorientation of the magnetization vector. However, generally speaking, the angle between the field and magnetization vector can be any and not only 0 and π. When inclined at any angle, the magnetization vector begins to precess around the magnetic field with the frequency of paramagnetic resonance, which is proportional to the difference of energies of oppositely oriented magnetic dipoles $\omega_0 = 2\mu H$. However, it is impossible to bring in correspondence the precessing magnetic dipole with a concrete energy level. Here we meet a new degree of freedom presented by the transverse component on magnetization. It is clear that the presence of precession means that the medium is not in equilibrium state, but this state is quite different from population inversion. It is possible to create such a state putting the paramagnetic sample in the magnetic field oscillating with frequency ω_0.

We can generalize all what has been said about paramagnetic to a two-level system of any origin. When placed in the resonance electromagnetic field, the quantum system can be converted to the state analogous to the state of precession, which is not characterized completely by the populations of the energy levels. In this case we speak about coherent superposition of states with nonzero values of nondiagonal elements of density matrix. Being in coherent state, the two-level system is capable to emitting at the transition frequency after the excited field is switched off. This phenomenon does not have any practical significance: the energy is extracted in the same form as spent but in smaller value.

Let us turn to more complicated situation with the three-level system. We can come to such a system, in particular, splitting the lower level into two close spaced sublevels (Fig. 1.1). The coherent superposition can be created between these sublevels, 1 and 2. Since $\omega_{21} \ll \omega_{32}$ and $\omega_{21} \ll \omega_{31}$, it can be called the low-frequency coherency. Such kind of nonequilibrium drastically changes the situation in the tree-level system, including the transition probabilities between the lower sublevels and the upper level. Atoms in a certain superposition state become uanble to pass to the upper level. It is similar to light diffraction on two slits when some areas in the space remain dark. The same is in Λ-scheme: due to existence of two transition routes (1→3 and 2→3), the upper level remains empty in spite of the presence of the resonance field! This means that absorption at frequencies of the optical transitions 1→3 and 2→3 (Fig. 1.1) is absent. But if the atoms in the lower states do not absorb, then even a small amount of upper level population is enough to make the media amplifying ones. The advan-

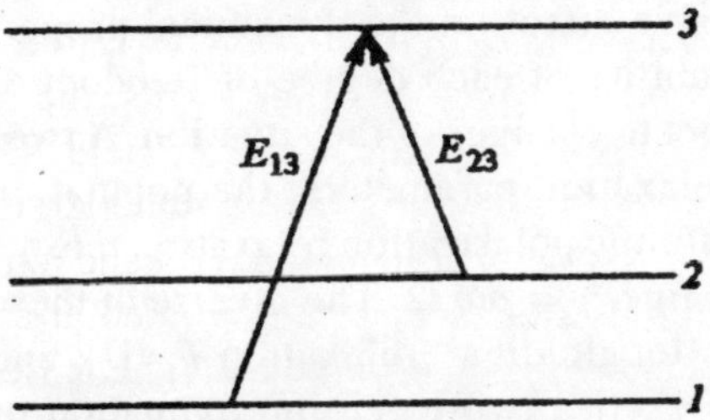

Fig. 1.1. Three-level Λ-scheme: monochromatic fields E_{13} and E_{23} create a coherent superposition of states *1* and *2*, the transitions from which to the level *3* under the action of these fields are completely suppressed. This establish the prerequisite for lasing without inversion.

tage of this method is that existence of low-frequency coherence leads to weakening of request to population of the upper level.

This idea was suggested and substantiated in Ref. [15]. Similar ideas were made independently in Refs. [16, 17]. In spite of the existence of forerunners, to which Refs. [18, 19] belong, the real progress in the field of inversionless amplification was stimulated by three fundamental studies [15–17] published in 1988 and 1989.

It is unlikely to think that lasers without population inversion, though they are realised in laboratories [20, 21], can compete with the traditional lasers in the applied sphere. But sometimes the inversionless amplification is of practical interest, in particular when it is difficult in principle to create population inversion. First, this is related to the problem of lasers in the domains of very short wavelength: vacuum ultraviolet, X-rays and gamma-rays [22].

1.2. Time-Dependent Processes in Quantum Oscillators

In addition to the three bases for laser classification discussed above, there is one more classification scheme, which is directly related to the dynamic properties.

1.2.1. Dynamic Properties of Lasers and Their Relation to Relaxation Rates

Induced transition under the action of an electromagnetic field is not the only way to make a quantum system to change its state. The lifetime of an excited state has a natural limit, defined by a spontaneous downward radiative transition. Meanwhile, there are nonradiative transitions caused by intermolecular collisions, interaction with lattice vibrations, etc. Nonradiative processes and spontaneous emission can be viewed as the result of the interaction of a limited number of isolated degrees of freedom (dynamical system) with all remaining degrees of freedom (thermal reservoir). Such processes are often called *relaxation processes* since they promote the establishment of an equilibrium state of a dynamical system.

Relaxation processes have a decisive effect on the dynamical properties of a laser since they define the ability of each degree of freedom of this complex oscillation system to react to changes of the situation. A two-level medium is described by two relaxation parameters: the population difference relaxation rate $\gamma_{\|}$, and the atomic polarization relaxation rate $\gamma_{\perp}$, which is the halfwidth of the spectral line, $\gamma_{\perp} = \delta\omega_0/2$. The inverse of these quantities are often used: the time of 'longitudinal' relaxation $T_1=1/\gamma_{\|}$ and that of 'transverse' relaxation $T_2=1/\gamma_{\perp}$[1]. If a two-level approximation of the medium is insufficient, then the description should involve the relaxation parameters of other transitions. For the complete set of laser characteristics we should add so-called 'photon lifetime', which is related to the cavity Q-factor by the relation $T_c= Q/\omega_c$. By analogy with the laser medium, we can make use of the decay rate $\kappa=1/2T_c$, which coincides with $\delta\omega_c/$, the half of the energy passband of the cavity.

When speaking about multimode lasers, the set of time parameters should be supplemented with a characteristic period of intermode beats $T_0 = 1/\Delta\omega$, which is equal to the round trip time in the case of longitudinal cavity modes.

We will call the processes in the laser the slow processes or the fast processes and, correspondingly, we will call the dynamics the low-frequency or the high-frequency depending on whether their spectrum matches the cavity passband. Of course, the fast processes are due to mode interference. The slow (low-frequency) processes depend on relationship between the relaxation parameters $\gamma_{\|}$, $\gamma_{\perp}$ and κ.

To make the following description more pictorial we turn, running ahead, to one of the basic models of the dynamical laser theory – the set of Lorenz–Haken equations:

$$\frac{dE}{dt} = \kappa(P-E)$$

$$\frac{dP}{dt} = \gamma_{\perp}(nE-P) \qquad (1.3)$$

$$\frac{dn}{dt} = \gamma_{\|}(A-n-PE)$$

The variables of this set of equations are the amplitude of the electric field E, the amplitude of the medium polarization P, and the population difference (inversion) n. For the purpose of simplification the notations

[1]The terms and symbols are taken from the magnetic resonance theory where the analogies for the population difference and polarization are the longitudinal and transverse components of magnetization with respect to the static magnetic field.

all the variables are normalized. The set of coefficients include, besides of three relaxation constants, also the pumping parameter *A*. The Lorenz–Haken model corresponds to a single-mode traveling-wave laser with a homogeneously broadened two-level laser.

Four important cases of relations between the coefficients of the Lorenz-Haken model can be identified [23, 24] that correspond to four dynamical classes of lasers.

Class A: $$\kappa \ll \gamma_{\parallel}, \gamma_{\perp} \tag{1.4}$$

The active medium follows without delay all the field variations. Both material variables, the inversion and the polarization, can adiabatically be eliminated from the set of equations.[1] The dynamical phase space is one-dimensional and the family of attractors can be represented only by the *fixed points*. The transient processes in the laser are aperiodic.

Class B: $$\gamma_{\perp} \gg \kappa \gg \gamma_{\parallel}, \tag{1.5}$$

Only the polarization follows the field without delay and it is the only variable eliminated. The phase space is two-dimensional and allows, besides fixed points, for the existence of closed (periodic) trajectories termed *limit cycles*. The transient process can be oscillatory.

Class C: $$\kappa \sim \gamma_{\perp} \tag{1.6}$$

All the variables have the equal rights. The presence of more than two dimensions means that the phase space can contain more complex attractors including strange attractors [25, 26].

The systems that satisfy the condition (1.4) are sometimes called *adiabatic systems*, and those meeting the condition (1.6) are called *nonadiabatic systems*. Meanwhile, the terms 'class *A*', 'class *B*' and 'class *C*' lasers have often been used recently following Arecchi. We should add one more class.

Class D: $$\kappa \gg \gamma_{\parallel}, \gamma_{\perp} \tag{1.7}$$

---

[1]The concept and procedure for adiabatic elimination of the variables will be discussed in Section 3.1.2. But, running ahead, it should be noted that the relations between the relaxation constants give only the necessary conditions for adiabatic elimination of this or that variables. Generally speaking, one should take into account also the rate of the induced transition, which can not exceed the largest of the relaxation constants. The last condition limits from above the intensity of field, interacting with the laser medium, and also the pumping parameter *A*.

In this case the field is an inertialess variable, which follows the state of the atomic system and should, therefore, be adiabatically eliminated. The main data characteristic for the mentioned above four classes are presented in the Table.

Dynamical class	Relations between the relaxation rates	Adiabatically eliminated variables	Dimension of the model
A	$\gamma_\perp, \gamma_\parallel >> \kappa$	P, n	1
B	$\gamma_\perp >> \kappa >> \gamma_\parallel$	P	2
C	$\gamma_\perp \sim \kappa$	—	3
D	$\kappa >> \gamma_\perp, \gamma_\parallel$	E	2

In the multimode lasers, we need to complete the list of important parameters by the intermode frequency space $\Delta\omega$. The large enough value of $\Delta\omega$ means that the populations of the working levels are not capable of reacting efficiently to the beats and this fact justify the use of the ordinary rate equations. If the mode frequencies are close to each other, it is necessary to take into account the oscillations of population with the intermode beat frequencies and use more complicated laser models with phase-sensitive interaction.

1.2.2. Widespread Types of Lasers

There are few representatives of *class D*. They include only beam masers among which ammonia and hydrogen masers are best known. The operating wavelength 1.25 cm corresponds to the strongest line in the inversion ammonia spectrum, related to the state with the quantum numbers of rotational angular momentum $J = 3$ and its projection onto the symmetry axis of the molecule $K = 3$. The molecules are sorted by states when passed through an inhomogeneous electric field.

The spontaneous emission transition probability, proportional to λ^{-3}, is negligibly small for centimetre wavelengths. The collisions between the beam molecules in a vacuum are also rare. Hence, the inertial properties of the active medium are defined by the transit time of the molecules through the interaction space, *i.e.* the maser cavity. The transit time is of the order of 10^{-4} s, *i.e.* it exceeds noticeably the photon lifetime in a microwave cavity (10^{-6} s or less) [27]. Thus, inequality (1.7) is satisfied.

Recently, the question has been discussed concerning the possibility of creating lasers operating in the super-radiation regime [28, 29]. It is necessary to satisfy the condition $\kappa \gg \gamma_\perp$, which means that these lasers belong formally to class *D*. However, the enormously high probability of radiative processes, which must exceed the rates of all relaxation processes, including κ, is also important.

Class C. The approximate equality of the cavity passband and the homogeneously broadened gain linewidth, which is implied by the relation (1.6), is not typical for lasers. Using the relation

$$\delta\omega_c = \frac{\omega}{Q} = \frac{c\Pi_{loss}}{L}$$

we obtain the passband $\delta\omega_c/2\pi = 5$ MHz, for the typical parameters of the cavity: length $L = 100$ cm and single-pass losses $\Pi_{loss} = 0.1$. Such a small homogeneously broadened width of the spectral lines is possible only in rarefied gases. Meanwhile, laser action will require that the gain in these lines be large enough. These are reasons for which the class *C* was considered to be empty for a long time. Noble gas-discharge lasers using infrared transitions were among the first ones to fit into this class.

The 3.51 μm xenon line is notable from the viewpoint of the attainable amplification. Its natural linewidth is 4.6 MHz, which corresponds to the lifetime of the laser levels $t_a = 1.2\cdot10^{-6}$ s and $t_b = 3.3\cdot10^{-6}$ s. The spontaneous decay of the upper level is mainly accompanied by a transition to the lower laser level. The additional homogeneous broadening due to collisions of xenon atoms (11.5 MHz/Torr Xe) with the partial pressure of tens of millitorr is negligibly small. For controlled action on the homogeneous linewidth, helium is introduced to the discharge tube to ensure the line broadening to 18.5 MHz/Torr He. In a finite range of pressures the relation (1.6) can be satisfied by sacrifice of the cavity *Q*-factor, which requires additional amplification.

The situation with laser transitions is similar for various noble gases [30]. It will be considered below in more detail using neon as an example. We only note that both laser levels are high enough. Since the excitation in a gas discharge is not a selective process, the upper levels can be occupied by atoms with different velocities. This is the reason why the gain line broadening is inhomogeneous. The Doppler width of the 3.51 μm Xe line is 110 MHz.

The interatomic collisions result both in dephasing of the atomic interaction with the radiation field and in variation of the atom velocity. The first factor is exhibited through the homogeneous broadening while the second through spectral cross-relaxation. The cross-relaxation rate is estimated to be 10^{-6} s^{-1} [31]. The remaining data on xenon are taken from Refs. [31, 32].

The relation (1.6) can be realized in the He–Ne laser for the wavelength 3.39 μm, although it is more difficult than in the 3.51 μm Xe line because of weaker amplification [33].

Other class *C* lasers are the molecular gas FIR (far infrared) lasers. The active media in such lasers are HCOOH [34], CH_3F [35], CH_3OD [36], CH_2F_2 [37] and some other gases. Ammonia again comes first [38–41].

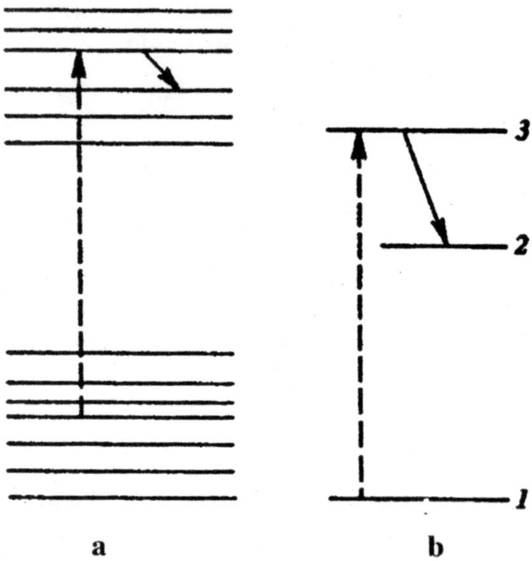

Fig. 1.2. Vibrational–rotational transitions in a molecular spectrum (*a*) and the three-level scheme explaining the principle of operation of a laser-pumped molecular FIR laser (*b*).

The ammonia FIR laser is optically pumped through a vibrational transition matched with the operating frequencies of a CO_2 or N_2O laser. The amplification is achieved on the rotational transitions. Fig. 1.2 shows a fragment of the ammonia-like energy spectrum. A three-level scheme is apparent if the fine structure is neglected.

For FIR lasers, the generated wavelength greatly exceeds the pumping wavelength. For example, the rotational transition $aR(7,7)$ in the $^{14}NH_3$ molecule corresponds to $\lambda \approx 81$ μm, while the vibrational transition $a(8,7)$ with $\lambda = 10.8$ μm, matched with the N_2O laser frequency, is used for pumping. The difference is still greater in the $^{15}NH_3$ laser: the laser transitions have the wavelength 374 μm ($aR(2,0)$ line) and 153 μm ($aR(4,4)$ line) while CO_2 lasers with the wavelength 10 μm are used for pumping.

The use of monochromatic pumping in the form of the travelling wave permits the selective excitation of molecules with definite velocity projection onto the wave vector to save the gain line from Doppler broadening. The homogeneous broadening at such long wavelength depends solely on intermolecular collisions. The linewidth of 1 MHz, realized under experimental conditions, allows the relation (1.6) to be satisfied even in a high Q-cavity. There are no reliable data on the relaxation times of the level populations.

The ability to provide a very high gain in a narrow homogeneous line under laser pumping is inherent not only to rotation transitions in chosen molecules. Similar effects can also be achieved also on some vibrati and electronic transitions and the total number of such transitions in d ferent gases is estimated as 10^6 [42].

Class B is represented by solid-state, semiconductor and some types low-pressure molecular gas lasers. Among the latter the most popular the carbon dioxide laser.

Ruby (Cr^{3+}:Al_2O_3) rivals ammonia in fame in the history of quantu electronics. The transitions between the Cr^{3+} ion spin levels are used paramagnetic masers. Ruby was the material in which visible laser actic was demonstrated for the first time in history. A system of the energy le els of Cr^{3+} impurities present in an Al_2O_3 crystal is shown in Fig. 1.3. Omi ting the details, we are dealing essentially with a three-level scheme.

The spontaneous emission spectrum of ruby exhibits two narrow in tense lines that correspond to transition from the metastable ^{2}E state to th ground state $^4A_2$. At the room temperature (300 K) the R_1 line has a maxi mum at λ = 0.6943 μm and the R_2 line at λ = 0.6927 μm. Cooling the crystal shifts the *R* lines towards shorter wavelength. The metastable ^{2}E levels can be excited through the wide bands, $^4F_1$ and $^4F_2$, connected to ^{2}E by radiationless transitions. Electrons pass from the $^4F_2$ band to the ^{2}E levels in a time of $t_{32} = 5\cdot10^{-8}$ s or emit spontaneously, returning to the ground state in $t_{31} = 2\cdot10^{-7}$ s. The spontaneous decay of the ^{2}E state is characterized by a time constant $t_{21} = 3\cdot10^{-3}$.

The $\bar{E}$ and $2\bar{A}$ sublevel populations are established in accordance with the Boltzmann distribution for thermal equilibrium. Since the sublevels are spaced by 29 cm^{-1}, their successive population difference is 15% for 300 K. The relaxation time between $\bar{E}$ and $2\bar{A}$ does not exceed 10^{-7} s. Thus, typically the laser does not operate on the R_2 line since the R_1

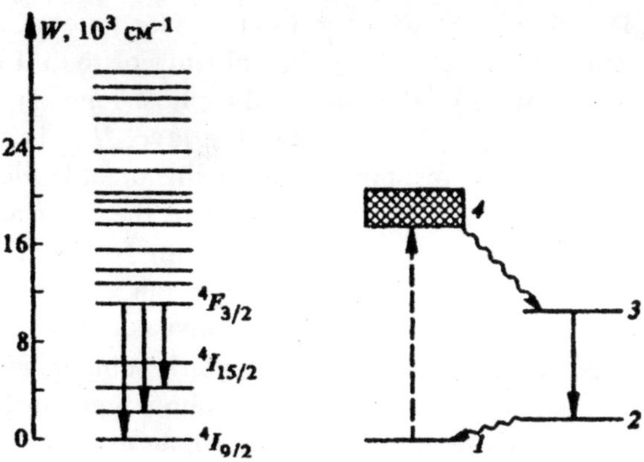

Fig. 1.3. Energy-level diagram of Cr^{3+} ion in ruby (*a*) and the scheme explaining the principle of operation of an optically pumped three-level solid-state laser (*b*).

lasing reduces the $2\overline{E}$ level population to a value below the laser threshold.

The R_1 linewidth depends on the number of impurity ions in the host. In a pink ruby with the Cr^{3+} concentration of the order of 10^{19} cm^{-3} for 300 K the R_1 linewidth is nearly 10 cm^{-1} or $3\cdot10^{11}$ Hz. This line narrows as the crystal is cooled. It is well known that the sapphire intracrystalline field splits the Cr^{3+} ground state into two components spaced by 0.38 cm^{-1}. Consequently, the R_1 line is split even for 77 K, for which of a separate component is 0.3 cm^{-1}. At high temperatures, the main contribution to linewidth is thermal oscillations of the crystal lattice, and the line can be considered a homogeneously broadened one. The cross-section of the transition, corresponding to the R_1 line, is $\sigma_{tr} = 10^{-20}$ cm^2.

It is an obvious drawback of the three-level ruby scheme that the laser transition uses the ground state as a basis. Half of the chromium ions are launched upwards with the only purpose of making the populations equal. Only the additional excited ions produce the laser effect.

Active media, which operate on a four-level scheme, are free of this drawback. They include *neodymium doped crystals and glasses.* The optical ion spectra of rare-earth elements are due to the transitions within nonfilled inner shells, screened by the outer shell from the external effects. Introducing the same ion into different materials, it is possible to vary the fine structure but the line position will not undergo any noticeable changes.

The energy level diagram of a Nd^{3+} ion is presented in Fig. 1.4. The $^4F_{3/2}$ level is metastable. Spontaneous transitions from this level to the ^4I levels are exhibited as four spontaneous emission lines. The most intense line, $^4F_{3/2}\rightarrow{}^4I_{11/2}$, has its maximum near λ = 1.06 μm. The $^4I_{11/2}$ level is elevated over the ground level by about 2000 cm^{-1}, which considerably exceeds k_BT.[1] This line is most favourable for laser action. Like the other lines, it has a structure stipulated by the splitting of the initial and final states by the electric field of the surrounding host ions. The metastable level $^4F_{3/2}$ is split into two sublevels, and the level $^4I_{11/2}$ is split into six Stark sublevels. Without a resonator comprising tunable elements, only the strongest spectral component is involved in the laser action. In the case of yttrium aluminium garnet with neodymium for 300 K this component has the following characteristics: $\delta\nu_0$ = 6 cm^{-1}, $T_1 = 2\cdot10^{-4}$ s, $\sigma_{tr} \approx 10^{-18}$ cm^2. Actually, the gain line 1.06 μm consists of two close components that correspond to different pairs of Stark sublevels [43]. Strictly speaking, this line should be considered as inhomogeneously broadened and in some contexts this feature is become apparent.

---

[1]The temperature 300K corresponds to $k_BT/\hbar$ of the order of 200 cm^{-1}.

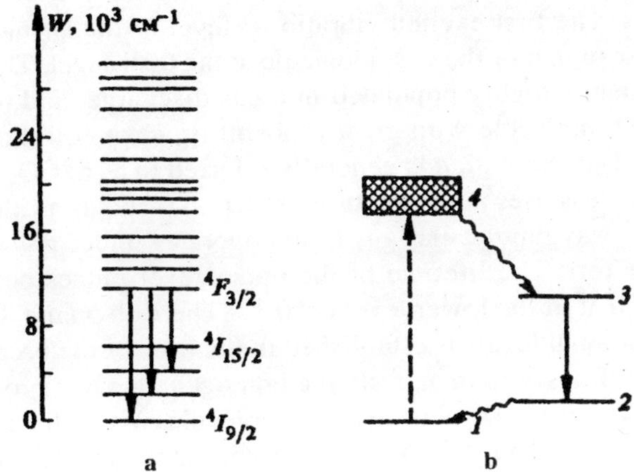

Fig. 1.4. Energy-level diagram of Nd^{3+} ion (*a*) and the scheme explaining the principle of operation of an optically pumped four-level solid-state laser (*b*).

In glasses, the ordered structure is absent and the implanted ions interact differently with their local environments. That is why the Stark components vary slightly in frequency from ion to ion and, consequently, the fluorescent line becomes inhomogeneously broadened. In silicon glass, a widely used laser material, the transition ${}^4F_{3/2} \rightarrow {}^4I_{11/2}$ is characterized by the cross-section $\sigma_{tr} \approx 10^{-20}$ cm², the metastable level has a lifetime $T_1 = 2 \cdot 10^{-4}$ s and the inhomogeneous fluorescent linewidth $\delta\nu_{inh} = 300$ cm⁻¹. The fraction of homogeneous broadening is an order of magnitude less.

A carbon dioxide laser is the next class *B* representative. The linear symmetrical CO_2 molecule has three basic internal modes of vibration: symmetric stretch with the distance between levels ν_1=1388.17 cm⁻¹, binding with ν_2= 667.4 cm⁻¹ and antisymmetric stretch with ν_3= 2349.16 cm⁻¹. The state of the molecule is generally described by a set of quantum numbers referring to the normal modes: $V_1 V_2^l V_3$. The quantum number l indicates the double degeneracy of the bending vibrations.

The main role in the population redistribution over the vibrational and rotational levels of CO_2 is played by collisions. The possibility of obtaining an excess population of the 00^01 state is due to the fact that the latter decays slower than the 10^00 or 02^00 states. This is because the energy transfer from one vibrational degree of freedom to another is slow compared to the time of onset each of them. The fast decay of the 10^00 state is contributed by the Fermi resonance, *i.e.* the coincident positions of 10^00 and 02^00 states. Under these conditions, the decay rates of the 01^00 state are a decisive factor. Helium can be introduced into the discharge to increase this decay rate.

The pumping of the upper laser level 00^01 is most effective with nitrogen molecules. The first excited vibrational level of the N_2 molecule has an energy close to that of the CO_2 molecule at the 00^01 level. This nitrogen level is metastable, highly populated in a gas discharge, and transfer energy to the CO_2 molecule with great probability when colliding with the N^*_2 molecule. Thus, though it is generally referred to as a 'CO_2 laser', this device nevertheless uses a gas mixture of CO_2–N_2–He, as a rule.

Continuous-wave molecular gas lasers operates under pressure of the order of 1–10 Torr. The lifetime of the upper level ranges between 10^{-3} and 10^{-4} s and that of the lower level is 10^{-5} s. The Boltzmann distribution among the rotational levels is established in the time of the order of 10^{-7}–10^{-8} s, which defines simultaneously the homogeneous line broadening of the rotational-vibrational spectrum of the molecule ($\delta\nu_0 \approx 10^{-3}$ cm^{-1}). Since the linewidth is less than the distance between lines (about 2 cm^{-1}) or the frequency range between longitudinal modes (usually $\Delta\nu \approx 10^{-2}$ cm^{-1}), single-frequency operation is readily provided in the CO_2 laser. The rotational occupation has its maximum for the levels with the quantum number J =20–30. Therefore, the laser action starts earlier for the rotational–vibrational transitions between the levels of this order. The lines belonging to the P branch ($\Delta J = -1$) of the 00^01–10^00 transition have an advantage. The operation at the 00^01–02^00 transition is suppressed since the 02^00 level is slightly below 10^00 thus being more occupied. The discrimination of the R branch ($\Delta J = 1$) can be explained in a similar fashion. Thus, the prevailing operating wavelength of the CO_2 laser is 10.6 mm. For tuning within this range or switching to 9.6 mm one should use a selective resonator.

Class B lasers also include *semiconductor lasers*. The main feature of this type of laser is that we deal with wide energy bands of the host crystal instead of the narrow energy levels of impurity atoms. Nevertheless, it is feasible to obtain the population inversion needed for the laser action. This should be problematic inside the band since the intraband relaxation rate is high enough and the interband transitions are generally involved. Hence, the frequency range is governed by the gap energy.

The optical properties of semiconductors are defined by the mutual arrangements and population of the two upper energy bands. The electrons obey the Fermi–Dirac statistics according to which the probability that an electron state with energy W is occupied is given by the law

$$h = \frac{1}{1+\exp[(W - W_{\mathrm{Fc,Fv}})/k_B T]}$$

If the sample is assumed to be in thermal equilibrium, then $W_{\mathrm{Fc}}=W_{\mathrm{Fv}}=W_{\mathrm{F}}$ and there is a single Fermi level with the occupation probability 0.5 instead of two quasi-Fermi levels W_{Fv} and W_{Fc}. This concept is valid because the carrier thermalization time within a band is much shorter than the interband relaxation time.

The population inversion condition in a semiconductor looks like $W_{Fc} - W_{Fv} > W_0$, where W_0 is the band-gap energy. This means that the quasi-Fermi levels are situated inside the bands and therefore the states near the bottom of the conduction band are occupied, and those near the top of the valence band are empty.

There are many ways to achieve the inversion in semiconductors: optical pumping, excitation by a fast electron beam, avalanche breakdown caused by an external electric field and carrier injection through the *p-n* junction. The last method has become most widespread due to its simplicity end effectiveness.

In an intrinsic semiconductor, which is in thermal equilibrium, the Fermi level is situated exactly in the middle of the band gap (Fig. 1.5*a*). If the crystal is alloyed with a dopant (the *n*-type semiconductor), then the Fermi level is nearer to the conduction band, and is inside this band in degenerate case. A similar shift, but towards the valence band, is obtained by an acceptor admixture (the *p*-type semiconductor). We can dope differently the different sites of one crystal and make a sharp boundary between the *n*-type and *p*-type domains. The boundary layer is called the *p–n* junction.

The position of the Fermi level is the same in different parts of a nonuniformly doped crystal if an external electric field is not applied. A relevant energy diagram is given in Fig. 1.5*b*. Carrier migration through the *p–n* junctions impeded by a potential barrier. A voltage applied to the junction in a forward direction (from the electron to the hole part) lowers the barrier (Fig. 1.5*c*) thus enabling the electron to penetrate into the *p*-region and the holes in the *n*-region. Consequently, layers with increased concentration of minority carriers are produced on both sides of the junction. The thickness of these active layers does not exceed a few microns.

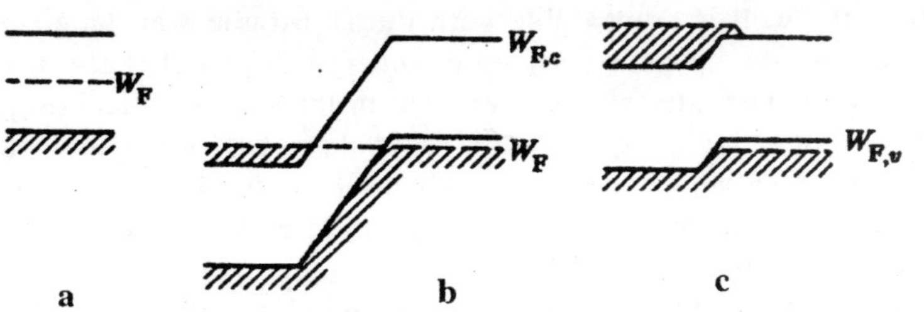

Fig. 1.5. Energy diagram and the position of the Fermi level (a) of the intrinsic semiconductor; (*b*) of the crystal with *p-n* junction in the absence of bias; (*c*) under positive bias. The electron levels in the hatched areas are occupied.

Carrier diffusion over great distances from the boundary is impossible because of recombination processes.

In GaAs lasers, a major role is played by electron injected into the *p*-region. The population inversion is produced near the boundary of the *p*–*n* junction. The cavity mirrors are normal to the *p*–*n* junction plane. Usually, these mirrors are two opposite facets of the cleaved crystal. By virtue of the high index of refraction of the semiconductor crystal, the Fresnel reflection is tens of per cent, so that the *Q*-factor of the cavity is high enough for lasing action even without additional reflecting coating. The gain can reach extremely high values because of the high density of states in the conduction and valence bands, which can be used to create a large population inversion.

It is an important factor for the operation of diode lasers that the index of refraction in the *p*–*n* junction is somewhat higher than in the reminder of the crystal. Such an index variation promotes localization of the radiation field near the *p*–*n* junction plane. This is the only place where amplification occurs, while outside the active region the absorption for the laser wavelength is very high.

In a laser with a homostructure, as discussed above, all of its properties are the same on the two sides of the *p*–*n* junction and, therefore, waveguide features are minor. This and other drawbacks are not seen in lasers with heterostructures. In this case, the layers adjoin the *p*–*n* junction, differ from the active region in the composition ($Ga_{1-x}Al_xAs$) and, consequently, in the optical and electrical properties. Besides the fact that an index jump at the boundaries of the active region contributes to field localization, the potential barrier between the layers prevents injected carriers from escaping from the active layer. Moreover, in heterostructures the losses are decreased outside the active layer, since the band gap in $Ga_{1-x}Al_xAs$ (adjoining layers) is wider than in GaAs (active layer).

The active layer of the laser diode confined to the both sides by layers with different chemical composition constitutes a potential well. If the width of the well is comparable with the DeBroglie wavelength for an electron, then the distance between the energy (levels of space quantization) exceeds the width of the levels that influence the optical properties of the system and, consequently, the laser characteristics. The devices of such type are called the quantum well lasers. What is the advantage of these lasers in comparison with the ordinary laser diodes? One of them is even not related to space quantization. The simple decreasing of the thickness of the active layer leads to the proportional decreasing of the threshold injection current, which corresponds to the transparency of a semiconductor. The further lowering of required current accompanies the qualitative changing of the distribution function of the energy states density, which becomes stepwise. Near the bottom of the potential well there are no levels and this fact is delivering from the necessity of pumping in the conduction band of great number of electrons with the aim to reach of needed localization of the quasi-Fermi levels.

A very important factor, which influences the threshold value and the energy characteristics of a laser, is the cavity Q-factor. It is possible to compensate the lowering of Q-factor due to decreasing of the cavity length by increasing the mirror reflectivity. But this is difficult to achieve in the framework of traditional design that assumes propagation of the emitted radiation in the plane of the active layer with the output through the crystal ends normal to this plane. The principal new result has been occurred when the direction of generation was chosen perpendicular to the active layer. This allowed decreasing noticeably the transverse dimensions of the cavity and growing the multilayer dielectric mirrors using unified for all a laser structure technology of epitaxial alternate deposition. The term VECSEL's is the abbreviature of Vertical Cavity Surface Emitting Lasers. The pictorial view of the construction of the lasers of the latter type is given in Fig. 1.6. A very important feature of such a laser is the absence of the preferential direction of polarization.

A distinctive feature of semiconductor lasers is their very short cavity, which does not exceed hundredths of a centimeter. That is why the intermode beat frequency is very large: $\Delta\omega \sim 10^{12}$ s^{-1}. The short cavity length together with the losses from transmission of the cleaved facets and absorption yields $\kappa \sim 10^{11} - 10^{12}$ s^{-1}. The rate at which the quasi-equilibrium distribution of carrier inside the bands is established significantly exceeds the intraband relaxation rate. The characteristic time of the first process is of the order of 10^{-13} s. This is coherence time T_2 in a semiconductor. The intraband relaxation mechanism is the spontaneous electron–hole recom-

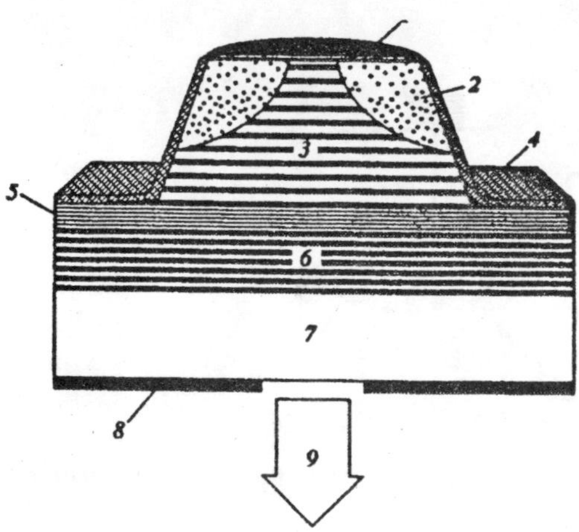

Fig. 1.6. Design of the Vertical Cavity Surface Emitting Laser (VECSEL) [44]. The small cavity volume is combined with the high mirror reflectivity, which leads to very low laser threshold: *1*–upper contact; *2*–area with irregular structure; *3*–Bragg reflector of *p*-type; *4*–silicon nitride; *5*–multilayer active area; *6*–Bragg reflector of *n*-type; *7*–substrate from GaAs; *8*–lower contact; *9*–output radiation.

bination. The time constant for this process depends on the carrier density but usually it is not less than 10^{-9} s and coincides with T_1.

Organic dye lasers fall into *class A*. The characteristic band-like structure of the energy spectrum of organic molecules is shown in Fig. 1.7. Owing to the even number of electrons, the ground state of the molecule is a singlet (the spins are antiparallel in pairs). The lowest singlet state $S_{(0)}$ and the first excited singlet $S_{(1)}$ play the primary roles in laser action. Meanwhile, triplet states, corresponding to the parallel spin orientation of optical electrons, are also important for the properties of the active medium.

The intercombination transitions between the levels of different multiplicity are spin-forbidden by the selection rules. Hence optical pumping can transfer the molecule from $S_{(0)}$ to $S_{(1)}$. If these states are purely electronic, amplification in a system of organic molecules would be impossible. Each electronic level, however, has a complex structure due to molecular vibrations and rotation. Thus, a version of a four-level scheme is realized [45].

The distance between the vibrational sublevels range is several thousands of inverse centimeters. The rotational structure has discrete steps of the order of 10–100 cm^{-1}. At a temperature 300 K only the ground sublevel $S_{(00)}$ is populated. When illuminated by pumping light, the molecules pass to one of the vibrational sublevels $S_{(1\nu)}$ of the first excited singlet state. Then over a short time compared with the lifetime of state $S_{(1)}$ the molecule enters the lover vibrational sublevel $S_{(10)}$ through the radiationless transition $S_{(1\nu)} \rightarrow S_{(10)}$. The molecule stays here for 10^{-8}–10^{-9} s and then returns to the lower singlet either directly or through the triplet state. The absorption line is formed by the transitions $S_{(00)} \rightarrow S_{(1\nu)}$, and luminescence line by the transitions $S_{(10)} \rightarrow S_{(0\nu)}$. The latter are shifted towards the longer

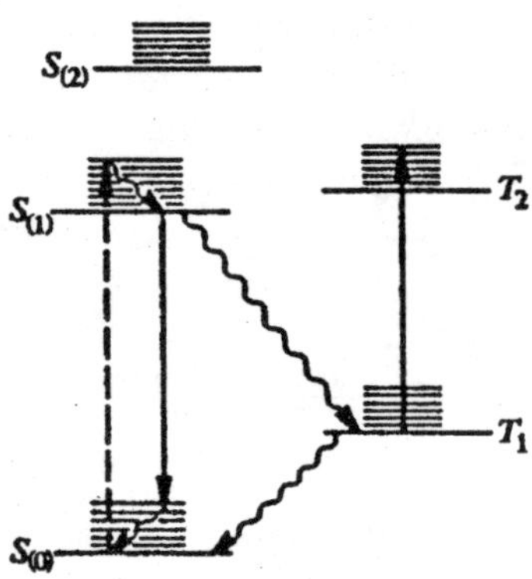

Fig. 1.7. Energy-level diagram of an organic dye laser. Short lines show the rotational-vibrational structure.

wavelength with respect to the former, though they partly overlap.

The lifetime of the state $S_{(1)}$ is short and, therefore, the population under the action of optical pumping is relatively small. Nevertheless, the population is greater than that of the excited vibrational sublevels of the ground state. Thus, inversion, accompanied by gain, is produced in a number of electron–vibrational transitions.

Some of the molecules come, without emission, from the singlet state $S_{(1)}$ to the triplet state $T_{(1)}$. The transition $T_{(1)} \rightarrow S_{(0)}$ is forbidden so that the state $T_{(1)}$ is metastable. The existence of such a state prevents the laser action. In particular, induced transitions $T_{(1)} \rightarrow T_{(2)}$ produce an additional absorption at the laser wavelength, (the so-called triplet absorption). The rate of accumulation of molecules in the triplet state is defined by the probability of intercombination transitions $S_{(1)} \rightarrow T_{(1)}$ and has a range of 10^7–10^9 s^{-1}. Due to this, in pumping by a giant pulse from a *Q*-switched solid-state laser triplet absorption is not observed, while the use of a flashlamp as a pumping source is possible only if there is an effective depletion of the lower triplet levels, which can be achieved by the choice of a special admixture in the dye solution.

If the triplet absorption problem is solved, primary attention should be given to the losses by spatially nonuniform heating of the medium during the pumping. The refractive index gradients arising in a cell with a dye solution are sufficient to distort noticeably the cavity geometry and even to make the cavity configuration unstable. This form of induced losses is the main obstacle to the continuous wave mode of operation of dye lasers. The CW lasing is achieved only in an active medium (such as a rhodamine 6G solution) having the form of a jet, with laser pumping (e.g. by the radiation of an argon ion laser).

The band-like energy spectrum is a common feature of organic molecules and semiconductor crystals. The relaxation times for these two cases are nearly the same: $T_1=10^{-8}$–10^{-9} s and $T_2=10^{-12}$–10^{-13} s. Nevertheless, the cavities of these two types of laser have nothing in common. In liquid lasers $T_c \sim 10^{-7}$ s, which ensures their belonging to class *A*.

The popularity of dye lasers is mainly due to the possibility of tuning within the broad luminescence line. A small collection of dyes is sufficient to cover the whole visible range. We can move to the IR, up to a wavelength of 3.5 μm, using active media of another type – ion crystals with intrinsic colour centres (*F* centres). The electron localized on an anion vacancy in the crystal is the simplest example of an *F* centre. Owing to its strong coupling with the crystal lattice, we should not consider the *F* centre as an isolated formation of a hydrogen atom type. The electron energy levels assume, because of the immediate environment, a vibrational band-like structure resembling the level structure of complex molecules.

The relaxation times (10^{-12} s within one electron state and 10^{-8} s between the different electron states) have the same order of magnitude as those in dye solutions.

A spectroscopic situation like this occurs in some doped laser crystals, including alexandrite (chromium chrizoberyllium) Cr^{3+}:$BeAl_2O_4$ [46] and Ti:Cr_2O_3 [47]. Here again the broad bands in the spectrum are caused by the effect of the crystal lattice on the doped ions. Unlike the rare-earth ions such as Nd^{3+}, the ions of the ferric transition metals have unfilled shells and are shielded by external electrons. The corresponding states are subject to environmental perturbation and thus are sensitive to the lattice vibrations. Consequently, the spectra show a vibrational structure that corresponds to transitions with simultaneous variations in the electron state of the impurity ion and in the vibrational state of the crystal matrix. A particular form of the spectrum depends on the host in which the ion is embedded. The lifetime of the upper level in alexandrite is 1.5 μs. That is why this laser falls into class *B*.

Class A includes most of atomic gas lasers. It was noted above that in lines with a very high gain it is possible to provide the conditions typical of class *C* lasers at the expense of the cavity *Q*-factor. Moreover, only a few transitions like these are known so far.

The helium–neon laser is the most used atomic gas laser. Neon serves as the lasing component of the mixture. The most intense CW generation is achieved for 0.6328 μm, 1.1523 μm and 3.39 μm. The ground state of neon atoms corresponds to the np^6 configuration of the outer shell electrons. Following Pashen's notation, the excited state with the $np^5(n+1)s$ configuration is denoted as $1s$ and its four sublevels are numbered in the order of decreasing energy from $1s_2$ to $1s_5$.The next excited state $np^5(n+1)p$, denotes as $2p$, consists of 10 sublevels from $2p_1$ to $2p_{10}$. The electron s-configuration levels have a greater lifetime than those of the p-configuration. Owing to this, as well as because of predominant population of the $2s$ and $3s$ levels of neon in collisions with metastable helium atoms, the medium gain is ensured for the frequencies of the $s \to p$ transitions.

The spectral line broadening in gas lasers is due to spontaneous emission, collisions and the Doppler effect. Any of these mechanisms can make the main contribution depending on the gas temperature and pressure, the atom mass, the transition type and the wavelength. Under the conditions typical of a helium–neon laser the neon line $3s_2$–$2p_4$ ($\lambda = 0.63$ μm) has the Doppler width $\delta\nu_D = 1700$ MHz and the line $3s_2$–$3p_4$ ($\lambda = 3.39$ μm) has the Doppler width $\delta\nu_D = 320$ MHz. In the literature, the homogeneous contribution to the linewidth $3s_2$–$3p_4$ is described by an empirical formula [48]

$$\gamma_\perp / 2\pi = 200\,\text{MHz} + 42\,\text{MHz/Torr} \quad (0.32\,\text{MHz/Pa})$$

which holds for the optimal relationship between the partial pressures of helium and neon 5:1. For the transition $3s_2$–$3p_4$ the analogous dependence is [49]

$$\gamma_\perp / 2\pi = 8.5\text{ MHz} + 59.5\text{ MHz/Torr} \quad (0.45\text{ MHz/Pa})$$

$$\gamma_\perp / 2\pi = 9.75\text{ MHz} + 14.9\text{ MHz/Torr} \quad (0.11\text{ MHz/Pa})$$

1.2.3. Some Experimental Facts

There is extensive experimental material on laser dynamics. We choose only the main facts without the knowledge of which it is difficult to discuss the theoretical problems. They refer to free running modes of operation.

Single-mode class A lasers show the simplest behaviour. The low inertia (short relaxation times) of the active medium makes the transients aperiodic. The time of onset of a stable steady state depends on the cavity Q-factor and the excess of the pumping above the laser threshold. In a helium–neon laser this time ranges in 10^{-6}–10^{-4} s [50–52]. To ensure that only one longitudinal mode falls within the Doppler linewidth, the length of the cavity should not exceed 10 cm for $\lambda = 3.39$ μm. In longer cavities single-mode operation requires the use of mode selectors. Single-mode operation of dye-lasers, the gain lines of which are extremely broad, can never been achieved without wavelength-selective elements.

The frequency dependence of the output power of a gas laser has a characteristic feature: fine adjustment of the laser mode to the centre of the gain line produces a local minimum in the power output [53]. This feature, called the Lamb dip, is typical of active media with Doppler broadening.

A frequency shift was discovered when the beat spectrum of lasers was investigated [54]. Laser frequencies are shifted from the cavity eigenfrequencies towards the line centre, and the beats between the longitudinal modes are less than $qc/2L$ (L is the length of the cavity and q is an integer). When more than two longitudinal modes are excited, the beat spectrum exhibits a fine structure indicating that the modes are not equally spaced in the laser spectrum. The spectrum nonequidistancy is sensitive to pumping power and cavity length.

In Ref. [55] it is found that the radiation process in the operation of the helium-neon laser is regular so long as not more than three longitudinal modes are involved. The degree of nonequidistancy a three-mode spectrum depends on the detuning of the strongest mode from the line centre. The laser action with rigorously equidistant frequencies arises continuously if the frequencies ω_c and ω_0 converge with $\omega_c > \omega_0$ and discontinuously if these frequencies converge from $\omega_c < \omega_0$.

Chaotic modulation of the envelope of laser emission can appear when a fourth mode is involved. The laser with $\lambda = 0.63$ μm, investigated in [55], exhibited the three well-known scenarios of transitions from regular to chaotic self-modulation of intensity as a control parameter was changed: through a sequence of period doubling bifurcations (of the secondary beat period in this particular case), through the appearance of incommensurate frequencies (again in the secondary beat frequency) and through intermittency [25,26].

In Refs. [56–58] one can find different statements concerning the behaviour of multimode He-Ne lasers: three modes are enough for the realization a chaotic regime. However, the talk is about the atomic transition with a wavelength of 3.39 μm, which is distinguished by a much larger gain. Under the experimental conditions realized in [56–58] the laser can be attributed to class *C* rather than to class *A*.

Another type of phenomenon is demonstrated by *dye lasers*. Figure 1.8 shows the scheme of the laser used in Refs. [56–58]. The ring cavity of the laser is supplemented with a retroreflecting mirror to ensure unidirectional lasing. The active medium in the form of a jet of Rhodamine 6G solution is pumped by an ion argon laser. The dispersive prism is intended for rough tuning of the cavity. It is seen from Fig. 1.9 that immediately above the threshold the laser emits in a narrow line. With an increase of the pumping a bifurcation point is achieved where the spectrum splits into two lines the distance between which is proportional to the square root of the dye laser power. The cavity detuning makes it possible to observe bifurcations leading to increasing complexity of the line spectrum. The position of the bifurcation points is also dependent on whether a travelling wave or a standing wave is used in the laser.

In Refs. [59-61] the modal composition of a CW dye laser output was not investigated. Meanwhile, the investigations concerned with the real sensitivity of the intracavity laser spectroscopy method indicate that hundreds of longitudinal modes participate when the laser with a nonselective cavity is in a steady state. A high-resolution scan of the laser spectrum has

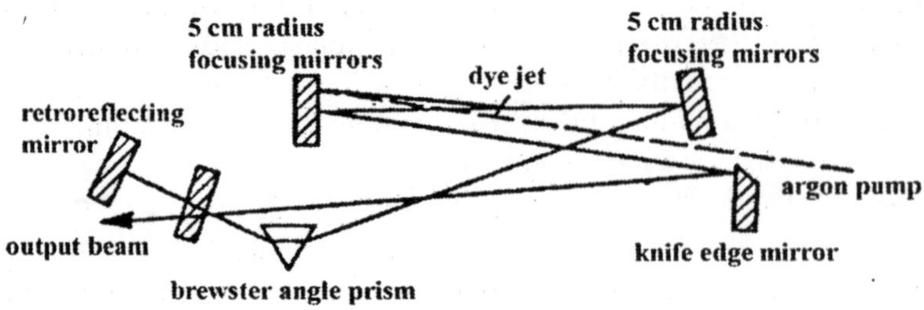

Fig. 1.8. Schematic diagram of CW ring dye laser [60]. When the retroreflector mirror is properly aligned the laser operate unidirectionally

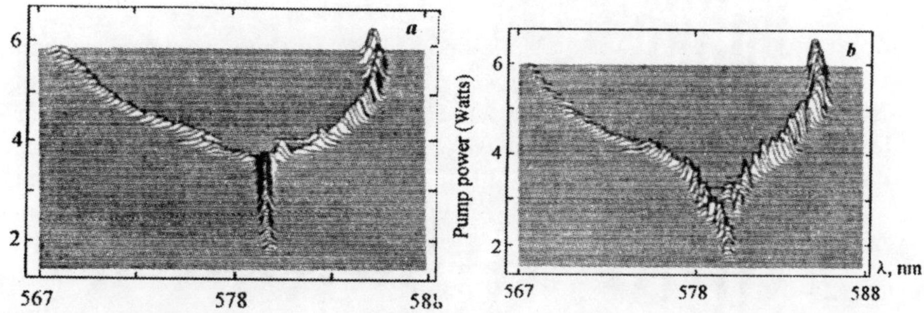

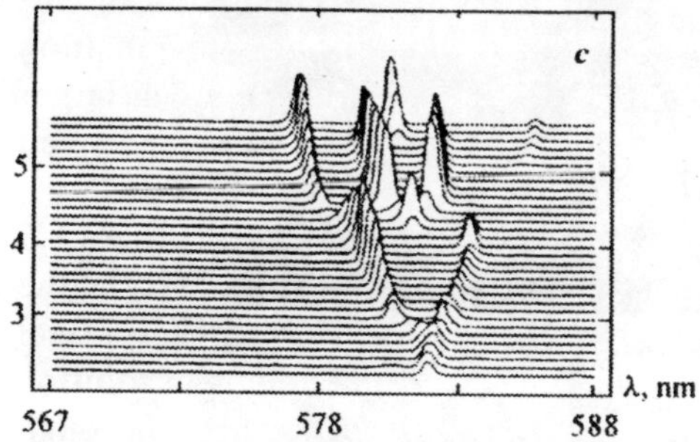

Fig. 1.9. Spectrum of the laser field as a function of the pump power for various cavity alignments [60]: (*a*) bidirectional operation at the gain line centre; (*b*) the laser is as in (a) except the retroreflector is aligned to produce unidirectional operation; (*c*) unidirectional operation at the frequency shifted by 20 A from the centre to the red.

shown that nearly periodic chirping occurs in the case of travelling wave [62]. The instantaneous width of the spectrum is small but the centre of gravity of the mode packet drifts slowly towards the red (Fig. 1.10). The drift ends in jump-like return of the spectrum to the initial position, followed by a new cycle. The period of these spectral chirps is inversely proportional to the laser power. In this experiment travelling wave operation was achieved by means of a Faraday cell.

It is possible to observe a regular self-sweeping of the spectrum in a standing wave laser with a short concentric resonator if the dye jet is slightly moved relative to the position of the beam waist [63]. It is shown also in the work [63] that the period of the mode amplitude pulsations can be noticeably increased at the expense of the cavity dispersion compensation. As a rule, in a standing wave jet laser is established the regime of irregular mode amplitude self-modulation (Fig.1.11) with constant total intensity of radiation [64, 65].

The different behaviour of the total intensity and spectral density of dye laser emission is also exhibited in the stage of onset of a steady state after pumping switching. The total intensity reaches the steady state very quickly, over a time of order $T_1 \sim 10^{-8}$ s while the characteristic time of spectral narrowing is 10^{-3} s or more [64]. Therefore, the dye lasers with pulsed pumping operate under nonsteady-state conditions.

Periodic antiphase sells-modulation of the amplitudes of counter-running waves is possible in a ring dye laser with two retroreflecting mirrors

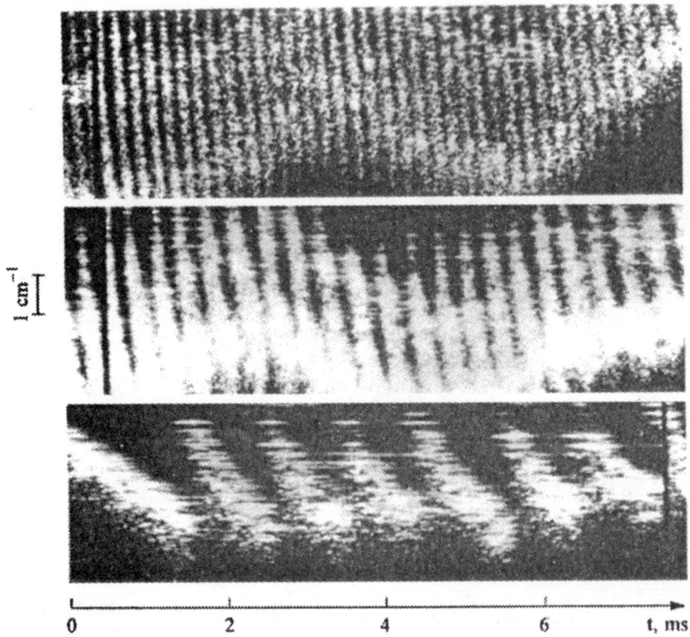

Fig. 1.10. Quasiregular self-sweeping of the spectrum of a unidirectional ring dye laser [62]: A = 1.9 (*a*); 1.3 (*b*); 1.15 (*c*).

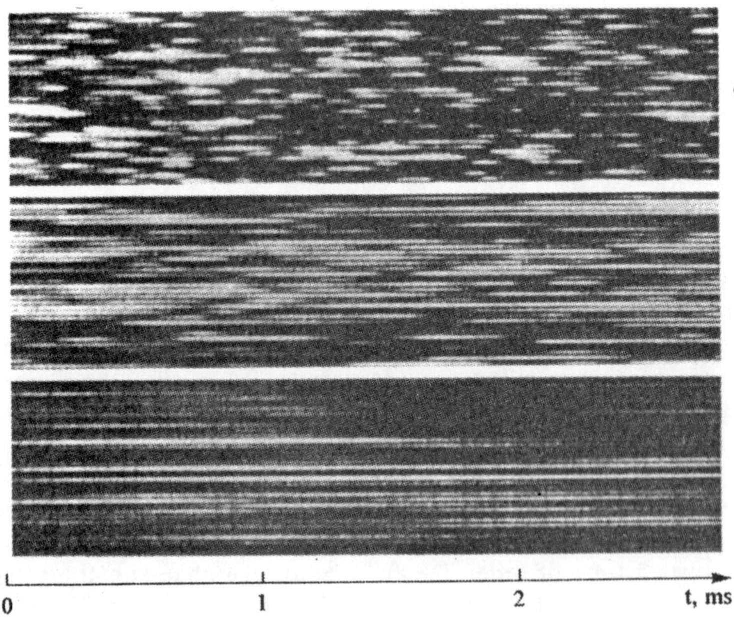

Fig. 1.11. Irregular kinetics of the dye laser spectrum [64, 65]: A = 1.30 (a); 1.13 (b); 1.02 (c).

[66]. Both periodic and chaotic self-modulation was observed in the helium-neon laser under similar conditions [67, 68].

Class B lasers exhibit diverse forms of behaviour. A characteristic feature is the phenomenon of 'spike generation', in which the laser emits a sequence of isolated bursts. This phenomenon was first discovered in the experiments with paramagnetic masers [69, 70] but thorough investigation of it came with the advent of solid-state lasers.

Solid-state lasers operate differently depending on the spectroscopic properties of the laser medium, the cavity type and the pumping parameter. These lasers are subject to mechanical vibrations and the temperature variation of the laser element. The basic elements of the design of lasers belong to first generation are a module containing a laser rod and pumping lamps, and a laser cavity is formed by mirrors mounted outside the module. A cooling system for the active element and pumping lamps are necessary in high power systems.

Free-running oscillation process of solid-state lasers depends on the spectroscopic properties of the active medium, on the cavity type and the pumping. A great influence can be provided by mechanical vibration and by temperature variation of the laser elements. The basic element of a first generation laser is a lighter with the laser rod and the flash lamps inside it. Sometimes this laser head block is called the quantron. The laser cavity more often is formed by mirrors mounted outside the quantron. The forced cooling system for the laser rod and the flash lamps is used when it is necessary.

The emission from a ruby laser with flash lamps pumping at the room temperature in a cavity formed by parallel plane mirrors is a series of spikes. No regular features are observed in the variation of either the amplitude of the spikes or the interval between them, nor is there a tendency for the pulsation damping. The laser start-up is delayed from the time of the pump switch-on by a fraction of a millisecond necessary to reach the threshold inversion. The average duration of the spike is about $5 \cdot 10^{-7}$ s while the peak power of the laser emission can reach 10^{-3}–10^{-4} W.

If the excess pumping over the laser threshold is small, then only a small part of the crystal, where the maximum inversion accumulates through focusing of the pump radiation, is involved in the laser action [71]. The emitting spot slightly enlarges with an increase in pumping energy. The radiation density is distributed nonuniformly within this domain. When film is exposed to the emission for a long time, simple structures, identified with the optical cavity modes, are recorded only if the excess over laser threshold is not great [72]. However, the dynamical behaviour can be more adequately recorded by the methods enabling one to follow the evolution of the optical pattern with time.

Processes, which last not more than 10^{-7} s, can be investigated by a streak camera, which uses the principle of mechanical scanning of the image along a fixed piece of film [73]. The shutter of the camera is opened when required or remains open as long as the exposure. A near field chronogram of a laser beam, recorded in this way, is shown in Fig. 1.12. Only a vertical strip of the output beam, limited by the entrance slit of the streak camera, is shown. The discrete pattern in the horizontal direction indicates that spikes are generated while that in the vertical direction indicates that individual transverse modes are excited. The structure varies from spike to spike meaning that a change of transverse mode occurs. The fact that there is a high probability of exciting only one transverse mode in a separate spike was mentioned by many authors [73–78]. Investigating the nature of the high angular divergence of the laser emission and of small diameter of the transverse modes the authors of Ref. [78] inferred that these features correspond to a spherical rather than a plane-parallel cavity. Optical inhomogeneities due to crystal imperfections, which arise when the crystal is heated inhomogeneneously by the pumping light, have the properties of a lens. With a lens between the plain mirrors, the cavity is equivalent to a spherical one. The sphericity is inconsequential only for very large focal lengths ($F \geq 10^5$ cm). The required high optical quality of the laser rod is seldom reached to avoid this effect, so that the term 'plane-parallel resonator' has a purely nominal meaning when applied to solid-state lasers. When the lensing property of the laser rod is compensated, the mode dimension approximates the crystal diameter and the beam divergence decreases to the diffraction limit [79].

Investigation of the spatial structure of the emission of a ruby laser with a plane-parallel cavity confirms that the transverse mode pattern is unstable; study of the optical spectrum offers a similar conclusion concerning the longitudinal modes. The total width of the spectrum is as large as a few tenths of an angstrom so that high-resolution Fabry–Perot interferometers can be used to observe its structure. The light transmitted through the interferometer forms a ring interference pattern in the far field zone (in the focal plane of a lens). The number of rings within free spectral range coincides with the number of longitudinal modes involved in the laser ac-

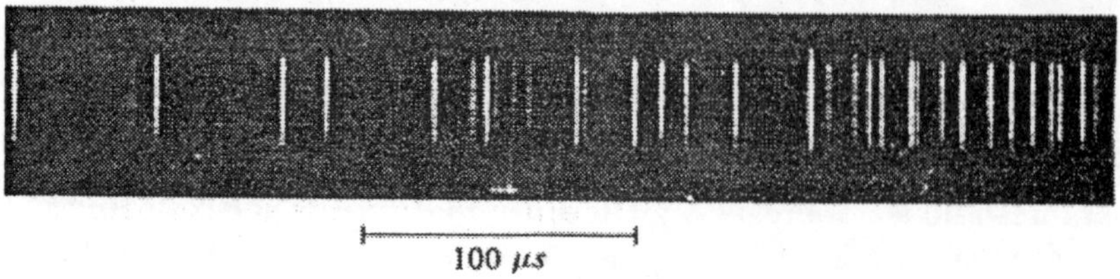

Fig. 1.12. The beam profile in the near-field zone of a ruby laser with plane-parallel cavity.

tion. Generally, tens of longitudinal modes are excited during each flash.

Selection of the longitudinal modes is evident in all cases where the cavity has a few parallel reflecting surfaces moderately inclined to the mirrors. The ends of the laser rod, the surfaces of the mirror substrates are examples of such surfaces. The parasitic selection is absent when the cavity mirrors are fixed directly to the ends of the active element. To avoid parasitic selection in the case of external mirrors the ends of the laser rod are cut at Brewster's angle and the mirrors are deposited on wedged substrates. The laser mode spectrum is also sensitive to the position of the active element and to the pumping distribution over its length [80].

A chronogram of the interferogram also can be obtained using a streak camera [73, 81–84]. Specifically, only a narrow band of the ring pattern of the interferogram (within the entrance slit of the camera) is recorded. The analysis of spectrochronograms like those in Fig. 1.13 has shown that the spectrum of an individual spike is formed by fewer modes than observed overall during the flash, but the collection of modes changes from spike to spike.

Variations in the pumping power do not influence the process and change only the peak power and the number of spikes in a flash. Lengthening the cavity has greater impact. When the cavity length exceeds 10 m the spiking become ordered [85] and remains regular up to the maximum achieved length of 400 m [86]. Such a base can be obtained when the delay optical line is inside the cavity. The spike duration increases with the cavity length. When the latter is extremely large only a single spike can be emitted; its envelope displays modulation with the intermode beat frequency.

Ruby lasers are very sensitive to cavity misalignment. An increase in the angle between the mirrors raises the laser threshold [87, 88], increases the time interval between spikes and increases their amplitudes [89–91]. The process seems to be more orderly than the spike generation in a plane-parallel cavity.

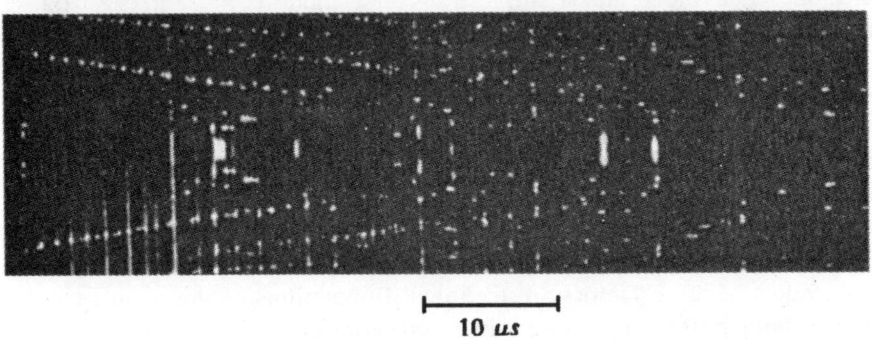

Fig. 1.13. Spectrochronogram of a ruby laser with a plane-parallel cavity.

An orderly laser action can be achieved in a ruby laser with a spherical cavity. Regular damped pulsations, which lead to eventual constant power output, are observed in confocal, concentric and similar cavities [83, 92–96]. The damping rate of regular pulsations is sensitive to the pumping power [85, 97] and the cavity geometry [95, 98]. Being maximal with concentric arrangement of the mirrors, the damping rate reduces as the ratio of the cavity length to the curvature of the mirrors is reduced. The absence of a visible transverse beam structure in the near field zone indicates that a large number of transverse modes are excited. The spectral width is reduced from a fraction of Angstrom in the first spike to a small value retained thereafter, but the centre of gravity of this narrowed spectrum is continuously shifted. Although the overall output power is constant, the laser action is not rigorously a time-independent process since each individual mode amplitude is subject to chaotic modulation variations [96], resembling the dynamic behaviour of dye laser.

In many experiments it is found that undamped regular pulsations can occur [85, 97, 99–104]. They feature a wide (tenth of an Angstrom) optical spectrum, which does not change during lasing. Any interruption of the pulsation regularity is accompanied by a narrowing of the spectrum [105]. The regular pulsations correspond to a homogeneous distribution of intensity over the beam cross-section, which is reproduced from spike to spike [93, 97, 102, 105].

The correlation between the regularity of pulsations and the number of lasing modes is not casual. This is proved by the experiments with mode selecting elements inside the cavity. A decrease of the number of excited modes breaks unambiguously the regular process regardless of whether transverse [106] or [107] longitudinal modes are selected. However, if the mode discrimination is very strong, then the laser dynamics becomes ordered again as in a single-mode laser, see for example [108, 109].

The laser action is rather regular when parasitic selection of longitudinal modes is avoided and a single transverse mode is present [97, 102, 105].

The wide range of possible behaviours of a ruby laser is apparent when one uses a concentric cavity misaligned by a few angular minutes [110, 111]. The lasing may have the form of spikes that follow in regular time intervals. The beam structure in the near-zone indicates that a few high-order transverse modes are excited. The emission frequency is fairly stable and the spectrum is narrow.

The laser emission process bears an imprint of many physical factors. The crudest of these factors are technical fluctuations of the parameters. In flash-lamp pulsed lasers mechanical vibrations of the rod are caused by switching the lamps. Vibrations of other cavity elements can be produced by external sources. The way the coolant passes over the laser rod is also important. The effects of bending vibrations of the rod in the form of amplitude modulation with a frequency about 10 kHz can be significantly reduced by careful choose of the size and position of the diaphragms on both sides of the active element [112, 114, 115]. This is also used to avoid the effect of variation of thermal lens in the rod. Vibrations of the mirrors and other elements can be reduced through mechanical decoupling and increased rigidity of the joints. An important role is played by careful alignment of the mirrors and elimination of the intracavity reflecting surfaces parallel to them. Therefore the ends of the laser rod are oriented at Brewster's angle, or they are antireflecting coated for a small angle of incidence.

The result obtained after taking these measures of passive stabilization of laser is apparent in Fig. 1.14. The spiking is retained but the degree of order is noticeable increased. Although the emission spectrum is initially broad, it is soon contracted so that thereafter one longitudinal mode is involved in each spike. The change of modes from spike to spike corresponds to the change of laser frequency by discrete steps depending on the cavity filling factor and the position of the laser rod [112, 113]. The self-frequency tuning rate can exceed several times the thermal drift of the ruby gain line.

The laser operation is also influenced by the spatial modulation of the inversion caused by the laser field. This produces a stronger difference in gain of the modes and therefore a higher discrimination among some of them. For longitudinal modes this factor is absent in travelling-wave lasers. However, if the modes are standing waves then their equal rights can be ensured only by continuously changing the localization of nodes and antinodes.

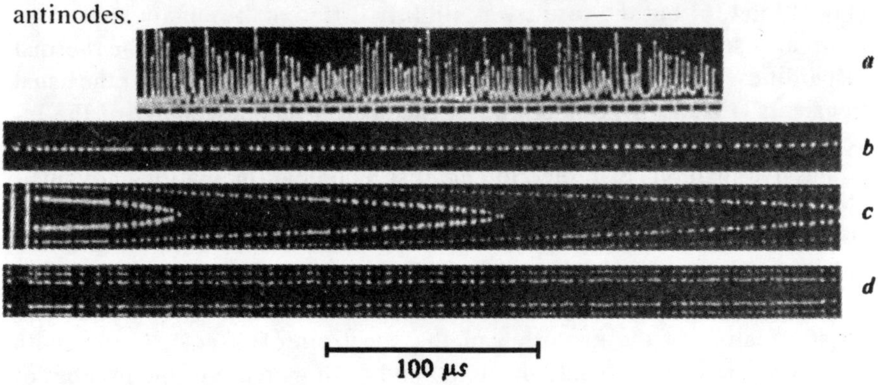

Fig. 1.14. Temporal characteristics of a ruby laser operated at the fundamental (TEM_{00q}) mode of a plane-parallel cavity under the conditions of passive stabilization of the device [114]: (*a*) oscillogram; (*b*) chronogram of the far-field zone of the laser; (*c, d*) spectrum chronograms without and with longitudinal mode selection.

One way to smooth the induced spatial inhomogeneities of inversion is to provide movement of the laser rod. Experiments with a 'running medium' [116–123] showed that the spike-free operation of a ruby laser is possible when the ruby rod moves with a velocity exceeding 20 cm/s. Single-frequency emission is ensured when the velocity is greater than 50 cm/s. A side effect of the moving medium is a 'kinematic modulation' with a frequency proportional to the velocity of the rod. The point is that the end of the rod parallel to the mirrors occupies equivalent positions at intervals equal to $\lambda/2$ as it moves along the cavity axis. The Q-factor of a composite cavity oscillates with the Doppler frequency $2\omega U/c$. This is exhibited as an amplitude modulation of the radiation in the spike-free mode of operation.

The standing wave pattern also can be moved along the active medium by displacement of one of the cavity mirrors [124, 125], but again there is a side effect: a scanning of the cavity eigenfrequencies. However, the compensated phase modulation is free of such drawbacks [114, 126–134]. The optical length of the gap between the active element and the mirror is controlled with an electrooptical phase modulator. When identical KDP or DKDP modulators are located on each side of the rod and control voltages are antiphased. Unlike the 'running medium' method, this procedure requires Brewster's angles on all interfaces inside the cavity.

Using the methods of smoothing of the longitudinal inhomogeneitiy together with passive methods of cavity stabilization, a ruby laser can be switched to damped regular pulsations (Fig. 1.15). During the transient process the spectrum is reduced to one longitudinal mode, then the modes alternate with the thermal drift velocity without any pulsations.

The same methods of laser stabilization were applied to *Nd:YAG lasers* [114, 135, 136] but the result was slightly different. To obtain spike-free lasing it is sufficient to take steps to avoid the mechanical and the thermal instabilities. Steady-state multimode operation is established after the usual transients (Fig. 1.15). Smoothing of the spatial inhomogeneity of the inversion does not produce any radical changes.

Passive methods of protection against technical fluctuations are sufficient to cancel the intensity pulsations in neodymium glass lasers as well (Fig. 1.16). However, the glass laser spectrum has a different structure compared with lasers that use crystals. This is explained by the inhomogeneous broadening of the spectral lines of impurity ions in glass. A characteristic feature of the evolution of the spectrum of laser operation with Nd:glass is that it splits into discrete bands. An increase of the number of these bands is accompanied by the growth of the pumping power. This was first noted in Refs. [138, 139]. A similar behaviour in dye lasers has been mentioned above.

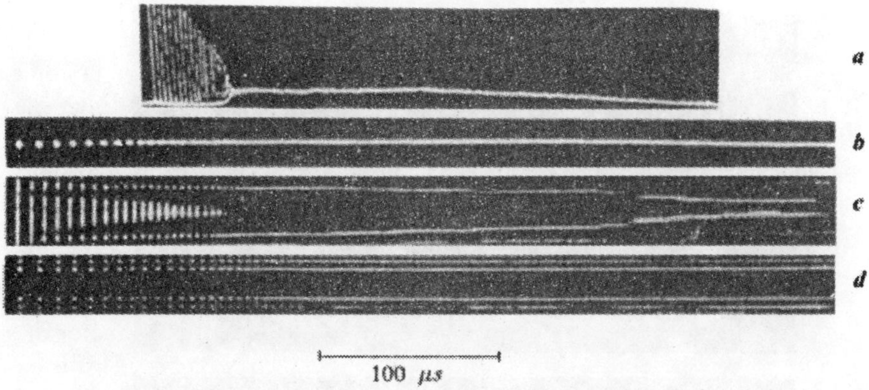

ig. **1.15.** Same as Fig. 1.14 but with the compensated phase modulator switched on.

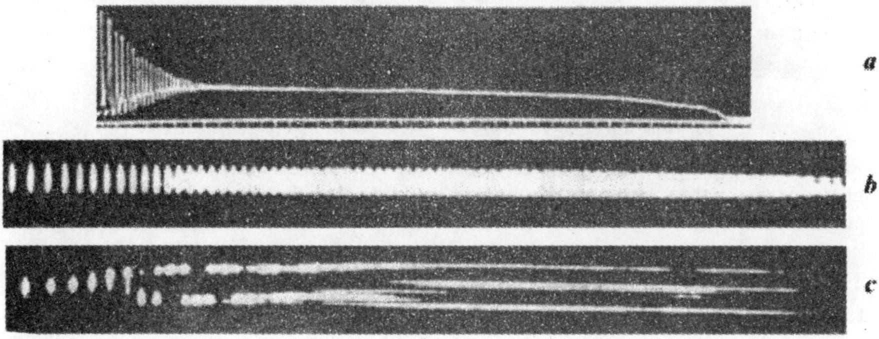

Fig. 1.16. Temporal characteristics of silicate neodymium glass laser with a plane-parallel cavity (TEM_{00q} mode) under the conditions of cut-off of the short-wave component of pumping light: (*a*) oscillogram; (*b*) chronogram of the far-field zone; (*c*) spectrochronogram.

We should mention one more feature of the dynamic behaviour of a silicon neodymium glass laser, namely the dependence on the spectrum of the pumping light [140–144]. The process shown in Fig. 1.17 occurs when the pumping spectrum components are cut off by short-wavelength light filters ($\lambda < 420$ nm). Such filtering prevents the formation of some colour centres in the glass which play the role of saturable absorber. If the whole spectrum of a xenon flash lamp, including the ultraviolet wing penetrates the active element, then the lasing takes a form of undamped oscillations (Fig. 1.17).

Spike emission is also characteristic of *semiconductor lasers* to the same extent as it is of the solid-state lasers mentioned above. The fact that such spikes were discovered several years after the advent of diode lasers [145] can be explained by the higher frequency of pulsations. The main features of the spiking of GaAs homo-structured lasers were found in Refs. [145-

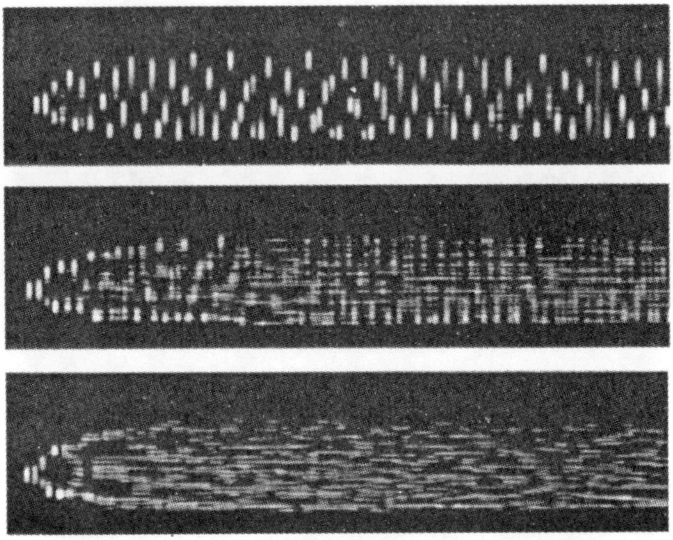

Fig. 1.17. Dependence of the dynamics of silicate neodymium glass laser with a spherical mirror cavity on the spectral composition of pumping. The optical density of the filter that absorbs the short-wave part of the pumping spectrum increases from (*a*) to (*c*) [137].

152].[1] Amplitude oscillations were discovered under both pulsed and continuous pumping. The modulation depth and frequency, and the orderliness of the pulsations depend on the diode temperature and the excess of the injection current over threshold pump level. Near the lasing threshold the modulation is shallow or absent. The most regular pulsations appear for a 1.1 to 1.4-fold excess over the threshold [148]. Increase in the injection current leads to a sacrifice of regularity.

The degree of the orderliness of the pulsations correlates with the field structure in the near-field zone. Because of crystal inhomogeneities the light in the active region of an injection laser can break up into filaments. The intensity pulsations of individual filaments, which act as independent laser, are not synchronized [145]. The experimental data concerning the relation between the form of the optical spectrum and the dynamic mode of operation, proposed by different groups of researchers, are in strong disagreement. This is an indication that different pulsation mechanisms exist.

Undamped pulsations are also observed in hetero-structure lasers [147]. For these lasers the injection current must exceed the threshold value by a factor of 1.5–2.0. Damped pulsations are observed for lower pumping levels. The period decreases as the injection current density and temperature

[1]A much more detailed list of references is given in [152, 153].

are increased, and the period growth as the cavity length increases. Typical spike durations are $(1–5)\times10^{-10}$ s. The ratio of the duration of individual spike to the spike repetition period is 0.2–0.4 [148].

Carbon dioxide lasers, which also belong to class *B*, do not show the same tendency towards spontaneous spiking. Observation of instability in a standing wave laser was reported only in [155, 156].

Other specific features are observed in the behaviour of *class B lasers with ring cavities*. Time-dependent processes of two types can be observed in a CW Nd:YAG ring laser provided spikes are avoided. The process of a first kind is the antiphase harmonic self-modulation of counter-running waves, whose frequency is close to the relaxation oscillations frequency [157–161]. A slower process of the second kind (Fig. 1.18) is a repetitive reversal of the laser direction [157, 158, 161]. Switching from one mode of operation to another can be achieved by cavity realignment. Both of these self-modulation processes occur for a single-frequency laser spectrum. If several frequencies occur in the spectrum, then the spike generation tends to be there.

The self-modulation regime of the second kind was also recorded in experiments with a CO_2 laser [162, 163]. This process could be obtained by detuning the cavity eigenfrequency from the gain line centre. Meanwhile, depending on the gas pressure, the discharge current and the detuning, other time-dependent processes can be observed in a CO_2 laser including synchronous intensity pulsations of both waves, having the form of short pulses.

The solid-state lasers of new generation differ from their predecessors mainly by the pumping scheme. The laser light sources are used for this purpose: semiconductor [164] or sometimes ion gas lasers [165] rather then flash lamps. A scheme of a solid-state laser with longitudinal pump-

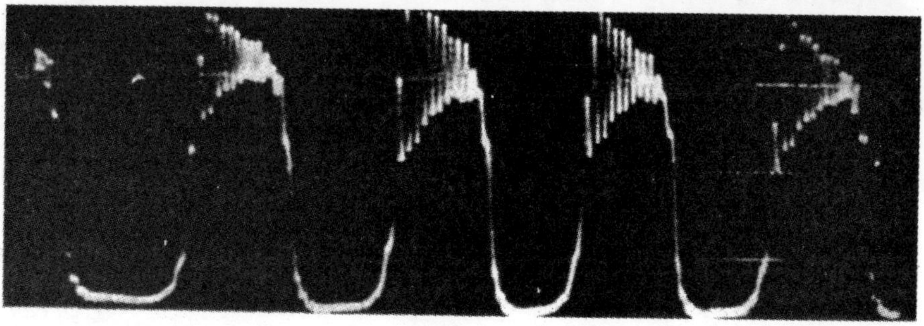

1 μs

Fig. 1.18. Intensity oscillogram of one of the waves generated by an Nd:YAG ring laser in the self-modulation regime of the second kind.

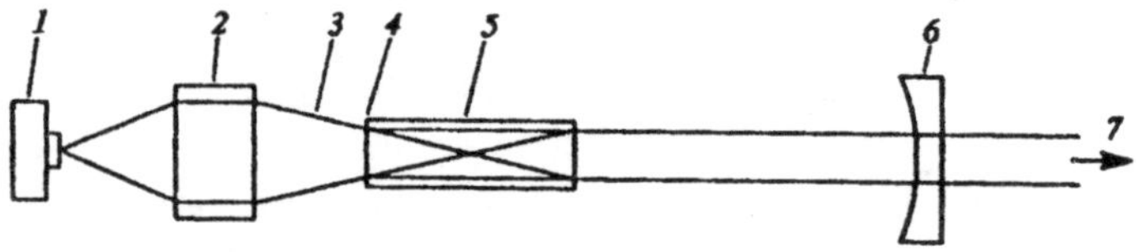

Fig. 1.19. Scheme of a Nd:YAG laser pumped by a semiconductor diode laser along the Fabri–Perot cavity axis [164]: *1*–laser diode; *2*–matched optics; *3*–pumping beam; *4*–mirror; *5*–laser rod; *6*–mirror; *7*–output radiation.

ing by means of a diode laser is presented in Fig. 1.19. The advantage of laser pumping is a narrow spectrum, which can be fit to the absorption line of the active medium. Due to this high efficiency of the pumping and weak heating of the active element are achieved. One more feature of the modern solid-state lasers is monoblock construction of the laser with mirrors at the ends of the active rod. The miniaturization of the laser results in a small number of lasing modes that is very convenient for dynamic investigation. Also the choice of the laser crystals is now much wider.

Diode pumping provides a much more stable laser operation than the flash-lamp pumping even without application of special measures. The CW spikeless oscillation is characterized by the level of fluctuations close to its natural limit. The dynamic features are exhibited as the resonance peaks in the power spectrum (spectrum of intensity fluctuations). The spectra of fluctuations of both the total intensity and individual modes of a laser with a very short cavity are shown in Fig. 1.20. The spectrum of laser oscillations consists of three longitudinal modes. There are three resonance peaks in the power spectra of the individual modes, while only one peak on the highest resonance frequency is seen in total intensity spectra. These experimental data have been confirmed in other works [166, 167]. They testify to existence of relaxation oscillations, the number of which is equal to the number of excited cavity modes, at least, if the number of the modes is not too large. All the relaxation oscillations except the high-frequency one belong to antiphase dynamics. That is why they do not exhibit themselves in the total intensity. Various types of relaxation oscillations can be seen also in a direct experiment by monitoring of the intensity of generation.

From the point of view of the spectrum, the fibre lasers are antipodes to the miniature solid-state lasers. The large cavity length and broad inhomogeneous gain lines promote the oscillation of a very large number of longitudinal modes. However, the most striking features of the fibre lasers are connected with the random orientation of the active centres in the fibre host and with always present birefringence. These features are exhibited in *polarization laser dynamics*. By separating at the laser output a compo-

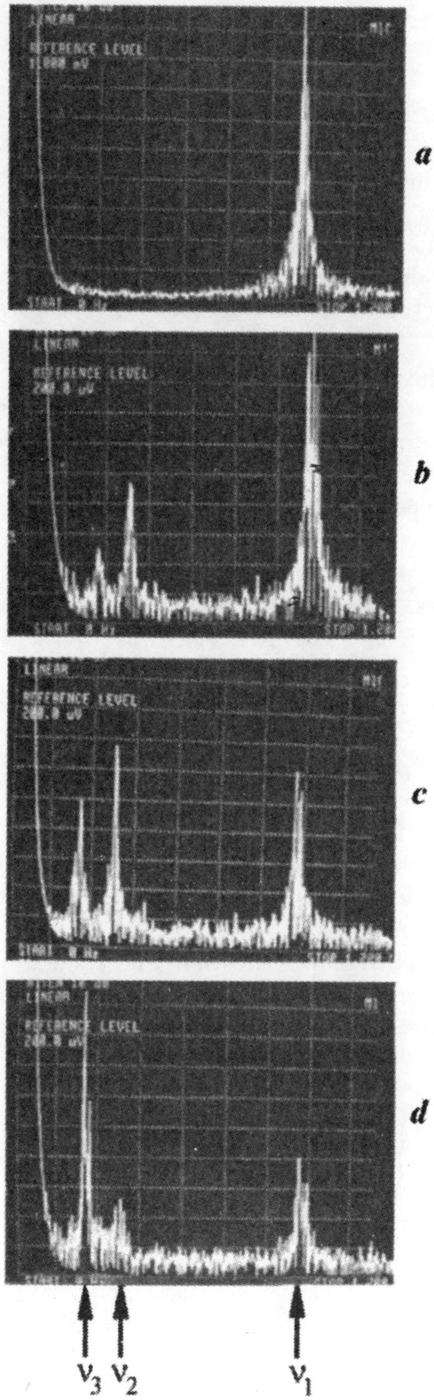

Fig. 1.20. Power spectra of total intensity (*a*) and of individual modes in the order of decreasing intensity (*b*,*c*,*d*) of LNP laser pumped by an argon ion laser [165].

nent with one of eigenpolarizations and analyzing its power spectrum, it is easy to see two resonance peaks while in the total intensity there is only one peak. The situation is similar to that in a two-mode laser. So, we can speak about the 'polarization modes' in spite of the fact that each field component with a fixed polarization contains a lot of longitudinal modes [168–170]. The dynamical effects of switching of the polarization states have been observed in semiconductor lasers, in particular, in VECSEL's [171, 172].

Comparing the experimental results for the lasers with the ring and two-mirror cavities, it is possible to note a sufficient feature: the number of resonance peaks in the power spectra of the latter does not exceed, with rare exception, the number of lasing modes, whereas in ring lasers we have three different types of relaxation oscillations though the number of lasing modes is only two [173, 174]. In lasers with the Fabry–Perot cavities such situation is possible only when there are modes with close enough eigenfrequencies, for example, in fibre lasers with a weak birefringence [169, 170].

Turning to *class C lasers* we should note that a ring version they admit of the same types of competition between counter-running waves as class *B* lasers. This was demonstrated using, as an example, a $^{15}NH_3$ laser with a CO_2 laser as the pumping source [175]. Besides the amplitude measurements, the phase measurements are also presented in [175]. The result related to self-modulation of the second kind is given in Fig. 1.21. The change in slope of the phase indicates that the laser frequency is switched between three discrete values as the mode intensity rises, falls and is kept at its maximum, respectively.

The experimental setup is shown in Fig. 1.22. The ring cavity is formed by three reflectors of different types. A concave opaque mirror M_3 is mounted on a piezoceramic to control the cavity perimeter within a small

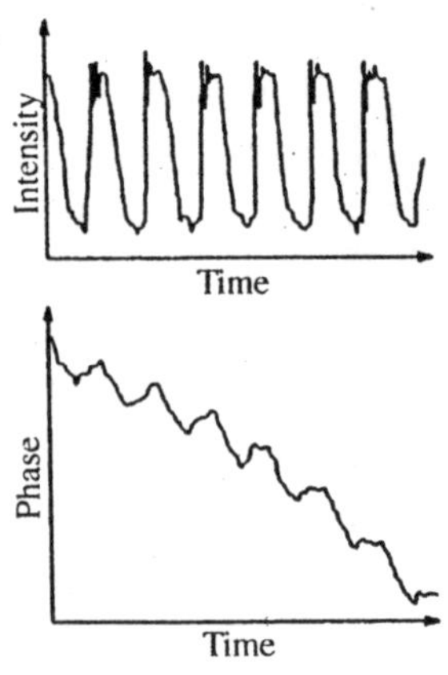

Fig. 1.21. Time dependence of the intensity and phase of one of the waves of a ring ammonia laser (λ= 153 μm) in the second-order self-modulation regime. The phase scale reads to π. The frequency shift in the areas of intensity rise and fall with respect to frequency at the maximum is +28 and –38 kHz, respectively [175].

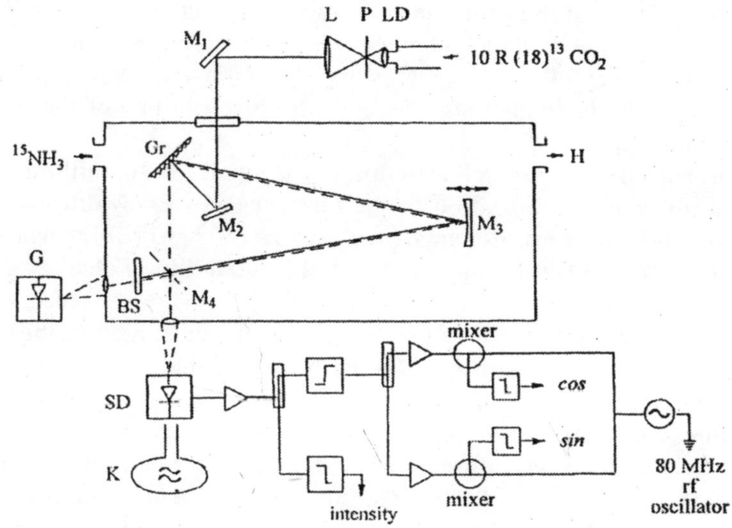

Fig. 1.22. Experimental setup showing an ammonia ring laser [175]: M–mirror; Gr–grating; P–aperture of CO_2 spatial filter; BS–pumping CO_2 beam stop; L–lens; D–diaphragm; SD–Schottky barrier diode detector; K–klystron.

range. The pump radiation is introduced into the resonator by the first order diffraction of a 10 μm grating, which simultaneously plays the role of the ring cavity mirror M_2. The far-infrared laser radiation outcoupling is by means of a partially transmitting gold mesh M_1. Recording of laser radiation generated both in the pumping wave and opposite direction is provided. The cavity is placed into a vacuum chamber, in which a required ammonia pressure is maintained.

The feature of basic importance is the heterodyne detection of the laser field carried out by mixing with the high frequency harmonics of a klystron. This method gives information on the field amplitude and phase dynamics while homodyne detection is informative only of laser intensity.

Unlike the other types of lasers, class C lasers are able to demonstrate instabilities due to the coherent interaction of the laser field with the atomic system, not associated with mode coupling. Ideally, the model of a single-mode, homogeneously broadened, travelling-wave laser, as Haken showed [176], is isomorphic to the famous Lorenz model [177], which initiated the study of deterministic chaos in the nonlinear dynamics of dissipative systems. The tenacious attempts to reach this ideal have been crowned by the experiments with far infrared ammonia lasers pumped by shorter-wavelength molecular gas lasers [23, 37–39, 178–186]. The unique feature of this type of lasers is that a large excess over the laser threshold can be achieved when the gain linewidth is less than the cavity passband. The spectral homogeneity of the gain line is ensured by selective excitation of the gas molecules in the field of a monochromatic travelling pumping wave [187]. If the frequency of the pump does not coincide with the absorption line centre, then the laser medium can interact only with one of the two counter-running waves.[1]

Since probability of spontaneous emission is small for submillimetre wavelength, the main contribution to the homogeneous linewidth is the gas pressure. There is a pressure range of composite (10^{-1}–10^{-2} Torr), where the linewidth (10^5–10^6 Hz) is combined with the possibility of obtaining a high gain.

Besides the fact that they proved the existence of a dynamic instability threshold, the experiments with an $^{15}NH_3$ laser showed that (a) this threshold is high enough, (b) finite-amplitude regular pulsations are established once this threshold is exceeded, (c) chaotic behaviour is achieved through a series of period-doubling bifurcations with an increase in pumping intensity, (d) the spectrum recorded by heterodyne detection of the laser field may or may not have a carrier frequency depending of combination of parameters (gas pressure, pumping power and pumping frequency), *i.e.* amplitude modulation or beats can occur.

A slightly different behaviour is exhibited by an $^{14}NH_3$ operated at λ = 81.5 μm and pumped by an N_2O laser. When the cavity is tuned exactly to the gain line centre and the pumping exceeds the instability threshold, the laser exhibits chaotic pulsations, which are typical for the Lorenz system. When the detuning between the cavity mode and the gain line is large enough the pulsations are regular. The route to chaos as the cavity tunes towards the line centre depends on the gas pressure. There is period doubling (Fig. 1.23) in the domain of 5.5–9 Pa [182, 183, 186], while one can observe intermittency if the pressure is lesser [184, 185].

Discussion of the experiments with ammonia FIR lasers will be continued in Section 3.5.4.

The instability of a steady-state operation was observed in other gas molecule FIR lasers with optical pumping such as a CH_2F_3 laser with λ = 117 μm [35], a CH_3F laser with λ = 496 μm [188], and a HCOOH laser with λ = 742 μm [189]. The self-modulation regimes, including the chaotic ones, were obtained also in a pulsed ammonia (λ = 12 μm) laser pumped by a TEA CO_2 laser [181, 190–193]. Unlike the CW FIR lasers, in this case the pressure in the ammonia cell was kept at a higher level of 1 to 10 Torr.

Inhomogeneous line broadening lowers the instability threshold [194], making it easier to achieve undamped pulsations in a free-running laser.

[1]Therefore, it does not matter what cavity, either ring or Fabry-Perot, is used.

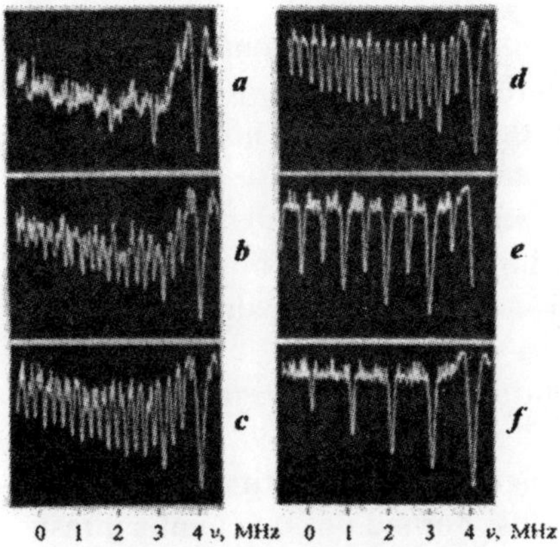

Fig. 1.23. Transition to chaos through period doubling with decreased detuning of an NH_3 laser operated at a wavelength of 81.5 μm. The detuning diminished from (*a*) to (*f*) [178].

This was first found in the operation at a xenon laser at a wavelength 3.51 μm [1,195,196]. Meanwhile, more freedom is given to experimental dealing with a helium–xenon laser due to the possibility of controlling the relationship between the homogeneous and inhomogeneous broadening of Xe, as mentioned above. On the transition in a He-Xe laser the instability and other bifurcations were established in Refs. [197, 198]. The dynamic instability threshold was reached also at λ = 3.39 μm in a He–Ne laser [31].

Some regular features were revealed by investigation of a ring He–Xe laser, in which the travelling-wave regime is ensured using a Faraday isolator [32, 199]. The results of the experiment are given in Fig. 1.24. Our attention is engaged by the low instability threshold and the similar behaviour of the laser when the pumping is changed and when the cavity is tuned. On the route to chaos, a subharmonic component, indicative of the

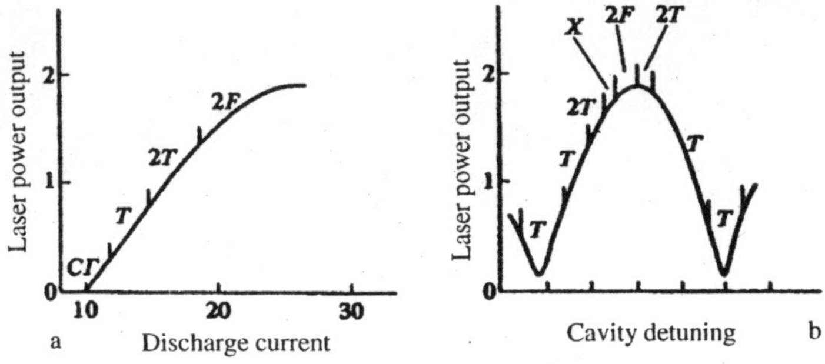

Fig. 1.24. Laser power output vs (*a*) relative excitation for laser tuning near the peak power output and (*b*) cavity detuning for gas fillings at 70 mTorr of xenon and 380 mTorr of helium [32]. Qualitatively different regions of constant intensity output (CW), of periodic pulsations with fundamental (*T*) and double (2*T*) periods, of quasiperiodic (2*F*) and chaotic (*C*) pulsations are shown.

period doubling, first appears in the spectrum and then a component with incommensurate frequency is involved. In [32] it is also found that an increase of the partial helium pressure and, therefore, of the homogeneous broadening of the gain line in the mixture leads to an increase of all bifurcation points of the discharge current as a control parameter. Using optical heterodyning it has been found that pulsations are possible both when the field spectrum has a pronounced central component and when the side components dominate.

Experimentalists have paid much attention to the investigation of the dynamic behaviour of standing-wave He–Xe lasers [197, 198, 200–202]. The bifurcation pattern in this case shows increased complexity, especially near the Doppler line centre where the holes, burned out by counter-running waves, overlap. The main tendencies revealed in these experiments can be formulated as follows:

–the steady-state instability threshold is lowered as the degree of inhomogeneous broadening increases;

–two types of time-dependent processes are observed; there are either smooth pulses or repeated damped pulse trains;

–the diverse scenarios of transition to chaos are possible; the transition from one regular regime to another through an intermediate strip of chaos is natural.

All these experimental facts concern only the free running lasers. They give an outline of each dynamic class. Missing details will be added as required. Knowledge of the general pattern of behaviour shows which methods of controlling the laser behaviour are reasonable in each particular case and help prediction of the results. Through this discussion we have indicated theoretical approaches to laser description that are needed for different phenomena.

Chapter 2

Basic Equations for the Dynamical Behaviour of Lasers

A self-consistent set of laser equations includes equations for the electromagnetic field and the equations describing the state of the medium, which interacts with this field. The complete set is often called the Maxwell-Bloch equations. Roughly the same form, now used in laser dynamics, of this semiclassical set was first written in 1957 by Fain [203] and in 1959 by Oraevsky [204]. In their most general form, these equations are too complicated for all but numerical simulations so that in particular situations one has to use radical simplifications.

There are many versions of the laser equations and some of them will appear in the next chapters. This chapter must give an idea of the main principles of simplifications of the Maxwell–Bloch equations, which make it possible to obtain the dynamical models of specific lasers.

2.1. Equations for the Electromagnetic Field

It is well known that a classical description of the electromagnetic field is fully justified for the dynamics of most phenomena in macroscopic lasers. Therefore, we will use Maxwell's equations as a basis:

$$\text{(a) } \nabla\times\mathbf{E} = -\frac{1}{c}\frac{\partial\mathbf{B}}{\partial t}, \quad \text{(b) } \nabla\times\mathbf{H} = \frac{4\pi\mathbf{j}}{c}+\frac{1}{c}\frac{\partial\mathbf{D}}{\partial t}, \tag{2.1}$$

$$\text{(c) } \nabla\cdot\mathbf{D}=0, \qquad \text{(d) } \nabla\cdot\mathbf{B}=0$$

The material equations will be written now in the usual form:

$$\text{(a) } \mathbf{j}=\sigma\mathbf{E}, \quad \text{(b) } \mathbf{B}=\mathbf{H}+4\pi\mathbf{M}, \quad \text{(c) } \mathbf{D}=\mathbf{E}+4\pi\mathbf{P}. \tag{2.2}$$

Conductivity σ is indicative of the bulk losses in the medium. Magnetization $\mathbf{M}$ and polarization $\mathbf{P}$ both split, in general, into two parts. The first part accounts for the nonresonant contribution of all the molecules (atoms) of the medium (the host) and can be presented as

$$\mathbf{P}_1 = \frac{\varepsilon - 1}{4\pi}\mathbf{E}, \quad \mathbf{M}_1 = \frac{\mu - 1}{4\pi}\mathbf{H}. \tag{2.3}$$

In what follows we will consider only nonmagnetic materials for which $\mu = 1$. Generally speaking, the dielectric constant of the medium can depend on the total intensity but the nonlinearity will not be taken into account for a while. The second part, at least for the polarization component, is solely due to the resonant interaction of the field with the active medium. This term is of great interest and the governing equations for it will be discussed below in Section 2.2.

2.1.1. Wave Equation

Consider the electromagnetic field in a weakly absorbing dielectric with active impurity atoms, which have transitions between the energy levels, which are assumed to be allowed under electric dipole approximation.

Differentiating Eq. (2.1b) with respect to time, we substitute the values $\partial\mathbf{H}/\partial t$ from Eq. (2.1a) and $\mathbf{j}$ from Eq. (2.2a). Bearing in mind that $\mathbf{B}=\mathbf{H}$, we arrive at

$$\nabla\times\nabla\times\mathbf{E} = \nabla(\nabla\cdot\mathbf{E}) - \nabla^2\mathbf{E} = -\frac{4\pi\sigma}{c^2}\frac{\partial\mathbf{E}}{\partial t} - \frac{1}{c^2}\frac{\partial^2\mathbf{D}}{\partial t^2} \tag{2.4}$$

Using Eq. (2.2c) we can write electric induction $\mathbf{D}$ as

$$\mathbf{D} = \mathbf{E} + 4\pi(\mathbf{P}_1 + \mathbf{P}) = \varepsilon(\mathbf{E} + 4\pi\mathbf{P}/\varepsilon)$$

and transform Eq. (2.4) to

$$\nabla^2\mathbf{E} - \nabla(\nabla\cdot\mathbf{E}) = \frac{4\pi\sigma}{c^2}\frac{\partial\mathbf{E}}{\partial t} + \frac{\varepsilon}{c^2}\frac{\partial^2}{\partial t^2}\left(\mathbf{E} + \frac{4\pi\mathbf{P}}{\varepsilon}\right). \tag{2.5}$$

If the medium is spatially homogeneous, then $\nabla(\nabla\cdot\mathbf{E})$ by virtue of Eq. (2.1c). Saturation effects slightly complicate the matter since the field inhomogeneities lead to an inhomogeinity of the medium. However, quantum electronics operates with wave beams of the form

$$\mathbf{E}(\mathbf{r},t) = \mathbf{E}_0(\mathbf{r},t)\exp[-i(\omega t - kz)],$$

$$\mathbf{P}(\mathbf{r},t) = \mathbf{P}_0(\mathbf{r},t)\exp[-i(\omega t - kz)]. \tag{2.6}$$

The complex amplitude is a slowly varying function of space coordinates and time, so that Eqs. (2.6) describe a monochromatic wave beam of limited, although large compared to the wavelength, cross-section, which propagates along the z-axis. Saying 'slowly varying', we mean that the scale of time for variation in phase and amplitude greatly exceeds the oscillation period and the characteristic scale of the spatial structure of the beam is much larger than the wavelength. This last fact and the fact that the wave is transverse suggest an inequality

$$\nabla(\nabla \cdot \mathbf{E}) << \nabla^2 \mathbf{E}\,.$$

The possibility of neglecting the term $\nabla(\nabla \cdot \mathbf{E})$ in the wave equation, describing the propagation of a wave beam in a weakly nonlinear medium, was shown in [205, 206]. The slow variations of the wave amplitude and phase lead to the inequalities

$$\left|\frac{1}{\mathbf{E}_0}\frac{\partial \mathbf{E}_0}{\partial t}\right| << \omega, \quad \left|\frac{1}{\mathbf{E}_0}\frac{\partial \mathbf{E}_0}{\partial z}\right| << k\,,$$

which make it possible to discard, by differentiation, a few terms:

$$\nabla^2 \mathbf{E} \approx \left(\nabla_\perp^2 \mathbf{E}_0 - k^2 \mathbf{E}_0 + 2ik\frac{\partial \mathbf{E}_0}{\partial z}\right)\exp[-i(\omega t - kz)]\,,$$

$$\frac{\partial \mathbf{E}}{\partial t} \approx -i\omega \mathbf{E}_0 \exp[-i(\omega t - kz)]\,,$$

$$\frac{\partial^2 \mathbf{E}}{\partial t^2} \approx -\left(\omega^2 \mathbf{E}_0 + 2i\omega\frac{\partial \mathbf{E}_0}{\partial t}\right)\exp[-i(\omega t - kz)]\,, \tag{2.7}$$

$$\frac{\partial^2 \mathbf{P}}{\partial t^2} \approx -\omega^2 \mathbf{P}_0 \exp[-i(\omega t - kz)]\,.$$

Inserting Eqs. (2.7) into the wave equation (2.5), we arrive at a parabolic equation

$$\nabla_\perp^2 \mathbf{E}_0 + 2i\left(k\frac{\partial \mathbf{E}_0}{\partial z} + \frac{\omega}{c'^2}\frac{\partial \mathbf{E}_0}{\partial t}\right) + \frac{4\pi i\omega}{c^2}\sigma \mathbf{E}_0 + \left(\frac{\omega^2}{c'^2} - k^2\right)\mathbf{E}_0 = -\frac{4\pi\omega^2}{c^2}\mathbf{P}_0\,. \tag{2.8}$$

The latter assumes its simplest form when the field and the medium can be considered homogeneous in a plane perpendicular to the wave propagation direction and when the dispersion law $\omega = c'k$ is valid:

$$\frac{\partial \mathbf{E}_0}{\partial z} + \frac{1}{c'}\frac{\partial \mathbf{E}_0}{\partial t} + \frac{2\pi\sigma}{c'\varepsilon}\mathbf{E}_0 = 2\pi i\frac{\omega}{c'\varepsilon}\mathbf{P}_0\,. \tag{2.9}$$

Here c' is the velocity of light in the medium.

The fact that ε is different from unity is disregarded quite often in the laser theory. From Eq. (2.9) it is seen that a transition from the case of an atomic system in vacuum ($\varepsilon = 1$) to the case $\varepsilon \neq 1$ is possible through the substitutions:

$$c \to c', \ \sigma \to \sigma/\varepsilon, \ \boldsymbol{P}_0 \to \boldsymbol{P}_0/\varepsilon \tag{2.10}$$

Also, the wave equation retains its form if the active molecules have a relevant magnetic dipole transition. In Eq. (2.5) one should substitute $\boldsymbol{E}$ by $\boldsymbol{H}$ and $\boldsymbol{P}$ by $\boldsymbol{M}$ and remember that magnetization is independent of ε.

2.1.2. Modal Decomposition

The field expansion in eigenfunctions (modes) of the laser cavity (see, for example, [207, 208]) aims at reducing Eqs. (2.1) to a set of ordinary differential equations. Rigorously speaking, the degree of the resultant equations is infinite. Almost always, however, it is possible to restrict oneself to a finite number of equations since there are a finite number of excited modes.

If the eigenfrequencies and eigenfunctions of the cavity are known, the problem is easy in principle. However, the eigenvalues problem is usually solved not for a real but for a similar ideal cavity. For a closed cavity the approximation means that the losses are fully neglected. Actually, the real boundary conditions on the metallized (S_1) and the open (S_2) areas of the surface are replaced by the ideal ones:

$$(\mathbf{n}\times\mathbf{E})_{S_1} = 0, \quad (\mathbf{n}\times\mathbf{H})_{S_2} = 0. \tag{2.11}$$

Here $\mathbf{n}$ is a unit vector normal to the surface. The conductance of the medium in the cavity is represented as a superposition of the eigenfunctions

$$\mathbf{E}(\mathbf{r},t) = \sum e_\lambda(t)\mathbf{E}_\lambda(\mathbf{r}), \quad \mathbf{H}(\mathbf{r},t) = \sum h_\lambda(t)\mathbf{H}_\lambda(\mathbf{r}). \tag{2.12}$$

The functions $\boldsymbol{E}_\lambda(\boldsymbol{r})$ and $\boldsymbol{H}_\lambda(\boldsymbol{r})$ are orthogonal and satisfy the normalization conditions

$$\int_{V_c} \mathbf{E}_\lambda \mathbf{E}_\mu \mathrm{d}V = V_c \delta_{\lambda\mu}, \quad \int_{V_c} \mathbf{H}_\lambda \mathbf{H}_\mu \mathrm{d}V = V_c \delta_{\lambda\mu}. \tag{2.13}$$

The ideal cavity functions themselves satisfy the equations

$$\nabla^2 \mathbf{E}_\lambda + k_\lambda^2 \mathbf{E}_\lambda = 0, \quad \nabla^2 \mathbf{H}_\lambda + k_\lambda^2 \mathbf{H}_\lambda = 0, \tag{2.14}$$

and the time-dependent expansion coefficients satisfy the equations

$$\frac{\mathrm{d}^2 e_\lambda}{\mathrm{d}t^2} + \omega_{c\lambda}^2 e_\lambda = 0, \quad \frac{\mathrm{d}^2 h_\lambda}{\mathrm{d}t^2} + \omega_{c\lambda}^2 h_\lambda = 0. \tag{2.15}$$

The latter oscillate harmonically with frequencies $\omega_{c\lambda} = k_\lambda c$. Bearing this in mind, it is easy to obtain from Eq. (2.1) another pair of equations, which links the eigenfunctions:

$$\nabla\times\mathbf{E}_\lambda = ik_\lambda\mathbf{H}_\lambda, \quad \nabla\times\mathbf{H}_\lambda = -ik_\lambda\mathbf{E}_\lambda. \tag{2.16}$$

We now turn to free oscillations in a real cavity. In the absence of appreciable bulk losses, the field in such a cavity can be approximated by a series (2.12) over the functions of an appropriate ideal cavity. The distinguishing features due to the difference of the boundary conditions from Eq. (2.11) will be seen when the other functions are expanded. In order to find the expansion for $\nabla\times\mathbf{E}$ we make use the vector identity

$$\nabla(\mathbf{E}\times\nabla\times\mathbf{E}_\lambda) = (\nabla\times\mathbf{E})(\nabla\times\mathbf{E}_\lambda) - \mathbf{E}\cdot\nabla\times\nabla\times\mathbf{E}_\lambda = ik_\lambda\mathbf{H}_\lambda\cdot\nabla\times\mathbf{E} - k_\lambda^2\mathbf{E}\mathbf{E}_\lambda.$$

Integrating this identity over the volume and using the Gauss divergence theorem we arrive at

$$\int_S \mathbf{n}(\mathbf{E}\times\mathbf{H}_\lambda)\mathrm{d}S = \int_V \mathbf{H}_\lambda(\nabla\times\mathbf{E})\mathrm{d}V + iV_c k_\lambda e_\lambda.$$

Bearing in mind that only the metallized areas of the boundary contribute to the surface integral we find

$$\nabla\times\mathbf{E} = \frac{1}{V_c}\sum\mathbf{H}_\lambda\int_V \mathbf{H}_\lambda(\nabla\times\mathbf{E})\mathrm{d}V = \sum\mathbf{H}_\lambda\left[-ik_\lambda e_\lambda + \frac{1}{V_c}\int_{S_1}(\mathbf{n}\times\mathbf{E})\mathbf{H}_\lambda\mathrm{d}S\right]. \tag{2.17}$$

In the similar way, it is easy to see that the relation

$$\nabla\times\mathbf{H} = \sum\mathbf{E}_\lambda\left[ik_\lambda h_\lambda + \frac{1}{V_c}\int_{S_1}(\mathbf{n}\times\mathbf{H})\mathbf{E}_\lambda\mathrm{d}S\right] \tag{2.18}$$

is valid. In the transition to an ideal cavity the surface integrals in Eqs. (2.17) and (2.18) are zero since the boundary conditions (2.11) are true.

Introducing the expansions (2.12), (2.17) and (2.18) into Eqs. (2.1) and making use of the orthogonality relations (2.13) we get

$$\frac{\mathrm{d}e_\lambda}{\mathrm{d}t} + 4\pi\sigma e_\lambda - ick_\lambda h_\lambda = \frac{c}{V_c}\int_{S_2}(\mathbf{n}\times\mathbf{H})\mathbf{E}_\lambda\mathrm{d}S,$$

$$\frac{\mathrm{d}h_\lambda}{\mathrm{d}t} - ick_\lambda e_\lambda = -\frac{c}{V_c}\int_{S_2}(\mathbf{n}\times\mathbf{E})\mathbf{H}_\lambda\mathrm{d}S. \tag{2.19}$$

Then, differentiating the first equation with respect to time we eliminate the function $\mathrm{d}h_\lambda/\mathrm{d}t$ with the aid of the second equation and arrive at

$$\frac{d^2 e_\lambda}{dt^2} + 4\pi\sigma \frac{de_\lambda}{dt} + c^2 k_\lambda^2 e_\lambda = -i\frac{c^2 k_\lambda}{V_c}\int_{S_1}(\mathbf{n}\times\mathbf{E})\mathbf{H}_\lambda dS + \frac{c}{V_c}\frac{d}{dt}\int_{S_2}(\mathbf{n}\times\mathbf{H})\mathbf{E}_\lambda dS\,. \tag{2.20}$$

We now transform the integral over the surface S_1, where Leontovich's boundary condition [209] is fulfilled,

$$\mathbf{n}\times\mathbf{E} = Z\mathbf{H}\,, \tag{2.21}$$

where Z denotes a surface impedance equal to $(\mu/\varepsilon)^{1/2}$ for a metal. Expressing the field $\mathbf{H}$ in the form of series (2.12) we substitute Eq. (2.21) into Eq. (2.20). Accurate to negligible terms, from Eqs. (2.19) it follows: $h_\lambda = -(i/ck_\lambda)de_\lambda/dt$ and, therefore,

$$\frac{ic^2 k_\lambda}{V_c}\int_{S_1}(\mathbf{n}\times\mathbf{E})\mathbf{H}_\lambda dS = \frac{ck_\lambda}{V_c} Z\sum_\mu\left(\frac{de_\mu/dt}{k_\mu}\int_{S_1}\mathbf{H}_\lambda\mathbf{H}_\mu dS\right). \tag{2.22}$$

The term on the right-hand side with $\mu = \lambda$ yields the damping of free oscillations of the λ-th mode, which is caused by the absorption in the cavity walls. The other terms have the sense of driving forces acting on a given mode from caused by all of the other modes. This mode coupling through the absorbing boundaries results from the use of the set of eigenfunctions of an ideal cavity for the field expansion in a real cavity.

A linear coupling between the modes of a real cavity is absent if the basis functions possess the orthogonality property on the walls:

$$\int_{S_1}\mathbf{H}_\mu\mathbf{H}_\lambda dS = I\delta_{\lambda\mu}\,. \tag{2.23}$$

The condition (2.23) is satisfied, for example, for the modes of a rectangular cavity. Even without an orthogonality relation, however, the decoupling of Eqs. (2.20) is promoted by the fact that the eigenfrequencies are not equal. If the frequency spacing between the modes exceeds the passband of the individual mode then the mode interaction can usually be neglected. Hence, we can retain one term with $\mu = \lambda$ in Eq. (2.22). When the mode coupling is appreciable, the basis system of functions is inadequate for the cavity in question. The quantity

$$Q_S = \frac{V_c k_\lambda}{Z\int_{S_1}|\mathbf{H}_\lambda|^2 dS}$$

represents the cavity Q-factor due to the absorption by the walls of the cavity. Using the Q-factor, we write the right part of Eqs. (2.22) as $(\omega_{c\lambda}/Q_{S\lambda})de_\lambda/dt$.

In exactly the same way we can make use of the integral over the surface S_2, which enters Eqs. (2.20), and write

$$-\frac{c}{V_c}\frac{d}{dt}\int_{S_2}(\mathbf{n}\times\mathbf{H})\mathbf{E}_\lambda dS = \frac{\omega_{c\lambda}}{Q_c}\frac{de_\lambda}{dt}.$$

The Q-factor of mode coupling, Q_c, describes the cavity losses from emission through the holes in the cavity walls.

The only thing that remains to be done is to introduce the notion of Q-factor for the bulk losses in the medium within the cavity,

$$Q_V = \frac{\omega_{c\lambda}}{4\pi\sigma},$$

and the notion of the net Q-factor,

$$Q = \frac{1}{Q_S^{-1} + Q_c^{-1} + Q_V^{-1}}. \tag{2.24}$$

and then to rewrite Eqs. (2.20)

$$\frac{d^2 e_\lambda}{dt^2} + \frac{\omega_{c\lambda}}{Q_\lambda}\frac{de_\lambda}{dt} + \omega_{c\lambda}^2 e_\lambda = 0. \tag{2.25}$$

Its solutions are the oscillations damped at a rate

$$\kappa = \frac{\omega_c}{2Q}. \tag{2.26}$$

Also, we should follow this scheme to obtain an equation describing the oscillations forced by the laser medium in the cavity. Using the expansions (2.12), (2.17) and (2.18), Eqs. (2.1) can be reduced to

$$\frac{d^2 e_\lambda}{dt^2} + \frac{\omega_{c\lambda}}{Q_\lambda}\frac{de_\lambda}{dt} + \omega_{c\lambda}^2 e_\lambda = -\frac{4\pi}{V_c}\int_{V_c}\frac{d^2\mathbf{P}}{dt^2}\mathbf{E}_\lambda dV. \tag{2.27}$$

The problem of excitation of open resonators is formulated somewhat differently [210]. The specific features are exhibited in the boundary conditions, which are defined not on a closed surface but only on the reflecting areas (mirrors). We should add the condition that specifies the field at infinity. An admissible approximation, which can be used when the eigenfunctions are sought, is to neglect the losses at the mirrors. The emission from the space between the mirrors should by no means be neglected, since for this reason all the eigenmodes of the optical cavity are damped, and they fall into two groups according to the damping power. There is a relatively small group of weakly damped modes with low diffraction losses. They are of greatest interest and should be distinguished in the field expansion

$$\mathbf{E} = \sum e_\lambda(t)\mathbf{E}_\lambda(\mathbf{r}) + \mathbf{E}_{damp}.$$

The term $\mathbf{E}_{damp}$ represents a group of strongly damped modes and this continuum can be omitted in what follows.

Following in the main the same procedure as in the case of a cavity resonator but remembering that the integration refers only to the surface S_1 and that radiation damping (diffraction losses) is taken into account through the complexity of the eigenfrequencies, we again arrive at Eqs. (2.27). By $\omega_{c\lambda}$ we denote the real part of the eigenfrequency, the imaginary part of which is included into Q_{diffr}. The net Q-factor of an open resonator consists of three parts describing the diffraction losses and the mirror losses. The latter include, besides the absorption in the mirror material, the removal of some of the field for its intended use. Decoupling of the equations is possible since the modes of open resonators, which have a different transverse structure (transverse modes), satisfy the condition of orthogonality at the mirrors while the modes differing in the number of half wavelengths between the mirrors (longitudinal or axial modes) are well spaced from each other in frequency.

Deriving Eq. (2.27) we take into account the corrections to the eigenvalues of the frequency of an ideal resonator. Corrections to the eigenfunctions are absent in this approximation.

Since the field damping in a cavity is small and the exciting sources are of moderate intensity, it is possible to single out a small parameter $\mu = \omega / Q$ and to present the equation in the form

$$\frac{\mathrm{d}^2 e_\lambda}{\mathrm{d}t^2} + \omega_{c\lambda}^2 e_\lambda = -\mu \left[\frac{\mathrm{d}e_\lambda}{\mathrm{d}t} - \frac{Q_\lambda}{\omega_{c\lambda}} \int_{V_c} \frac{\mathrm{d}^2 \mathbf{P}}{\mathrm{d}t^2} \mathbf{E}_\lambda \mathrm{d}V \right]. \tag{2.28}$$

Its solutions are nearly harmonic oscillations with a slowly varying amplitude and phase [211, 212]

$$e_\lambda = \frac{1}{2}[F_\lambda(t)\exp(-i\omega t) + F_\lambda^*(t)\exp(i\omega t)]. \tag{2.29}$$

Substituting Eq. (2.29) in Eq. (2.27), neglecting small terms of order μ^2 and averaging over the oscillation period $T = 2\pi/\omega$ we arrive at the abbreviated equations

$$\frac{\mathrm{d}F_\lambda}{\mathrm{d}\tau} + [\kappa + i(\omega_{c\lambda} - \omega)]F_\lambda = 4\pi i \omega_\lambda P_\lambda. \tag{2.30}$$

The complex amplitude of the m-th polarization component is introduced through the equality

$$\frac{1}{V_c} \int_{V_c} \mathbf{P}\mathbf{E}_\lambda \mathrm{d}V = P_\lambda(t)\exp(-i\omega t) + P_\lambda^*(t)\exp(i\omega t). \tag{2.31}$$

The reference frequency remains undetermined for the present. Its choice depends on the problem to be resolved. One should only bear in mind the obligatory requirement

$$|\omega - \omega_{c\lambda}| << \omega .$$

2.1.3. Ring-Cavity Field Equations

In the previous section we have considered the cavities, whose modes are standing waves. The cavity resonators and the open Fabry–Perot resonators are among them. The ring resonators, whose modes are counter-running travelling waves, are also often used in practice. The specificity of these cavities is connected with frequency degeneration of counter-running waves. Due to this the weak rescattering of the waves into each other by microinhomogeneities of the cavity optical elements or by the objects located in the beam outside the cavity can be critical. Clearly, a more delicate approach is needed for taking into account the coupling between counter-running waves than that which is used for the standing waves.

Let us return to Eq. (2.5). Averaging over time we obtain, in the one-dimensional case, the abbreviated equation

$$c^2 \frac{\partial^2 F}{\partial z^2} = -4\pi i\omega\sigma F - \varepsilon\left(\omega^2 F + 2i\omega\frac{\partial F}{\partial t}\right) - 4\pi\omega^2 P . \qquad (2.32)$$

We now represent the variables as a superposition of counter-running waves:

$$F = F_1 e^{ikz} + F_{-1} e^{-ikz}, \quad P = P_1 e^{ikz} + P_{-1} e^{-ikz} . \qquad (2.33)$$

In substituting Eq. (2.33) into Eq. (2.32) we bear in mind the smallness of the field amplitude variation over the perimeter of a high-Q resonator. Averaging over space we find

$$\frac{dF_{\pm 1}}{dt} + \left(\frac{\omega^2 - c^2k^2 L/L'}{2i\omega} + \frac{2\pi}{L'}\int_0^L \sigma dz\right) F_{\pm 1} = 2\pi i\omega P_{\pm 1} + F_{\mp 1}\frac{i\omega}{2L'}\int_0^L\left(\varepsilon + i\frac{4\pi\sigma}{\omega}\right)e^{\mp 2ikz} dz \qquad (2.34)$$

The effective length of the cavity is determined by

$$L' = L + L_a(\sqrt{\varepsilon} - 1), \qquad (2.35)$$

where L is the cavity length, L_a is the length of the active element. The wave number k belongs to the discrete set of eigenvalues defined by the cyclic condition $F(z+L,t) = F(z,t)$. Thus, $ckL/L' = \omega_c$ is the frequency eigenvalue. Then we notice that the terms

$$\kappa = \frac{2\pi}{L'}\int_0^L \sigma \mathrm{d}z, \quad \xi_{\pm} = \frac{\omega}{L'}\int_0^L \left(\varepsilon + i\frac{4\pi\sigma}{\omega}\right) e^{\pm 2ikz}\mathrm{d}z \tag{2.36}$$

have the meaning of the loss coefficient and the mode coupling coefficients. Eq. (2.34) can be rewritten in a more compact form

$$\frac{\mathrm{d}F_{\pm 1}}{\mathrm{d}t} + [\kappa - i(\omega - \omega_c)]F_{\pm 1} = 2\pi i\omega P_{\pm 1} + \frac{i}{2}\xi_{\mp}F_{\mp 1}. \tag{2.37}$$

The main contribution to $\xi_{\pm}$ is due to spatial inhomogeneities on the scale of order of the wavelength. The dielectric constant and conductivity inhomogeneities are responsible for phase shifts of the scattered waves, which differ by $\pi/2$. Consequently, the phase difference of the mode coupling coefficients ξ_+ and ξ_- is equal to zero in the first case and equal to π in the second case. Sometimes these are referred to as the cases of complex conjugate and anticomplex-conjugate coupling [160, 213–215].

The apparent generalization of Eqs. (2.37) extends their applicability to the case where the wave characteristics are not the same in the opposite directions. Putting a Faraday cell in combination with polarizers into the laser cavity cancels the loss degeneracy (amplitude nonreciprocity) and the phase velocity degeneracy (phase nonreciprocity) of waves. Cavity rotation with respect to the axis perpendicular to the cavity plane also leads to nonreciprocity. Therefore, the laser equations contain two loss coefficients, κ_+ and κ_-, and two eigenfrequencies, ω_c^+ and ω_c^-. Thus, instead of Eq. (2.37) we have

$$\frac{\mathrm{d}F_{\pm 1}}{\mathrm{d}t} + [\kappa_{\pm} - i(\omega - \omega_c^{\pm})]F_{\pm 1} = 2\pi i\omega P_{\pm 1} + \frac{i}{2}\xi_{\mp}F_{\mp 1}. \tag{2.38}$$

2.2. Equations for the Dynamics of the Material

It is a natural simplification in describing a medium interacting with a field to distinguish a dynamical system with a finite number of degrees of freedom. The latter is formed by only those molecules, which resonantly interact with the field (active molecules), and only those levels are specified in the energy spectrum, between which the transitions are induced. The effect on the dynamical system from the remaining environment (thermal reservoir) is considered as a perturbation tending to equilibrium in this system.

The dynamical system that interacts with the thermal reservoir, from the point of view of quantum mechanics, is, in fact, in a mixed state. The density matrix formalism is most useful to describe this system. The medium polarization is expressed through the density matrix as

$$\mathbf{P} = \frac{1}{\Delta V} \sum \mathrm{Tr}(\mathbf{d}_j \rho_j) .$$

The summation is taken over all the molecules with a physically infinitesimal volume ΔV. This summation can be treated as averaging $\mathrm{Tr}(\mathbf{d}_j \rho_j)$ over such a volume and it can be written, if the density matrix (statistical operator) $\rho(t,\mathbf{r})$ is considered as a continuous function of coordinates, in the form

$$\mathbf{P} = N_S \mathrm{Tr}(\mathbf{d}\rho) . \tag{2.39}$$

Here N_S denotes the number of molecules in a unit volume. The product ρN_S has the sense of the volume density of a statistical operator.

2.2.1. Master Equations

The differential equations describing the evolution of the density matrix elements, in which the relaxation terms are presented due to averaging over the reservoir states, are sometimes called the quantum kinetic equations. These equations can be written in the form [3,5,6]:

$$\frac{\mathrm{d}\rho_{mn}}{\mathrm{d}t} + (\gamma_{mn} + i\omega_0^{mn})\rho_{mn} = \frac{i}{\hbar}\mathbf{E}\sum_q (\mathbf{d}_{mq}\rho_{qn} - \mathbf{d}_{qn}\rho_{mq}) , \tag{2.40a}$$

$$\frac{\mathrm{d}\rho_{mm}}{\mathrm{d}t} + \sum_q (w_{mq}\rho_{mm} - w_{qm}\rho_{qq}) = \frac{i}{\hbar}\mathbf{E}\sum_q (\mathbf{d}_{mq}\rho_{qm} - \mathbf{d}_{qm}\rho_{mq}) . \tag{2.40b}$$

One should also bear in mind the conditions for the normalization and the Hermitian character of the density matrix:

$$\sum \rho_{mm} = 1, \quad \rho_{mn} = \rho_{nm}^* . \tag{2.41}$$

We have used the notation: ρ_{mn} is the density matrix element; $\boldsymbol{d}_{mn}$ is the matrix element of the dipole operator (dipole moments); γ_{mn} is the relaxation rate of the off-diagonal density matrix element; w_{mn} is the probability of a relaxation transition between the indicated energy levels; ω_0^{mn} is the frequency of this transition; $\boldsymbol{E}$ is the intensity of the electric component of the radiation field (we consider only electric dipole transitions).

Let us represent the field as a sum of quasi-monochromatic components, the frequency of which, ω_{mn}, is nearly the same as the molecular transition frequencies ω_0^{mn}:

$$\mathbf{E}(t,\mathbf{r})=\frac{1}{2}\sum[\mathbf{F}_{mn}(t,\mathbf{r})\exp(-i\omega_{mn}t)+\mathbf{F}^{*}_{mn}(t,\mathbf{r})\exp(i\omega_{mn}t)]. \quad (2.42)$$

The elements of the density matrix of a molecule in this field can be presented as

$$\rho_{mn}(t)=\sigma_{mn}(t)\exp(-i\omega_{mn}t), \quad (2.43)$$

provided that the inequalities

$$\frac{|\mathbf{dE}|}{\hbar}<<\omega, \quad \gamma_{mn}<<\omega \quad (2.44)$$

are satisfied. The complex amplitude of the matrix element σ_{mn} is a slow variable. Substituting Eqs. (2.42) and (2.43) into Eq. (2.40) and averaging over the oscillation period we arrive at

$$\frac{\mathrm{d}\sigma_{mn}}{\mathrm{d}t}=-[\gamma_{mn}-i(\omega_{mn}-\omega_0^{mn})]\sigma_{mn}+\frac{i}{2\hbar}\sum_q(\mathbf{d}_{mq}\mathbf{F}_{mq}\sigma_{qn}-\mathbf{d}_{qn}\mathbf{F}_{qn}\sigma_{mq}), \quad (2.45a)$$

$$\frac{\mathrm{d}\sigma_{mm}}{\mathrm{d}t}=-\sum_q\left[w_{mq}\sigma_{mm}-w_{qm}\sigma_{qq}+\frac{i}{2\hbar}(\mathbf{d}_{mq}\mathbf{F}_{mq}\sigma_{qm}-\mathbf{d}_{qm}\mathbf{F}_{qm}\sigma_{mq})\right], \quad (2.45b)$$

The variables and coefficients of Eqs. (2.45) depend on two indices and, generally speaking, they are sensitive to the sequence, in which these indices are written:

$$\sigma_{mn}=\sigma^{*}_{nm};\quad \mathbf{d}_{mn}=\mathbf{d}^{*}_{nm};\quad \mathbf{F}_{mn}=\mathbf{F}^{*}_{nm};\quad \omega_{mn}=-\omega_{nm};\quad w_{mn}=w_{nm}\exp(\hbar\omega_{mn}/k_B T).$$

The only exception is $\gamma_{mn}=\gamma_{nm}$.

2.2.2. Two-Level Medium

The approximation is such that we choose two levels from the full set of the energy levels of the medium and consider only the transitions between them. Then Eqs. (2.45) transform to

$$\frac{\mathrm{d}\sigma_{21}}{\mathrm{d}t}=-[\gamma_{21}-i(\omega-\omega_0)\sigma_{21}-\frac{i}{2\hbar}\mathbf{d}_{21}\mathbf{F}_{21}(\sigma_{22}-\sigma_{11}), \quad (2.46a)$$

$$\frac{\mathrm{d}\sigma_{11}}{\mathrm{d}t}=-w_{12}\sigma_{11}+w_{21}\sigma_{22}+\frac{i}{2\hbar}(\mathbf{d}_{12}\mathbf{F}_{12}\sigma_{21}-\mathbf{d}_{21}\mathbf{F}_{21}\sigma_{12}). \quad (2.46b)$$

$$\frac{d\sigma_{22}}{dt} = w_{12}\sigma_{11} - w_{21}\sigma_{22} - \frac{i}{2\hbar}(\mathbf{d}_{12}\mathbf{F}_{12}\sigma_{21} - \mathbf{d}_{21}\mathbf{F}_{21}\sigma_{12}). \quad (2.46c)$$

Instead of two parameters, w_{mn}, we can use a single time of relaxation $T_1 = \sigma_0^{mm} / w_{mn}$ [6], where σ_0^{mm} belongs to the thermal equilibrium state. Slightly changing the notation: $D = \sigma_{22} - \sigma_{11}$, $\gamma_{21} = \gamma_\perp$, $\sigma_{21} = \sigma$, taking into account Eqs. (2.41) and omitting the unnecessary transition indices we obtain, instead of Eqs. (2.46), a set of two equations

$$\frac{d\sigma}{dt} + [\gamma_\perp - i(\omega - \omega_0)]\sigma = -\frac{i}{2\hbar}\mathbf{dF}D, \quad (2.47a)$$

$$\frac{dD}{dt} + \frac{D - D^{(0)}}{T_1} = -\frac{i}{\hbar}(\mathbf{d}^*\mathbf{F}^*\sigma - \mathbf{dF}\sigma^*). \quad (2.47b)$$

Here $D^{(0)}$ is the unsaturated value of D, which corresponds to $\mathbf{F} = 0$.

2.2.3. Three-Level Medium: Coherent Pumping

If three levels are included, many variations are possible. Therefore, we specialize the equations to a particular situation. Let us consider the scheme shown in Fig. 1.2b, for which Eqs. (2.45) reduce to

$$\frac{d\sigma_{32}}{dt} + [\gamma_{32} - i(\omega_{32} - \omega_0^{32})]\sigma_{32} = \frac{i}{2\hbar}[\mathbf{d}_{31}\mathbf{F}_{31}\sigma_{12} - \mathbf{d}_{32}\mathbf{F}_{32}(\sigma_{33} - \sigma_{22})],$$

$$\frac{d\sigma_{31}}{dt} + [\gamma_{31} - i(\omega_{31} - \omega_0^{31})]\sigma_{31} = \frac{i}{2\hbar}[\mathbf{d}_{31}\mathbf{F}_{31}(\sigma_{11} - \sigma_{33}) + \mathbf{d}_{32}\mathbf{F}_{32}\sigma_{21}],$$

$$\frac{d\sigma_{21}}{dt} + [\gamma_{21} - i(\omega_{31} - \omega_{32} - \omega_0^{21})]\sigma_{21} = \frac{i}{2\hbar}[\mathbf{d}_{23}\mathbf{F}_{23}\sigma_{31} - \mathbf{d}_{31}\mathbf{F}_{31}\sigma_{23}],$$

(2.48)

$$\frac{d\sigma_{33}}{dt} + (w_{31} + w_{32})\sigma_{33} - w_{23}\sigma_{22} - w_{13}\sigma_{11} = \frac{i}{2\hbar}(\mathbf{d}_{31}\mathbf{F}_{31}\sigma_{13} - \mathbf{d}_{13}\mathbf{F}_{13}\sigma_{31} + \mathbf{d}_{32}\mathbf{F}_{32}\sigma_{23} - \mathbf{d}_{23}\mathbf{F}_{23}\sigma_{32}),$$

$$\frac{d\sigma_{22}}{dt} + (w_{23} + w_{21})\sigma_{22} - w_{32}\sigma_{33} - w_{12}\sigma_{11} = \frac{i}{2\hbar}(\mathbf{d}_{23}\mathbf{F}_{23}\sigma_{32} - \mathbf{d}_{32}\mathbf{F}_{32}\sigma_{23}),$$

$$\sigma_{11} + \sigma_{22} + \sigma_{33} = 1,$$

A specific feature of a FIR laser operated by this scheme is that $\hbar\omega_{31} >> k_B T$, whereas $\hbar\omega_{32} << k_B T$ and, therefore, $w_{m1} >> w_{1m}$, $w_{32} \approx w_{23}$. Thus, using the normalization condition and expressing the diagonal matrix elements through $D = \sigma_{33} - \sigma_{22}$, we rewrite Eqs. (2.48) in a somewhat simpler form:

$$\frac{d\sigma_{32}}{dt} + [\gamma_{32} - i(\omega_{32} - \omega_0^{32})]\sigma_{32} = \frac{i}{2\hbar}[\mathbf{d}_{31}\mathbf{F}_{31}\sigma_{12} - \mathbf{d}_{32}\mathbf{F}_{32}D],$$

$$\frac{d\sigma_{31}}{dt} + [\gamma_{31} - i(\omega_{31} - \omega_0^{31})]\sigma_{31} = \frac{i}{2\hbar}[\mathbf{d}_{31}\mathbf{F}_{31}(1 - 2D - 3\sigma_{22}) + \mathbf{d}_{32}\mathbf{F}_{32}\sigma_{21}]$$

$$\frac{d\sigma_{21}}{dt} + [\gamma_{21} - i(\omega_{31} - \omega_{32} - \omega_0^{21})]\sigma_{21} = \frac{i}{2\hbar}[\mathbf{d}_{23}\mathbf{F}_{23}\sigma_{31} - \mathbf{d}_{31}\mathbf{F}_{31}\sigma_{23}],$$

$$(2.49)$$

$$\frac{dD}{dt} + (w_{31} + 2w_{32})D - (w_{31} - w_{21})\sigma_{22} = \frac{i}{2\hbar}(\mathbf{d}_{31}\mathbf{F}_{31}\sigma_{13} - \mathbf{d}_{13}\mathbf{F}_{13}\sigma_{31} + 2\mathbf{d}_{32}\mathbf{F}_{32}\sigma_{23} - 2\mathbf{d}_{23}\mathbf{F}_{23}\sigma_{32})$$

$$\frac{d\sigma_{22}}{dt} - w_{32}D + w_{21}\sigma_{22} = \frac{i}{2\hbar}(\mathbf{d}_{23}\mathbf{F}_{23}\sigma_{32} - \mathbf{d}_{32}\mathbf{F}_{32}\sigma_{23}).$$

Deriving the maser equations, where all $\hbar\omega_{mn} << k_B T$ and the laser equations, where $\hbar\omega_{mn} >> k_B T$, is recommended as an exercise.

If the coefficients of a set of differential equations, such as (2.49) differ in order of magnitude, then it is possible to simplify them [211]. This was noticed first in the paper [216], devoted to the problem of stability of molecular beam maser. We therefore have the term 'Khokhlov's method', which is used in Russian scientific literature [217]. In the West, this problem originates from Haken [218, 219], being associated with the method of adiabatic elimination of variables [220–222]. The applicability of this method will be discussed in more detail in Section 3.1 in connection with the single-mode laser models. The subject matter is as follows.

When some coefficients are much greater than the others, the variables also may not be equivalently important in the dynamics. Fast-relaxing variables may respond instantly (inertialess reaction) to changes in the state of the system while slow variables cannot. Adiabatic elimination of fast variables means that corresponding derivatives are put equal to zero and that subset of equations, which has become algebraic, is resolved with respect to these variables.

In this particular laser problem the off-diagonal elements of the density matrix can be the fast variables since $\gamma_{mn} \geq w_{mn}$ [223]. Suppose that the coherency is decay rapidly in transitions 3→1 and 2→1, i.e., $\gamma_{31}, \gamma_{21} >> \gamma_{32}, w_{mn}$, and that γ_{m1} exceeds the velocity of any motion in the system including Rabi oscillations with frequencies $\omega_R = \mathbf{dF}/\hbar$ [224]. When the whole set of inequalities

$$\gamma_{31}, \gamma_{21} >> \gamma_{32}, w_{mn}, \omega_R \tag{2.50}$$

is satisfied, we can assume $d\sigma_{31}/dt = d\sigma_{21}/dt = 0$ to adiabatically eliminate σ_{31} and σ_{21} from Eqs. (2.49). We obtain a set of three equations:

$$\frac{d\sigma_{32}}{dt} = -\left[\gamma_{32} + \frac{|\mathbf{d}_{31}\mathbf{F}_{31}|^2}{4\hbar^2\gamma_{21}} - i(\omega_{32} - \omega_0^{32})\right]\sigma_{32} - \frac{i}{2\hbar}\mathbf{d}_{32}\mathbf{F}_{32}D,$$

$$\frac{dD}{dt} = -\left(w_{31} + 2w_{32} + \frac{|\mathbf{d}_{31}\mathbf{F}_{31}|^2}{4\hbar^2\gamma_{31}}\right)D + \frac{|\mathbf{d}_{31}\mathbf{F}_{31}|^2}{\hbar^2\gamma_{31}}$$

$$+\left(w_{21} - w_{31} - \frac{3|\mathbf{d}_{31}\mathbf{F}_{31}|^2}{2\hbar^2\gamma_{31}}\right)\sigma_{22} + \frac{i}{\hbar}(\mathbf{d}_{32}\mathbf{F}_{32}\sigma_{23} - \mathbf{d}_{23}\mathbf{F}_{23}\sigma_{32}). \tag{2.51}$$

$$\frac{d\sigma_{22}}{dt} = w_{32}D - w_{21}\sigma_{22} + \frac{i}{2\hbar}(\mathbf{d}_{23}\mathbf{F}_{23}\sigma_{32} - \mathbf{d}_{32}\mathbf{F}_{32}\sigma_{23}).$$

With a more rigid criterion, that there is no saturation in the pumping transition,

$$|\mathbf{d}_{31}\mathbf{F}_{31}/\hbar|^2 << \gamma_{31}w_{31}, \tag{2.52}$$

the first two equations can be simplified still further:

$$\frac{d\sigma_{32}}{dt} = -[\gamma_{32} - i(\omega_{32} - \omega_0^{32})]\sigma_{32} - \frac{i}{2\hbar}\mathbf{d}_{32}\mathbf{F}_{32}D, \tag{2.53}$$

$$\frac{dD}{dt} = -(w_{31} + 2w_{32})D + (w_{21} - w_{31})\sigma_{22} + \frac{|\mathbf{d}_{32}\mathbf{F}_{31}|^2}{2\hbar^2\gamma_{31}} + \frac{i}{\hbar}(\mathbf{d}_{32}\mathbf{F}_{32}\sigma_{23} - \mathbf{d}_{23}\mathbf{F}_{23}\sigma_{32}).$$

The equality of the population relaxation rates, $w_{21} = w_{31}$, means that Eqs. (2.53) do not contain σ_{22}, i.e., they become a closed set.

2.2.4. Three- and Four-Level Media; Transition to an Equivalent Two-Level Description

All the known ways of producing population inversion, except for laser pumping, are based on incoherent processes and, therefore, permit a probabilistic description. This is the case of a three-level medium, such as ruby, when it is pumped by a flash lamp. Since the level spacing is much larger than $k_B T$ it is possible to neglect upward relaxation transition. Since the laser levels are 1 and 2, in place of Eqs. (2.48) we have

$$\frac{d\sigma_{21}}{dt} + [\gamma_{21} - i(\omega - \omega_0^{21})]\sigma_{21} = \frac{i}{2\hbar}\mathbf{d}_{21}\mathbf{F}_{21}(\sigma_{11} - \sigma_{22})],$$

$$\frac{d\sigma_{33}}{dt} = W_{\text{pump}}(\sigma_{11} - \sigma_{33}) - (w_{31} + w_{32})\sigma_{33}, \tag{2.54}$$

$$\frac{d\sigma_{22}}{dt} = -w_{21}\sigma_{22} + w_{32}\sigma_{33} + \frac{i}{2\hbar}(\mathbf{d}_{21}\mathbf{F}_{21}\sigma_{12} - \mathbf{d}_{12}\mathbf{F}_{12}\sigma_{21}),$$

$$\frac{d\sigma_{11}}{dt} = w_{21}\sigma_{22} + w_{31}\sigma_{33} - W_{\text{pump}}(\sigma_{11} - \sigma_{33}) + \frac{i}{2\hbar}(\mathbf{d}_{12}\mathbf{F}_{12}\sigma_{21} - \mathbf{d}_{21}\mathbf{F}_{21}\sigma_{12}),$$

where $W_{\text{pump}} = |\mathbf{d}_{31}\mathbf{F}_{31}|^2 / 2\hbar^2\gamma_{31}$ is the rate of the transition between levels 1 and 3, induced by pumping.

This model can be reduced to an equivalent two-level one if the adiabatic elimination of σ_{33} is possible. A necessary condition is a high rate for radiationless decays of the third level. If we use ruby, which also has high quantum efficiency for luminescence, then we have

$$w_{32} >> w_{m1}, W_{\text{pump}}.$$

Thus, we can assume $d\sigma_{33}/dt = 0$ and obtain

$$\sigma_{33} \approx \frac{W_{\text{pump}}}{w_{32}}\sigma_{11} << \sigma_{11}, \tag{2.55}$$

which means there is only a small population in level 3. Substituting Eq. (2.55) into the other equations and using the relation $\sigma_{11} + \sigma_{22} \approx 1$ we reduce Eqs. (2.54) to

$$\frac{d\sigma_{21}}{dt} + [\gamma_{21} - i(\omega - \omega_0^{21})]\sigma_{21} = \frac{i}{2\hbar}\mathbf{d}_{21}\mathbf{F}_{21}D, \tag{2.56a}$$

$$\frac{dD}{dt} + \gamma_{\parallel}(D - D^{(0)}) = -\frac{i}{2\hbar}(\mathbf{d}_{12}\mathbf{F}_{12}\sigma_{21} - \mathbf{d}_{21}\mathbf{F}_{21}\sigma_{12}). \tag{2.56b}$$

The inversion relaxation rate

$$\gamma_{\parallel} = \frac{1 + T_1 W_{\text{pump}}}{T_1} \tag{2.57}$$

is defined not only by the decay of upper level but also by the pumping. Without the laser action the inversion is given be the unsaturated quantity

$$D^{(0)} = \frac{W_{\text{pump}}T_1 - 1}{W_{\text{pump}}T_1 + 1}. \tag{2.58}$$

We now turn to a four-level scheme characteristic of neodymium lasers (see Fig. 1.4). Using the same arguments as above we write

$$\sigma_{44} \approx \frac{W_{\text{pump}}}{w_{43}}\sigma_{11}.$$

The changes in populations of the upper and lower levels obey the equations

$$\frac{d\sigma_{33}}{dt} = W_{pump}\sigma_{11} - (w_{32} + w_{31})\sigma_{33} + \frac{i}{2\hbar}(\mathbf{d}_{32}\mathbf{F}_{32}\sigma_{23} - \mathbf{d}_{23}\mathbf{F}_{23}\sigma_{32}), \tag{2.59a}$$

$$\frac{d\sigma_{22}}{dt} = \frac{W_{pump}w_{42}}{w_{43}}\sigma_{11} + w_{32}\sigma_{33} - w_{21}\sigma_{22} - \frac{i}{2\hbar}(\mathbf{d}_{32}\mathbf{F}_{32}\sigma_{23} - \mathbf{d}_{23}\mathbf{F}_{23}\sigma_{32}), \tag{2.59b}$$

under the normalization condition

$$\sigma_{11} + \sigma_{22} + \sigma_{33} = 1.$$

If the rate of decay of the lower laser level greatly exceeds the rates of all processes that lead to its population, then we can assume $\sigma_{22} << \sigma_{11}; \sigma_{33}$ and $D_{32} = D \approx \sigma_{33}$. After such a simplification the laser equations can be written

$$\frac{d\sigma_{32}}{dt} = -[\gamma_{\perp} - i(\omega - \omega_0^{32})]\sigma_{32} - \frac{i}{2\hbar}\mathbf{d}_{32}\mathbf{F}_{32}D, \tag{2.60a}$$

$$\frac{dD}{dt} = -\gamma_{\parallel}(D - D^{(0)}) - \frac{i}{2\hbar}(\mathbf{d}_{23}\mathbf{F}_{23}\sigma_{32} - \mathbf{d}_{32}\mathbf{F}_{32}\sigma_{23}). \tag{2.60b}$$

The effective rate of population inversion relaxation is given by

$$\gamma_{\parallel} = \frac{1 + W_{pump}T_1}{T_1}, \tag{2.61}$$

and the population difference without the laser action, is proportional to

$$D^{(0)} = \frac{W_{pump}T_1}{1 + W_{pump}T_1}. \tag{2.62}$$

In a three-level medium, as it seen from Eq. (2.58), the inversion is achieved under the condition $W_{pump} > T_1^{-1}$, which corresponds to more than a two-fold excess of $\gamma_{\parallel}$ over $1/T_1$. In a four-level medium the condition $W_{pump} > 0$ is sufficient for inversion, and the laser operates at $W_{pump}T_1 << 1$ and $\gamma_{\parallel} \approx 1/T_1$.

Comparison of Eqs. (2.47), (2.56) and (2.60) shows that the only difference among them is due to the coefficient of the last term in the equation for the population difference. In a three-level medium the lower laser level is the ground state, and the emission of a photon is accompanied, as in a two-level medium, with a change of the inversion by two. In a four-level medium, in which the lower level is above the ground state and this level rapidly decays, photon emission corresponds to a change of the inversion by one. Introducing a coefficient β_a, equal to 1 for a four-level medium and to 2 for a three-level medium, we can write the equations in a form combining all the cases we have considered:

$$\frac{\mathrm{d}\sigma}{\mathrm{d}t} = -[\gamma_\perp - i(\omega - \omega_0^{32})]\sigma - \frac{i}{2\hbar}\mathbf{dF}D\,, \tag{2.63a}$$

$$\frac{\mathrm{d}D}{\mathrm{d}t} = -\gamma_\parallel(D - D^{(0)}) - \frac{i\beta_a}{2\hbar}\mathbf{d}(\mathbf{F}^*\sigma - \mathbf{F}\sigma^*)\,. \tag{2.63b}$$

It is assumed that the dipole moment matrix is real, i.e., $\mathbf{d}_{mn} = \mathbf{d}_{nm} = \mathbf{d}$.

The set of Eqs. (2.63) can be considered as a generalization of the well-known equations for paramagnetic resonant systems obtained by Bloch [225] to the case of a two-level system of arbitrary nature. In such an interpretation, the population difference and the polarization have the sense of the longitudinal and the transversal components of some vector (energy spin or the generalized Bloch's vector [223–226]) in a configuration space.

2.2.5. Material Equations Specialized to an Ensemble of Moving Atoms

The motion of atoms is essential to the form of the material equations provided that the atoms move, without changing in state, distances comparable to the wavelength. This condition is not satisfied in condensed media. However, the mean free path of gas atoms can exceed λ; if so, the spatial dispersion has to be taken into account.

The density matrix of a moving atom is a function of time, coordinates and velocities. Assuming that interaction with the electromagnetic field does not change the velocity we find

$$\frac{\mathrm{d}\rho}{\mathrm{d}t} = \frac{\partial\rho}{\partial t} + \mathbf{U}\nabla\rho\,, \tag{2.64}$$

where $\mathbf{U}$ is the velocity of the atom. This operator equation transforms to a set of abbreviated equations for density matrix elements [227]:

$$\frac{\partial\sigma}{\partial t} + \mathbf{U}\nabla\sigma = -[\gamma_\perp - i(\omega - \omega_0)]\sigma - \frac{id}{2\hbar}F(\sigma_a - \sigma_b)\,,$$

$$\frac{\partial\sigma_a}{\partial t} + \mathbf{U}\nabla\sigma_a = -w_a\sigma_a - \frac{id}{2\hbar}(F^*\sigma - F\sigma^*) + W_\mathbf{a}\,, \tag{2.65}$$

$$\frac{\partial \sigma_b}{\partial t} + \mathbf{U}\nabla\sigma_b = -w_b\sigma_b + w_{ab}\sigma_a + \frac{id}{2\hbar}(F^*\sigma - F\sigma^*) + W_b\,.$$

Here *a*, *b* are the indices of the upper and the lower laser levels, $w_{a,b}$ are the probabilities of decay of the atomic states, and w_{ab} is the probability of spontaneous transition from the upper to the lower laser level.

In the model for the laser medium being considered, both laser levels are assumed to be well above the ground state. The pumping that causes the atoms to make transitions from the ground state to the laser levels is represented by the terms $W_{a,b}$, which have the form of excitation rates. The relative population of the ground state remains close to unity. Assuming that an excitation event does not change the atomic velocity distribution function $h(U)$ we have

$$W_j = w_j \Lambda_j(t,\mathbf{r}) h(U)\,. \tag{2.66}$$

The parameter $\Lambda_j(t,\mathbf{r})$ that characterizes the rate of pumping to the j-th level may have a slow dependence on the spatial coordinates and on time.

In derivation of Eqs. (2.65) we used the assumption of none distinct orientation in the medium. This means that the interaction with radiation does not change the atomic dipole angular distribution either because of weak saturation of the medium or fast relaxation of the orientation. Therefore, the mean energy of the interaction can be represented as *dF*, where $d^2 = |\mathbf{d}|^2/3$.

The spatial inhomogeneity of the density matrix elements, at least its small-scale component, is due to the interaction of the gas with the electromagnetic beam. Eqs. (2.65) can be slightly simplified since the transverse structure of the wave beam is much greater than the longitudinal one: $k_\perp << k_z$.

If the characteristic scale of any inhomogeneity is $1/k$, then

$$U\nabla\sigma_j / w_j\sigma_j \approx Uk / w_j\,.$$

Substituting for *U* its most probable value U_0, we obtain the criterion for sufficient smallness of the term $U\nabla\sigma_j$ in the form $k << w_j/U_0$. Using typical values for atomic gases, $w_j \approx 10^8$ s^{-1}, $U_0 \approx 10^5$ cm/s, we come to the conclusion that it is reasonable to retain the gradient terms in Eqs. (2.65) if $k \geq 10^3$ cm^{-1}. The wave numbers that correspond to IR and, of course, to the optical range, satisfy this criterion. Outside the beam focusing region, the transverse wave numbers do not exceed 10^2 cm^{-1}, as a rule; thus, $U\nabla\sigma_j \approx U\partial\sigma_j/\partial z$. Introducing, instead of σ_a and σ_b, the new variables $D = \sigma_a - \sigma_b$ and $S = \sigma_a + \sigma_b$, we write Eqs. (2.65) in the form

$$\frac{\partial\sigma}{\partial t}+U\frac{\partial\sigma}{\partial z}=-[\gamma_\perp-i(\omega-\omega_0)]\sigma-\frac{id}{2\hbar}FD\,, \tag{2.67a}$$

$$\frac{\partial D}{\partial t}+U\frac{\partial D}{\partial z}=-\frac{1}{2}(w_a+w_b+w_{ab})D-\frac{1}{2}(w_a-w_b^*+w_{ab})S-\frac{i}{\hbar}d(F^*\sigma-F\sigma^*)+W_a-W_b \tag{2.67b}$$

$$\frac{\partial S}{\partial t}+U\frac{\partial S}{\partial z}=-\frac{1}{2}(w_a+w_b-w_{ab})S-\frac{1}{2}(w_a-w_b-w_{ab})D+W_a+W_b\,. \tag{2.67c}$$

If $w_b=w_a+w_{ab}$ the variable S is deleted from (2.67b) and first two equations form a closed system

$$\frac{\partial\sigma}{\partial t}+U\frac{\partial\sigma}{\partial z}=-[\gamma_\perp-i(\omega-\omega_0)]\sigma-\frac{id}{2\hbar}FD\,, \tag{2.68a}$$

$$\frac{\partial D}{\partial t}+U\frac{\partial D}{\partial z}=-\gamma_\parallel D-\frac{i}{\hbar}d(F^*\sigma-F\sigma^*)+W_a-W_b\,, \tag{2.68b}$$

which represents a two-level model of the active medium. In this case $\gamma_\parallel=w_b=w_a+w_{ab}$. Since $\gamma_\perp\geq(w_a+w_b)/2$ [228] (the equality corresponds to the natural linewidth), we have $\tilde{\gamma}=\gamma_\parallel/\gamma_\perp\leq 2w_b/(w_a+w_b)$ or $\tilde{\gamma}\leq 1$.

Equations (2.67) can be reduced to ordinary differential equations if the functions are expanded with respect to the basic set of travelling waves:

$$F=\sum_{-\infty}^{\infty}F_\lambda\exp(i\lambda\Delta kz),\quad \sigma_j=\sum_{-\infty}^{\infty}\sigma_{j\lambda}\exp(i\lambda\Delta kz), \tag{2.69}$$

where Δk is the discrete step in the wave number spectrum, which is equal to π/L for a Fabry–Perot resonator and $2\pi/L$ for a ring resonator. By substitution of relations (2.69) and by separation of the terms with equal coordinate dependence, the material equations (2.67) can be transformed to

$$\frac{\partial\sigma_\lambda}{\partial t}+[i(\lambda\Delta kU-\omega+\omega_0)]\sigma_\lambda=-\frac{id}{2\hbar}\sum_\nu F_\nu D_{\lambda-\nu}\,,$$

$$\frac{\partial D_\lambda}{\partial t}+\left[i\lambda\Delta kU+\frac{1}{2}(w_a+w_b+w_{ab})\right]D_\lambda=-\frac{1}{2}(w_a-w_b+w_{ab})S_\lambda-\frac{i}{\hbar}d\sum_\nu(F_\nu^*\sigma_{\nu+\lambda}-F_\nu\sigma_{\nu-\lambda}^*)+W_{a\lambda}-W_{b\lambda} \tag{2.70}$$

$$\frac{\partial S_\lambda}{\partial t}+\left[i\lambda\Delta kU+\frac{1}{2}(w_a+w_b-w_{ab})S_\lambda\right]=-\frac{1}{2}(w_a-w_b-w_{ab})D_\lambda+W_{a\lambda}+W_{b\lambda}\,.$$

In our previous considerations we ignored the fact that collisions between gas atoms change not only the internal states but also the velocities of the colliding particles. Thus, the atom collisions produce spectral cross-relaxation that tends to remove the distortion of the equilibrium inhomogeneous lineshape. To describe this process we introduce terms in the form of a collision integral, into the equations describing the population dynamics (see [31] and the reference cited):

$$\int [\Gamma_j(U',U)\sigma_j(U,z,t) - \Gamma_j(U,U')\sigma_j(U',z,t)]\mathrm{d}U' \qquad (2.71)$$

Here, $\Gamma_j(U',U)\mathrm{d}U'$ is the probability that a gas atom will pass over, due to collision, from the velocity range $U, U+\Delta U$ to the velocities $U', U'+\Delta U'$ unchanged in state.

For simplicity's sake, we will use the strong collisions model, in which the final velocity of the atoms assumes, with equal probability, any arbitrary value within the Doppler profile, and consider only two cross-relaxation parameters, $\Gamma_{a,b}(U',U) = \Gamma_{a,b}h(U)$, reducing Eqs. (2.71) to

$$\frac{1}{2}\Gamma_{a,b}\left[\sigma_{a,b}(U) - h(U)\int \sigma_{a,b}(U')\mathrm{d}U'\right].$$

Finally, disregarding the difference of Γ_a from Γ_b, we have a single parameter $\Gamma = \Gamma_a = \Gamma_b$ and write, instead of Eqs. (2.67), the set of equations

$$\frac{\partial\sigma}{\partial t} + U\frac{\partial\sigma}{\partial z} = -[\gamma_\perp - i(\omega-\omega_0)]\sigma - \frac{id}{2\hbar}FD,$$

$$\frac{\partial D}{\partial t} + U\frac{\partial D}{\partial z} = -\frac{1}{2}(w_a + w_b + w_{ab})D - \frac{1}{2}(w_a - w_b + w_{ab})S,$$

$$-\Gamma[D(U) - h(U)\int D(U')\mathrm{d}U'] - \frac{i}{\hbar}d(F^*\sigma - F\sigma^*) + W_a - W_b,$$

$$(2.72)$$

$$\frac{\partial S}{\partial t} + U\frac{\partial S}{\partial z} = -\frac{1}{2}(w_a + w_b - w_{ab})S - \frac{1}{2}(w_a - w_b - w_{ab})D - \Gamma[S(U) - h(U)\int S(U')\mathrm{d}U'] + W_a + W_b$$

Because of their complexity the use of equations containing cross-relaxation terms has been mainly limited to numerical simulations of the laser processes [31].

2.3. Self-Consistent Semiclassical Set of Laser Equations

In the previous sections we have discussed separately the equations describing the medium polarization in a given electromagnetic field and the radiation field ensured by a given polarization of the atomic system. Combining these equations, i.e., assuming that the polarization, which is the source of the field, is in turn, determined by the same field, we arrive at a self-consistent set of equations describing the laser. In general form this set of equations is written as

$$\nabla\times\nabla\times\mathbf{E}=-\frac{4\pi\sigma}{c^2}\frac{\mathrm{d}\mathbf{E}}{\mathrm{d}t}-\frac{1}{c^2}\frac{\partial^2}{\partial t^2}(\mathbf{E}+4\pi\mathbf{P}),$$

$$\mathbf{P}=N_S\langle\mathrm{Tr}(\mathbf{d}\rho)\rangle, \tag{2.73}$$

$$\frac{\mathrm{d}\rho}{\mathrm{d}t}=\frac{i}{\hbar}[H,\rho]+R.$$

This is often called a semiclassical or quasiclassical set of equations [229, 230] to emphasize that the field equations are of a classical and the laser equations are of a quantum origin.

Let us reduce the problem by assuming that the medium consists of homogeneously broadened identically oriented two-level atoms. We also assume that the field is linearly polarized along the direction of the dipole moment. To describe the propagation of a plane wave in such a medium we will turn to Eq. (2.8). If we speak about an unlimited medium, then the dispersion law $\omega = ck$ is valid. In the case of a ring cavity the cyclic boundary condition

$$E(t,z)=E(t,z+L) \tag{2.74}$$

determines the spectrum of the eigenvalues of propagation constant k_λ, which correspond to the cavity eigenfrequencies $\omega_{c\lambda}=ck_\lambda$. By L in Eq. (2.74) we mean the perimeter of the laser cavity.

Combining Eq. (2.9) with Eq. (2.63) and assuming $\mathbf{d}\parallel\mathbf{F}$ we arrive at a scalar model of a travelling wave laser:

$$c\frac{\partial F}{\partial z}+\frac{\partial F}{\partial t}+[\kappa-i(\omega-\omega_c)]F=2\pi i\omega dN_S\sigma, \tag{2.75a}$$

$$\frac{\partial\sigma}{\partial t}+[\gamma_\perp-i(\omega-\omega_0)]\sigma=-\frac{i}{2\hbar}dFD, \tag{2.75b}$$

$$\frac{\partial D}{\partial t}+\gamma_\parallel(D-D^{(0)})=-\frac{i\beta_a}{2\hbar}d(F^*\sigma-F\sigma^*), \tag{2.75c}$$

The self-consistent set of equations obtained by expanding the field in regard of the modes of an ideal resonator includes Eqs. (2.30) and (2.63):

$$\frac{\mathrm{d}F_\lambda}{\mathrm{d}t}+[\kappa-i(\omega-\omega_{c\lambda})]F_\lambda=4\pi i\omega dN_S\frac{1}{V_c}\int\sigma E_\lambda(\mathbf{r})\mathrm{d}V,$$

$$\frac{\partial\sigma}{\partial t}+[\gamma_{\perp}-i(\omega-\omega_0)]\sigma=-\frac{id}{2\hbar}D\sum E_{\lambda}(\mathbf{r})F_{\lambda}\,, \tag{2.76}$$

$$\frac{\partial D}{\partial t}+\gamma_{\|}(D-D^{(0)})=-\frac{i\beta_a d}{2\hbar}\sum(F_{\lambda}^{*}\sigma-F_{\lambda}\sigma^{*})E_{\lambda}(\mathbf{r})\,.$$

This set of equations will be widely used below to formulate the concrete laser models.

Chapter 3

Single-Mode Lasers

The simplest single-mode models play a special role in the dynamic laser theories. They possess extremely low dimensions and include only the most fundamental and unavoidable nonlinearity that accompanies the process of matter-field interaction but do not cover the mode interaction, additional nonlinear elements and external signals. The behaviour of single-mode lasers depends on the dynamical class they belong to.

3.1. Dynamical Models of Homogeneously Broadened Lasers

In what follows we often use equations written in dimensionless form. The advantage of using this form is its simplicity owing to which it is possible to 'hide' almost all the coefficients not defined in the experiment. Meanwhile, the normalization of the observed quantities has a clear physical meaning: the field amplitude is normalized to saturation value and the inversion is normalized to a value corresponding to the laser threshold.

3.1.1. Equations for the Quadratic Quantities

We now turn to the laser model expressed by Eqs. (2.75). If the losses for the separate cavity elements (mirrors) are small and, therefore, there are no areas of abrupt field increase or decrease inside the cavity, then we can assume $\partial F/\partial z = 0$. Let us introduce a table of dimensionless symbols:

$$\tau = \frac{t}{\hat{t}}, \quad f = \frac{F}{F_{\text{sat}}} = \frac{dF}{\hbar}\left(\frac{2}{\beta_a}\gamma_{\parallel}\gamma_{\perp}\right)^{-1/2}, \quad n = \frac{\pi\omega d^2}{\hbar\gamma_{\perp}\kappa}N_S D,$$

$$p = \frac{2\pi i\omega d^2}{\hbar\kappa}\left(\frac{2}{\beta_a}\gamma_{\parallel}\gamma_{\perp}\right)^{-1/2} N_S\sigma, \quad A = \frac{\pi\omega d^2}{\hbar\gamma_{\perp}\kappa}N_S D^{(0)},$$

$$\hat{\kappa} = \kappa\hat{t}, \quad \hat{\gamma}_i = \gamma_i\hat{t}, \quad \Delta_0 = (\omega-\omega_0)/\gamma_{\perp}, \quad \Delta_c = (\omega-\omega_c)/\kappa. \tag{3.1}$$

It is most convenient to choose a normalization factor $\hat{t}$ comparable with a time scale of the time-dependent process in a laser. The hard unification of this factor is not reasonable. The variety of possibilities dictates the individual choice in each concrete situation. We will use inverse relaxation rates γ_i^{-1} and κ^{-1} as the normalization coefficients. In notations (3.1) the equations become

$$\frac{df}{d\tau} - i\hat{\kappa}\Delta_c f = \hat{\kappa}(p - f), \tag{3.2a}$$

$$\frac{dp}{d\tau} - i\hat{\gamma}_\perp \Delta_0 p = \hat{\gamma}_\perp (nf - p), \tag{3.2b}$$

$$\frac{dn}{d\tau} = \hat{\gamma}_\parallel \left[A - n - \frac{1}{2}(f^* p + fp^*) \right]. \tag{3.2c}$$

The complex form of equations is most compact but not always convenient. Sometimes it is more reasonable to use real variables, for example, the real and imaginary parts of f, p or its modules and arguments. As real variables can serve also the quadratic quantities:

$$m = |f|^2, \quad r = |p|^2, \quad s = \frac{1}{2}(fp^* + f^* p), \quad q = \frac{i}{2}(fp^* - f^* p). \tag{3.3}$$

These variables as well as inversion n are coupled by means of five equations:

$$\frac{dm}{d\tau} = 2\hat{\kappa}(s - m), \tag{3.4a}$$

$$\frac{dn}{d\tau} = \hat{\gamma}_\parallel (A - n - s), \tag{3.4b}$$

$$\frac{dr}{d\tau} = 2\hat{\gamma}_\perp (ns - r), \tag{3.4c}$$

$$\frac{ds}{d\tau} = -(\hat{\gamma}_\perp + \hat{\kappa})s + \hat{\gamma}_\perp \Delta_{0c} q + \hat{\gamma}_\perp mn + \hat{\kappa} r, \tag{3.4d}$$

$$\frac{dq}{d\tau} = -(\hat{\gamma}_\perp + \hat{\kappa})q - \hat{\gamma}_\perp \Delta_{0c} s, \tag{3.4e}$$

that one can easy obtain from (3.2). Here $\Delta_{0c} = (\omega_0 - \omega_c)/\gamma_\perp$. Eqs. (3.4) are convenient, for example, for investigation the conditions of transfer to the second order rate equations.

3.1.2. Adiabatic Elimination of the Atomic Polarization; Single-Mode Rate Equations

The rate equations for the inversion N and the photon number M can be written using simple considerations based on the concept of transition probability per unit time:

$$\frac{dM}{dt} = BMN - 2\kappa M \ , \tag{3.5a}$$

$$\frac{dN}{dt} = \gamma_{\|}(N^{(0)} - N) - \beta_a BMN \ . \tag{3.5b}$$

These equation show that the number of photons in a cavity increases due to stimulated emission with the rate BMN (B is Einstein's coefficient defined relative to the total number of photons in the cavity), and decreases with the rate $2\kappa M$ due to the losses. The inversion N changes due to stimulated emission, relaxation and pumping processes; as a result, the inversion tends to the equilibrium value $N^{(0)}$. Eqs. (3.5) were published first by Statz and DeMars [231] and are sometimes mentioned in the literature under their names.

The transition probability basis for writing the rate equations is simple and illustrative but it does not provide information on the limits of their applicability. We can gain such information by considering more general set (3.2) or (3.4), from which Eqs. (3.5) follow as a limiting case [232-234].

Equations (3.5) have the same variables, m and n (although they are renamed), which enter the system (3.4). However, the latter is of a higher order and, correspondingly, it contains three additional variables, r, s and q, which should be omitted provided the dimension of the model is reduced properly. This is achieved by adiabatic elimination of variables mentioned in Section 2.2.3. A standard procedure consists of disregarding of 'extra' derivatives, **d**r/**d**τ, and to find an algebraic expression for r and through τ and eliminate them from the remaining two differential equations.

In general, it becomes possible to reduce the number of differential equations by adiabatic elimination of part of the variables when a subspace with a smaller number of dimensions, in which fast motions are absent, is distinguished in the phase space of the system [211]. Starting from arbitrary initial conditions, the representative point rapidly passes in this subspace and then it moves along the phase space trajectories localized in it. Following the terminology of Oppo and Politi [235], such a subspace is called the centre manifold.

Mathematically, the motion in the system described by

$$\mu\dot{x}_i = F_i(x,y), \quad \dot{y}_j = G_j(x,y),$$

where μ is the small parameter, is very close to the motion in the limiting system

$$F_i(x,y) = 0, \quad \dot{y}_j = G_j(x,y).$$

Thus, for adiabatic elimination of variables it is necessary to have a small parameter at some derivatives. This means that coefficients are very different in magnitude, and variables can be separated in two groups: 'fast' variables and 'slow' ones. The fast variables are able to reach the quasistationary values determined by instantaneous values of the slow variables and to keep track of evolution of the latter.

In application to Eqs. (3.4), where the variables r, s and q are subject to adiabatic elimination, it can be asserted that the role of a small parameter is played by a quantity proportional to $1/\widehat{\gamma}_\perp$ and that the first group of conditions providing a transition for validity to the rate equations looks like

$$\gamma_\perp >> \gamma_\parallel, \quad \gamma_\perp >> \kappa. \tag{3.6}$$

These inequalities are fulfilled for class B lasers. The typical values of parameters for solid-state lasers are $\gamma_\parallel = 10^3$–10^4 s^{-1}, $\gamma_\perp = 10^{12}$ s^{-1}, $\kappa = 10^8$ s^{-1}.

Localization of the phase trajectories in the limits of centre manifold is guaranteed due to [211] by the negative real parts of the roots of characteristic equation

$$\begin{vmatrix} \dfrac{\partial F_1}{\partial x_1} - \lambda & \dfrac{\partial F_1}{\partial x_2} & \cdots & \dfrac{\partial F_1}{\partial x_s} \\ \cdot & \cdot & \cdot & \cdot \\ \cdot & \cdot & \cdot & \cdot \\ \dfrac{\partial F_s}{\partial x_1} & \dfrac{\partial F_s}{\partial x_2} & \cdots & \dfrac{\partial F_s}{\partial x_s} - \lambda \end{vmatrix} = 0 . \tag{3.7}$$

As applied to subsystem (3.4c)–(3.4d) the characteristic equation (3.7) becomes

$$\lambda^3 + 4\lambda^2 + (5 + \Delta_{0c}^2 - 2n\widetilde{\kappa})\lambda + 2(1 + \Delta_{0c}^2) - 2n\widetilde{\kappa} = 0.$$

Note that the last equation contains the quantity $\widetilde{\kappa}$ that can pretend on the role of the small parameter. Using the Routh–Hurwitz criterion [236] we obtain the condition of negativity of Reλ:

$$n < \begin{cases} (1 + \Delta_{0c}^2)/\widetilde{\kappa} & \text{for } \Delta_{0c}^2 < 3, \\ (9 + \Delta_{0c}^2)/\widetilde{\kappa} & \text{for } \Delta_{0c}^2 > 3. \end{cases} \tag{3.8}$$

Supposing that the criteria (3.6) and (3.8) are fulfilled, we then use the standard procedure of adiabatic elimination of variables r, s and q from the Eq. (3.4), i.e., we mean a $\partial r/\partial\tau = \partial s/\partial\tau = \partial q/\partial\tau = 0$ and, therefore, we have algebraic expressions

$$q = -\Delta_{0c}s, \quad r = ns, \quad s = \frac{mn}{1+\tilde{\kappa}(1-n)+\Delta_{0c}^2(1+\tilde{\kappa})^{-1}}. \tag{3.9}$$

Substituting Eqs. (3.9) into Eqs. (3.4a) and (3.4b), we find

$$\frac{dm}{d\tau} = 2\tilde{\kappa}m\left(\frac{n}{1+\tilde{\kappa}(1-n)+\Delta_{0c}^2(1+\tilde{\kappa})^{-1}} - 1\right), \tag{3.10a}$$

$$\frac{dn}{d\tau} = \tilde{\gamma}_{\parallel}\left(A - n - \frac{mn}{1+\tilde{\kappa}(1-n)+\Delta_{0c}^2(1+\tilde{\kappa})^{-1}}\right). \tag{3.10b}$$

Limiting by zero approximation in the small parameter $\tilde{\kappa}$ we come to the rate equations in the traditional form:

$$\frac{dm}{d\tau} = 2\tilde{\kappa}m\left(\frac{n}{1+\Delta_{0c}^2} - 1\right), \tag{3.11a}$$

$$\frac{dn}{d\tau} = \tilde{\gamma}_{\parallel}\left[A - n\left(\frac{m}{1+\Delta_{0c}^2} + 1\right)\right]. \tag{3.11b}$$

The time scale of the processes in this system is determined by the parameters $\gamma_{\parallel}$ and k. It will be logical to use $\hat{t} = (\gamma_{\parallel}\kappa)^{-1/2}$, $\tilde{\kappa} = (\kappa/\gamma_{\parallel})^{1/2}$, $\tilde{\gamma}_{\parallel} = (\gamma_{\parallel}/\kappa)^{1/2}$. However, more often people choose two other possibilities: $\hat{t} = \kappa^{-1}$ and $\hat{t} = \gamma_{\parallel}^{-1}$. Below we will adhere to the latter and introduce the only fundamental parameter $\kappa/\gamma_{\parallel} = G/2$.

Inequalities (3.6) and (3.8) do not form a complete set of conditions for validity of the rate equations. The point is that inequalities do not give sufficient ground to assume $1/\gamma_{\perp}$ as a small parameter. In order to obtain the additional conditions we act as follows. We different the first of Eq. (3.11) with respect to time, insert, in place of $dm/d\tau$ and $dn/d\tau$ their respective values according to Eq. (3.4) and take the inequalities (3.6) into account. The desired conditions are equivalent to the smallness of the resultant expression compared to the right-hand side of the corresponding equation in (3.4). A term-by-term comparison shows that we should add two inequalities to (3.6):

$$|n| \ll 1/\tilde{\kappa}, m \ll 1/\tilde{\gamma},$$

where $\tilde{\gamma} = \gamma_{\parallel}/\gamma_{\perp}$

The inequalities

$$\gamma_{\perp} \gg \gamma_{\parallel}, \quad \gamma_{\perp} \gg \kappa, \quad \gamma_{\perp} \gg \gamma_{\parallel} m, \quad |n| \ll \gamma_{\perp}/\kappa \tag{3.12}$$

guarantee the existence of the domain of slow motions of the system (3.4), and the softest inequality (3.8) guarantee the stability of these motions. Note, that according to Eq. (3.1), the inequality $m \ll \gamma_{\perp}/\gamma_{\parallel}$ is equivalent to $W \ll \gamma_{\perp}$, where $W = (dF)^2/\hbar^2\gamma_{\perp}$ have the meaning of probability of the stimulated transition. The same inequality can be interpreted as $\omega_R \ll \gamma_{\perp}$ where $\omega_R = dF/\hbar$ is the frequency of the Rabi oscillations.

The rate equations can be obtained from (3.2) assuming $\mathrm{d}p/\mathrm{d}\tau = 0$ and $F = m^{1/2}e^{i\varphi}$. Adiabatic elimination of polarization followed by a transformation to real variables yields

$$\frac{\mathrm{d}m}{\mathrm{d}\tau} = Gm\left[\frac{n}{1+\Delta_0^2} - 1\right], \tag{3.13a}$$

$$\frac{\mathrm{d}n}{\mathrm{d}\tau} = A - n\left[\frac{m}{1+\Delta_0^2} + 1\right], \tag{3.13b}$$

$$\frac{\mathrm{d}\varphi}{\mathrm{d}\tau} = \frac{G}{2}\left[\frac{n\Delta_0}{1+\Delta_0^2} + \Delta_c\right]. \tag{3.13c}$$

Equations (3.13a) and (3.13b) differ from Eqs. (3.11) because the place of Δ_{0c}^2 is occupied by Δ_0^2. When $\gamma_{\perp} \gg \kappa$, this difference is less than the accuracy of the approximation.

3.2. Traveling-Wave Laser with Homogeneous Active Medium

The model considered below is based on the assumption that a unique cavity mode is excited and that the laser medium is homogeneous (spectrally and spatially). These assumptions are best satisfied by a unidirectional ring laser. However, the spatial uniformity of inversion is also provided if a large number of modes of a standing wave type under the approximately equal conditions are involved in the laser action. The rate equations for total radiation intensity and population difference in such a multimode laser look like those for a single-mode laser Eq. (3.11), which are considered in what follows.

3.2.1. Steady States and Relaxation Oscillations

With time normalized to $\gamma_{\parallel}^{-1}$ and with exact coincidence of the cavity eigenfrequency and the laser transition frequency, Eqs. (3.11) become

$$\frac{dm}{d\tau} = Gm(n-1)\,, \tag{3.14a}$$

$$\frac{dn}{d\tau} = A - n(m+1)\,. \tag{3.14b}$$

The fixed points of the set of rate equations (3.14) and the solutions in their vicinity have been considered by many authors [231, 232, 237–243]. The steady states

$$\begin{aligned} \overline{m}_a &= 0, & \overline{n}_a &= A, \\ \overline{m}_b &= A-1 & \overline{n}_b &= 1 \end{aligned} \tag{3.15}$$

can be readily found from (3.14) provided $d/d\tau = 0$. The type of the fixed points can be specified by linearizing the Eqs. (3.14) in the vicinity of each of them with respect to small deviations $\delta m = m - \overline{m}, \delta n = n - \overline{n}$.

The following linearized equations

$$\frac{d}{d\tau}(\delta m) = G(A-1)\delta m, \quad \frac{d}{d\tau}(\delta n) = -A\delta m - \delta n$$

hold near point a. Substitution of the solutions $\{\delta m, \delta n\} = \{\delta m', \delta n'\}e^{\lambda\tau}$ into these equations leads to the characteristic equation

$$(\lambda+1)[\lambda - G(A-1)] = 0\,. \tag{3.16}$$

One root of Eq. (3.16), $\lambda_1 = -1$ is negative, while the sign of the other, $\lambda_2 = G(A-1)$, depends on A. Where $A<1$ the second root is negative too and the fixed point is a stable node. Where $A>1$ the sign of λ is positive and a becomes a saddle point, i.e., the fixed point is no longer stable. The inequality

$$A > 1 \tag{3.17}$$

expresses the laser self-excitation condition.

The motion of the system in the vicinity of the fixed point b obeys the linearized equations

$$\frac{d}{d\tau}(\delta m) = G(A-1)\delta n, \quad \frac{d}{d\tau}(\delta n) = -\delta m - A\delta n\,. \tag{3.18}$$

The corresponding characteristic equation

$$\lambda^2 + A\lambda + G(A-1) = 0 \tag{3.19}$$

has roots

$$\lambda_{1,2} = -\frac{A}{2} \pm \sqrt{\frac{A^2}{4} - G(A-1)}\,.$$

The fixed point can be either a stable node if

$$A^2 - 4G(A-1) > 0\,,$$

or a stable focus if the inverse inequality is satisfied. For class *B* lasers $G = 2\kappa/\gamma_{\|} >> 1$. Therefore, the fixed point will almost always be a focus and Eqs. (3.18) describe the damped oscillations of laser intensity near stationary level $\overline{m}_b$ with the frequency

$$\Omega_1 = \sqrt{G(A-1)} \quad \left(\nu_1[\text{Hz}] = \frac{1}{2\pi}\sqrt{\gamma_{\|}\kappa(A-1)} \right) \tag{3.20}$$

and the decrement

$$\theta_1 = -A/2\,. \tag{3.21}$$

These oscillations are generally what is meant when the term *relaxation oscillations* is used.

Equations of the form (3.20) and (3.21) hold true also in the case where the cavity is not tuned to the line centre. One should only substitute $A/(1+\Delta_{0c}^2)$ for A.

According to Eq. (3.20) the relaxation oscillations frequency is found as a geometric mean of the inversion decay rate and the field damping rate. Since for the more dielectric laser crystals $\gamma_{\|} \sim 10^3 - 10^4\ \text{s}^{-1}$, and the excess excitation over the laser threshold ranges for solid-state lasers from tens to thousands of per cent, the relaxation oscillations frequency falls in the range of tens kilohertz. In semiconductor lasers where $\gamma_{\|} \sim 10^9\,\text{s}^{-1}$ and $\kappa \sim 10^{12}\,\text{s}^{-1}$ frequency ν_1 is shifted to the gigahertz range.

3.2.2. Phase Portrait of Laser; Spikes Characteristics

Rather full information of the transients in considered laser model could be obtained by use of approximate analytical methods.

We will obtain the phase space trajectories equations dividing Eq. (3.14a) by Eq. (3.14b):

$$\frac{\mathrm{d}m}{\mathrm{d}n} = G\frac{(n-1)m}{A-(m+1)n}\,. \tag{3.22}$$

Defining the motion in the immediate vicinity of the fixed point the linear approximation gives an insight into the structure of whole phase plane. To gain a better understanding of this structure we should use the general properties of Eq. (3.22). Since $G \gg 1$, the inclination of the phase space trajectories is strong on the whole plane except of the domains close to the straight lines

$$n = 1, \quad m = 0, \tag{3.23}$$

which are the isoclines (the lines of equal inclination of phase space trajectories) with horizontally arranged tangents. The isocline with vertically arranged tangents is given by

$$m = \frac{A}{n} - 1. \tag{3.24}$$

Figure 3.1a shows the structure of the phase plane of a system for which the self-excitation condition (3.17) is not satisfied. Figure 3.1*b* refers to the case $A>1$, which is considered below. Besides the phase space trajectories, Fig. 3.1 shows the isoclines (3.23) and (3.24). It is impossible to find exact solutions of Eq. (3.14) analytically. Therefore, some authors have tried numerical solutions [244, 245]. However the basic parameters of the nonlinear process can be found by the way of an approximate analysis [242, 246]. The approximate method is based on the large value of the parameter G, owing to which the phase space trajectories can be divided into segments of fast and slow motions. This procedure fails near the fixed points where, instead, the linear approximation is possible. The representative point passes slowly through the lower trajectory segments close to the abscissa axis. The stimulated emission probability is small and the velocity of the representative point depends solely on the rate of pumping. The representative point passes the trajectory segment, on which the stimulated emission dominates the pumping (the interval of emission), with high velocity. For each interval Eqs. (3.14) admit simplifications based on the neglecting of some terms and can therefore be integrated. Matching solutions is facilitated by the fact that in the transient areas the pumping is compensated by stimulated emission and the inversion is nearly constant.

The phase space trajectories in Fig. 3.1*b* are slowly convergent to point *b* on a spiral pass. One turn to the spiral corresponds to a spike in the

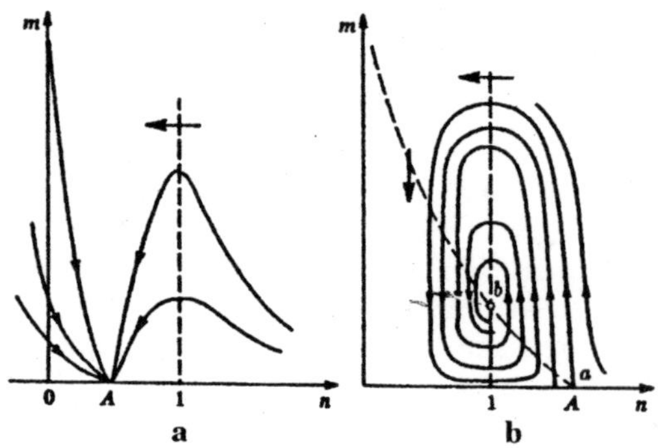

Fig. 3.1. Phase portrait of the rate equation model represented by Eqs. (3.14) at the pumping parameter values (*a*) below and (*b*) above threshold. Dashed lines indicate the isoclines.

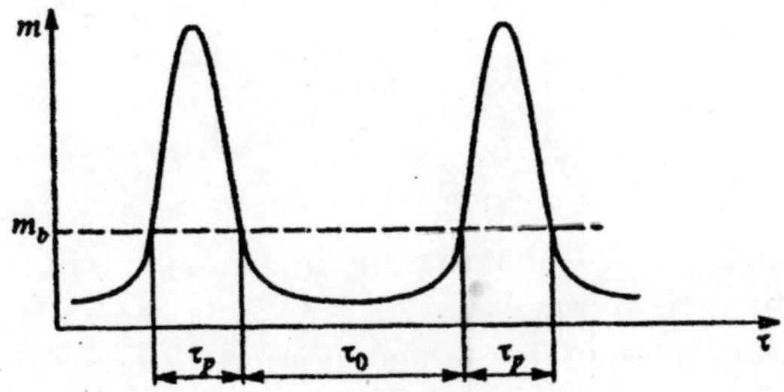

Fig. 3.2. Time-dependent solutions of the rate equations in a conservative approximation.

emission (Fig. 3.2). It is interesting to note that the minimum and maximum radiation intensities correspond to the same population difference $n = 1$. Since the damping rate of the spikes is slow we can solve the problem in two stages. First, disregarding the spike amplitude variation (the conservative approximation) we find the spike amplitude, the spike duration and the interval between spikes. Then, taking account of the spike damping as a perturbation we define the law by which the spike amplitude is diminishes.

When we find the spike amplitude and duration we bear in mind that the population difference varies little in a free running laser: $|n-1| << 1$. Such an assertion can be validated by a direct estimate. Consider Eq. (3.14*b*), which, under the condition $m \ll 1$ is simplified:

$$\frac{dn}{d\tau} = A - n \,. \tag{3.25}$$

The solution of Eq. (3.25) is apparent and is given by

$$n = A + [n(0) - A]e^{-\tau} \,. \tag{3.26}$$

The time τ, over which n is varied from $n(0) = 1$ to n_{max}, is small compared to unity; thus Eq. (3.26) reduces to

$$n - 1 = (A - 1)\tau \,. \tag{3.27}$$

Insertion of Eq. (3.27) into Eq. (3.14a) yields the law of field growth in the pumping interval

$$\ln\frac{m}{m_{min}} = \frac{1}{2}G(A-1)\tau^2 \,. \tag{3.28}$$

Eliminating the time from Eqs. (3.27) and (3.28) we obtain the relationship between n and m in the explicit form

$$\eta = \left[\frac{2}{G}(A-1) \ln \frac{m}{m_{\min}} \right]^{1/2} , \tag{3.29}$$

where

$$\eta \equiv n - 1 \ .$$

It is seen from Eq. (3.24) that that the inversion maximum is achieved at $m \leq \overline{m}_b$. At these field intensities Eq. (3.29) is generally inapplicable. Meanwhile, bearing in mind that the population difference is almost constant near $m = \overline{m}_b$ and m enters Eq. (3.29) under the logarithmic sign, we can adopt, without introducing a noticeable error

$$\eta_{\max} = \left[\frac{2}{G}(A-1) \ln \frac{\overline{m}_b}{m_{\min}} \right]^{1/2} . \tag{3.30}$$

The distance the representative point penetrates into domain $n > 1$ is the longer the lower its intersection with the line $n = 1$ is. Consequently, $\eta_{\max}$ must be the largest for the first spike after pumping is switched when $m_{\min}$ is defined solely by the fluctuation field in the cavity. In Section 3.2.3 it is shown that $\ln(\overline{m}_b / m_{\min}) \approx 25$ for all solid-state lasers, whence $\eta_{\max} \approx 0.1$.

The approximate conservative equations can be obtained by disregarding the small value of η in Eq. (3.14b)

$$\frac{dm}{d\tau} = Gm\eta , \tag{3.31a}$$

$$\frac{dn}{d\tau} = \overline{m}_b - m . \tag{3.31b}$$

These equations have the integral

$$\frac{1}{2} G(\eta^2 - \eta_0^2) = \overline{m}_b \ln \frac{m}{m_0} - m + m_0 , \tag{3.32}$$

which describes a family of closed trajectories in the phase plane.

To find the spike amplitude we should fix a trajectory assuming $\eta_0 = \eta_{\max}$, $m_0 = \overline{m}_b$ in Eq. (3.32). Assuming that the peak intensity $m_{\max} >> \overline{m}_b$ and knowing that this value is achieved at $\eta = 0$, we find

$$m_{\max} = \frac{1}{2} G \eta_{\max}^2 = (A-1) \ln \frac{\overline{m}_b}{m_{\min}} . \tag{3.33}$$

We now calculate the spike duration, defining it as the time the representative point moves over the upper part of the trajectory between the values $m_{\max} = \overline{m}_b$. On the segment $m_{\max} >> \overline{m}_b$ Eq. (3.31b) is convenient. Since the logarithmic term is small here, Eq. (3.32) is resolved with respect to m, and the time the representative point moves along the phase space trajectory between the points with equal ordinates $m = m_1$ is

$$\tau_p' = \int_{\eta_1}^{-\eta_1} \frac{d\eta}{\overline{m}_b - m(\eta)} \approx \frac{2}{G\eta_1} \ln \frac{2\eta_1}{(\eta_1^2 + 2G^{-1}m_1 - \eta_1)^{1/2}} .$$

If $m_1 << m_{max}$, then $\eta_1 \approx \eta_{max}$ and confining ourselves to the linear term in the expansion of the denominator of the logarithmic term we have

$$\tau_p' = \frac{2}{G\eta_{max}} \ln \frac{2G\eta_{max}^2}{m_1} .$$

The time the field rises from $m = \overline{m}$ to $m = m_1$ can be easily found from Eq. (3.31a) by letting $\eta = \eta_{max}$

$$\tau_p'' = \frac{1}{G\eta_{max}} \ln \frac{m_1}{\overline{m}_b} .$$

The desired spike duration is determined by

$$\tau_p = \tau_p' + 2\tau_p'' = \frac{2}{G\eta_{max}} \ln \frac{2G\eta_{max}^2}{A-1} . \tag{3.34}$$

The time between spikes, τ_0, is found from Eq. (3.27)

$$\tau_0 = \frac{2\eta_{max}}{A-1} . \tag{3.35}$$

We now propose two numerical examples.

Example 3.1
Ruby laser

$\gamma_\parallel = 10^3$ s^{-1};	$\eta_{max} = 4.5 \cdot 10^{-2}$; $m_{max} = 100$;
$\kappa = 5 \cdot 10^7$ s^{-1}; $G = 10^5$;	$\tau_p = 2 \cdot 10^{-3}$; $\tau_0 = 2.25 \cdot 10^{-2}$;
$A = 5$;	$t_p = 2$ μs; $t_0 = 22.5$ μs;
$\ln(\overline{m}_b / m_{min}) = 25$;	$\tau_p / \tau_0 = 0.08$.

Example 3.2
Nd:YAG laser

$\gamma_\parallel = 5\cdot 10^3\ \mathrm{s}^{-1};$	$\eta_{\max} = 5\cdot 10^{-2};\ m_{\max} = 25;$
$\kappa = 5\cdot 10^7\ \mathrm{s}^{-1};\ G = 2\cdot 10^4;$	$\tau_\mathrm{p} = 9.2\cdot 10^{-3};\ \tau_0 = 0.1;$
$A = 2;$	$t_\mathrm{p} = 1.8\ \mu\mathrm{s};\ t_0 = 20\ \mu s;$
$\ln(\overline{m}_b / m_{\min}) = 25;$	$\tau_\mathrm{p} / \tau_0 = 0.09.$

The damping rate of spikes in the nonlinear regime can be calculated if we do not use the conservative approximation. First of all, we find the $\eta_{\max}$ variation on one turn on the spiral (Fig. 3.3).

In the emission interval ($m >> \overline{m}_b$) the motion of the representative point obeys the approximate equation

$$\frac{\mathrm{d}m}{\mathrm{d}\eta} = -\frac{G\eta}{\eta + 1}, \tag{3.36}$$

the solution of which, if the limits of integration with respect to m are the same, has the form

$$\eta_2 - \eta_1 = \ln\frac{1+\eta_2}{1+\eta_1}. \tag{3.37}$$

In the pumping interval ($m \ll 1$) equations

$$\frac{\mathrm{d}m}{\mathrm{d}\tau} = G\eta m, \quad \frac{\mathrm{d}\eta}{\mathrm{d}\tau} = \overline{m}_b - \eta \tag{3.38}$$

are valid. Integrating Eq. (3.38) over the lower part of the spiral turn in the limits $\eta_2 \le \eta \le \eta_3$ that correspond to equal values of m we arrive at

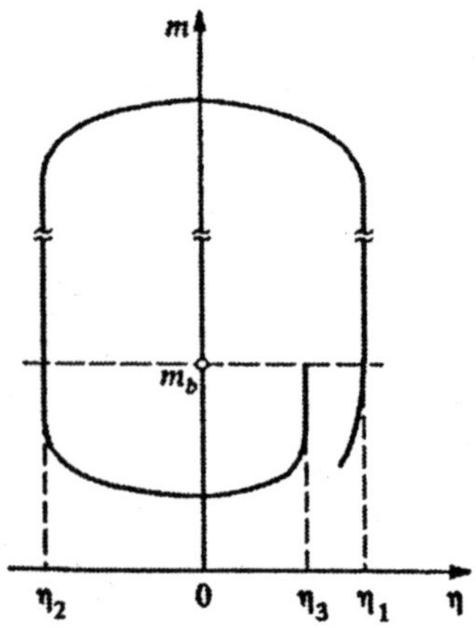

Fig. 3.3. Turn of a spiral phase space trajectory in a nonconservative rate equation model.

$$\eta_3 - \eta_2 + \overline{m}_b \ln \frac{\overline{m}_b - \eta_3}{\overline{m}_b - \eta_2} = 0 . \tag{3.39}$$

We stress once again that the population difference variation is the slowest near the points $\eta = \eta_{\max}$. Now it is necessary to estimate the real variation rate. To do this, we expand the function $\eta(m)$ in a series in the vicinity of the point $\eta = \eta_{\max}$. Since the first derivative in the extremum turns to zero,

$$\eta = \eta_{\max} - \frac{(m - \overline{m}_b)^2}{2G(A-1)\eta_{\max}} . \tag{3.40}$$

Assuming that the η variation in one turn is the small quantity of order η^2 we will try to match the solutions of Eqs. (3.37) and (3.39). This can be done if the values at the ends to be matched differ by less than η^2 . The criterion of validity of such an operation is obtained from (3.40):

$$(m - \overline{m}_b)^2 << 2G(A-1)\eta_{\max}^2 . \tag{3.41}$$

Using these considerations we assume in Eq. (3.37) $\eta_1 = \eta_{\max}, \eta_2 = \eta_{\min}$. Confining the accuracy in calculation of $\Delta\eta_1 = |\eta_2| - \eta_1$ to the terms of order η^2 , we thereby assign the number of terms in the expansion of the logarithm and obtain

$$\Delta\eta_1 = -\frac{2}{3}\eta_1^2 . \tag{3.42}$$

With the same accuracy we find from Eq. (3.39) the following expression for $\Delta\eta_2 = \eta_3 - |\eta_2|$:

$$\Delta\eta_2 = -\frac{2\eta_2^2}{3\overline{m}_b} . \tag{3.43}$$

Since the amplitude variation of the neighboring spikes is small, Eqs. (3.42) and (3.43) can be combined:

$$\Delta\eta_{\max} = -\frac{2}{3}\frac{A}{A-1}\eta_{\max}^2 . \tag{3.44}$$

Equations (3.33) and (3.44) yield the law for diminishing of the spike amplitude with time. From (3.33.) is follows that each spike is less that preceding one by

$$\Delta m = G\eta_{\max}\Delta\eta_{\max} .$$

Submitting $\Delta\eta_{\max}$ from Eq. (3.44.) we find

$$\Delta m = -\frac{4}{3}\frac{A}{A-1}\eta_{\max} m_{\max} . \tag{3.45}$$

Dividing this expression by the time interval between the spikes (3.35) we determine the spike envelope derivative

$$\frac{\mathrm{d}m_{\max}}{\mathrm{d}\tau} \approx \frac{\Delta m_0}{\tau_0} = -\frac{2Am}{3} .$$

The envelope itself, representing the law of diminishing spike amplitudes, is given by

$$m_{\max} = m_{\max}^{0} \exp(-2A\tau/3) . \tag{3.46}$$

The damping rates of intense spikes and the small relaxation oscillations appeared to be nearly the same.

The main physical result of this consideration is that system comes back to the equilibrium position with any arbitrary departure from equilibrium. The regimes of undamped oscillations are beyond the scope of this model and are not described by the simplest rate equations (3.14).

3.2.3. The Linear Stage of the Onset

After pumping is switched on, a definite time is needed for the population difference to reach the threshold value. This time can be found by use of Eq. (3.25) assuming in general $A = A(\tau)$. The solution of this equation has the form

$$n = e^{-\tau}\left[n(0) + \int_0^{\tau} A(\tau')e^{\tau'}\mathrm{d}\tau'\right]. \tag{3.47}$$

The time needed for the creation of threshold inversion ($n = 1$) is most easily expressed in the case A = const:

$$\tau_{\mathrm{d}} = \ln\frac{A - n(0)}{A-1} . \tag{3.48}$$

Initially, all the particles of the laser medium are at the ground level. Simultaneously this ground level is the lower laser level of materials like ruby, which operate on a three-level scheme, so that $n(0) = -n_S$. For threshold inversion to be achieved over a time $\tau_{\mathrm{d}} = 1$ ($t_{\mathrm{d}} = T_1$), the pumping power should be high enough to ensure that

$$A_{\min} = \frac{2.7 - n(0)}{1.7} .$$

A typical value for a ruby laser is $n_S = A_{\max} = 10$ and, therefore, $A_{\min} = 7.5$. Rather often the pumping pulse is not of rectangular form under the experimental conditions. Threshold inversion is achieved after the pumping power passes a maximum and laser is operated at values A less than the values obtained above. For estimates we have chosen $A = 5$.

If we deal with a four-level atomic system, then $n(0) = 0$ and the threshold value is achieved at $A = 1.5$ over a time $\tau_d = 1$.

The time τ_d we have found is not the total delay time of laser action, which is manifested in experiment. The point is that induced emission is absent at the time the self-excitation threshold achieved and starts to rise after that time. It should be noted that in a system without fluctuations the representative point, after passing the value $n = 1$, would proceed moving along the n-axes, since the latter is a phase space trajectory over its whole length. The fluctuating field (first of all, the spontaneous emission) in the cavity causes the representative point to enter one of upgoing trajectories. Thus, one must take into account the necessary fluctuations, in principle, since it qualitatively changes the laser dynamics. After a triggering mechanism when the laser passes the threshold, the fluctuating field in fact off the ground until the representative point is again close to the n-axes.

Since induced emission grows from the fluctuation level we should introduce a term taking into account (on the average) the spontaneous emission to estimate the time of the linear development of the first spike. The probability of the process can be written as

$$W_{sp} = BN_2 = \frac{1}{2}B(N + N_S)\,.$$

This value should be added to the right-hand side of rate equation in order to find the total number of photons in the cavity (3.5)

$$\frac{\mathrm{d}M}{\mathrm{d}t} = BMN - 2\kappa M + \frac{1}{2}B(N + N_S)\,. \tag{3.49}$$

The transition to dimensionless variables follows the pattern:

$$\tau = \gamma_{\parallel} t, \quad m = \beta_a BM / \gamma_{\parallel}, \quad n = BN/(2\kappa)\,.$$

Adding

$$\varepsilon_{\mathrm{sp}} = \frac{\beta_a B}{2\gamma_{\parallel}} = \frac{2\pi\beta_a \omega d^2}{\hbar V \gamma_{\parallel}\gamma_{\perp}} = \frac{\beta_a c\sigma_{\mathrm{tr}}}{2\gamma_{\parallel} V_{\mathrm{c}}}\,, \tag{3.50}$$

we rewrite Eq. (3.49) as

$$\frac{\mathrm{d}m}{\mathrm{d}\tau} - G(n-1)m = G\varepsilon_{\mathrm{sp}}(n + n_S)\,. \tag{3.51}$$

At the linear stage of lasing at A = const, the inversion variation obeys the Eq. (3.27). Substituting (3.27) into (3.51) we can integrate the resultant equation. In the right-hand side we can put $n = 1$, since a small-range variation in n is not essential. The solution of the linear Eq. (3.51) is given by

$$m=\exp\left[\frac{1}{2}G(A-1)\tau^2\right]\left\{m_0+G\varepsilon_{sp}\left[\int_0^{\tau}\exp\left[-\frac{1}{2}G(A-1)\tau'^2\right]d\tau'+(A-1)\int_0^{\tau}\tau'\exp\left[-\frac{1}{2}G(A-1)\tau'^2\right]d\tau'\right]\right\}$$

where m_0 is the field intensity at the time the laser threshold is achieved.

In what follows it will be shown that $G\tau_d^2(A-1)>>1$, and we use it as an assumption for the time being. In doing so we neglect the first and the third terms in the brackets compared to the second term, since these relate as $G^{-1/2}$. Owing to this simplification, the last relation is reduced to

$$m\approx\varepsilon_{sp}(n_S+1)\left[\frac{\pi G}{A-1}\right]^{1/2}\exp\left[\frac{1}{2}G(A-1)\tau_d^2\right]. \tag{3.52}$$

Comparing Eqs. (3.52) and (3.28) it can be asserted that m_{min} entering (3.28) is nothing but

$$m_{min}\approx\varepsilon_{sp}(n_S+1)\left[\frac{\pi G}{2(A-1)}\right]^{1/2}. \tag{3.53}$$

Example 3.3

Ruby laser

$\gamma_\parallel=10^3\ s^{-1}$; $\sigma_{tr}=10^{-20}\ cm^2$; $\varepsilon_{sp}=10^{-13}$; $m_{min}=2\cdot10^{-10}$;

$G=10^5$; $A=5$; $n_S=10$; $G\tau_d^2(A-1)/2=\ln(\overline{m}_b/m_{min})=23$.

$V_c=3\ cm^3$;

Example 3.4

Nd:YAG laser

$\gamma_\parallel=5\cdot10^3\ s^{-1}$; $\sigma_{tr}=10^{-18}\ cm^2$; $\varepsilon_{sp}=10^{-12}$; $m_{min}=3\cdot10^{-10}$;

$G=2\cdot10^4$; $A=2$; $n_S=1$; $G\tau_d^2(A-1)/2=\ln(\overline{m}_b/m_{min})=22$

$V_c=3\ cm^3$;

3.3. Single-Mode Standing-Wave Class *B* Laser

The single-mode model of standing wave laser is described by a set of Eqs. (2.76). Since we deal with class *B* laser, it is possible to eliminate adiabatically the atomic polarization, which leads, in the absence of detuning, to the rate equations

$$\frac{\mathrm{d}m}{\mathrm{d}\tau} = Gm\left(\int n\psi^2 \mathrm{d}v - 1\right), \tag{3.54a}$$

$$\frac{\partial n}{\partial \tau} = A - n(m\psi^2 + 1), \tag{3.54b}$$

where

$$n = \frac{2\pi d^2 \omega N_S}{\hbar \gamma_\perp \kappa} D, \quad A = \frac{2\pi d^2 \omega N_S}{\hbar \gamma_\perp \kappa} D^{(0)}, \quad m = \frac{\beta_a d^2}{2\hbar \gamma_\parallel \gamma_\perp} F^2,$$

$$\tau = \gamma_\parallel t, \quad v = V/V_c, \quad \psi = E_\lambda(\mathbf{r}). \tag{3.55}$$

3.3.1. Model with Extended Laser Medium

One steady state of Eqs. (3.54), namely $\overline{m}_a = 0, \overline{n}_a = A$, is easy to find. The others are determined from

$$\overline{n}_b = \frac{A}{\overline{m}_b \psi^2 + 1}, \tag{3.56}$$

$$\int \frac{A\psi^2}{\overline{m}_b \psi^2 + 1} \mathrm{d}v = 1. \tag{3.57}$$

The number of nontrivial steady states is equal to the number of the real positive roots of Eq. (3.57). In the simplest case, the population inversion is created over the entire volume of the laser rod, as assumed. At $A > 1$, the left-hand side of Eq. (3.57) is a monotonically decreasing function of $\overline{m}_b$ and, therefore, this equation has an unique root.

The zero branch of solutions becomes unstable if the self-excitation condition

$$\int A\psi^2 \mathrm{d}v > 1. \tag{3.58}$$

is satisfied. In order to investigate the stability of the nontrivial branch of solutions, we linearize Eq. (3.54) in the vicinity of the states $m = \overline{m}_b, n = \overline{n}_b$:

$$\frac{\mathrm{d}(\delta m)}{\mathrm{d}\tau} = G\overline{m}_b \int \psi^2 \partial n \mathrm{d}v, \tag{3.59a}$$

$$\frac{\partial(\delta n)}{\partial \tau} = -\frac{A}{\overline{m}_b}\delta n - \psi^2 \overline{n}_b \delta m. \tag{3.59b}$$

On assuming a perturbation of form $\{\delta m, \delta n\} = \{\delta m_0, \delta n_0\} e^{\lambda\tau}$, resolving

(3.59b) with respect to δn_0 and substituting the results into (3.59a), we arrive at a characteristic equation

$$\lambda = -G\overline{m}_b \int \frac{A\psi^4}{(\psi^2\overline{m}_b + 1)(\lambda + \psi^2\overline{m}_b + 1)} \mathrm{d}v \,. \tag{3.60}$$

Our further analysis is based on the large value of the parameter G. Thus, it can be inferred that the roots of Eq. (3.60) are divided into two groups. One includes the negative roots with modulus of order unity. The roots of the other group are of order $G^{1/2}$. There are only two of them and they are complex conjugate, and $\mathrm{Im}\,\lambda >> \mathrm{Re}\,\lambda$. If we make use of the large value $\mathrm{Im}\,\lambda = \Omega$ and expand the integrand in (3.60) into a series of Ω^{-1}, then we can easily find the approximate values

$$\Omega_1 = \left(G\overline{m}_b \int \frac{A\psi^4}{\psi^2\overline{m}_b + 1} \mathrm{d}v \right)^{1/2} \tag{3.61a}$$

$$\theta_1 = -\frac{1}{2} \frac{\int A\psi^4 \mathrm{d}v}{\int A\psi^4 (\overline{m}_b\psi^2 + 1)^{-1} \mathrm{d}v} \,. \tag{3.61b}$$

It is important to note that $\mathrm{Re}\,\lambda = \theta_1 < 0$ and, therefore, the spatial field structure does not disturb the stability of a time-independent solution.

It is interesting to compare Ω_1 and θ_1 for the cases of uniformly and nonuniformly saturated laser media with the comparable parameter values. We can make this comparison by specifying the form of the functions $\psi(\zeta)$ and $A(\zeta)$. Consider the mode of a plane standing wave type and assume the pumping is spatially homogeneous. Taking into account the normalization condition, the eigenfunction of the cavity is written as

$$\psi_q = \sqrt{2}\sin(\pi q \zeta) \,. \tag{3.62}$$

As a dimensionless coordinate, it is convenient to choose $\zeta = z / L_a$, where L_a is the length of the laser rod.

First of all, we determine the laser intensity in the steady state. Using Eq. (3.62) in (3.57) and integrating under the condition of constant A we obtain

$$\frac{A}{\overline{m}_b}\left[\zeta - \frac{1}{\pi q\sqrt{1+2\overline{m}_b}} \arctan\left(\sqrt{1+2\overline{m}_b}\,\tan(\pi q \zeta)\right)\right]_{\zeta_1}^{\zeta_2} = 1 \,, \tag{3.63}$$

where ζ_1 and ζ_2 denote the coordinates of the boundaries of the active element, $\zeta_2 - \zeta_1 = 1$. Shifting the integration limits from the active element boundaries to the nearest points at which $\tan(\pi q\zeta'_{1,2})=0$, we arrive at

$$\arctan\left[\sqrt{1+2\overline{m}_b}\,\tan(\pi q\zeta)\right]_{\zeta'_1}^{\zeta'_2}=\pi q\,. \tag{3.64}$$

Since $L_a >> \lambda$, such a substitution does not change essentially the result. After this it is easy to pass over to

$$\overline{m}_b=\frac{1}{4}\left(4A-1-\sqrt{8A+1}\right). \tag{3.65}$$

Taking the same approximation, from (3.61) we find the frequency of small oscillations

$$\Omega_1=\sqrt{G(A-1)}\,, \tag{3.66}$$

and their decrement

$$\theta_1=-\frac{3A\overline{m}_b}{4(A-1)}\,. \tag{3.67}$$

3.3.2. Model with the Sinusoidal Inversion Grating

In the above consideration we did not limit the shape of spatial distribution of saturated steady-state inversion. We only assumed that the pumping in the volume of active element is distributed uniformly, i.e., the pumping parameter A = const. However, it is clear that the real structure of the field mode leads to nonuniform saturation of the laser medium, or, as people often spoken, to inversion hole burning. Thus, the standing wave burns out the inversion grating with the period $\lambda/2$. Neglecting all of the higher harmonics of inversion we reduce the problem to a simple enough set of ordinary differential equations.

In this case, we present the inversion as a sum of the mean value and the first spatial harmonic:

$$n=n_0+2n_1\cos(2\pi q\zeta)\,, \tag{3.68}$$

where

$$n_0=\int_0^1 n\,\mathrm{d}\zeta,\qquad n_1=-\int_0^1 n\cos(2\pi q\zeta)\,\mathrm{d}\zeta,\quad \zeta=z/L, \tag{3.69}$$

L is the cavity length. Substituting Eqs. (3.69) into (3.54) we reduce the latter to three ordinary differential equations:

$$\frac{\mathrm{d}m}{\mathrm{d}\tau}=Gm(n_0+n_1-1)\,, \tag{3.70a}$$

$$\frac{dn_0}{d\tau} = A - n_0(1+m) - n_1 m, \tag{3.70b}$$

$$\frac{dn_1}{d\tau} = -n_1(1+m) - \frac{1}{2}n_0 m. \tag{3.70c}$$

The steady-state intensity of radiation is given by

$$\overline{m}_b = \frac{1}{2}\left(A - 4 + \sqrt{A^2+8}\right), \tag{3.71}$$

and other variables are expressed over the intensity as

$$\overline{n}_{0b} = A - \overline{m}_b, \quad \overline{n}_{1b} = 1 - A + \overline{m}_b. \tag{3.72}$$

Linear stability analysis in the vicinity of this steady state leads to a cubic characteristic equation

$$\lambda^3 + a_1\lambda^2 + a_2\lambda + a_3 = 0$$

with coefficients

$$\begin{gathered} a_1 = 2(1+\overline{m}_b), \\ a_2 = (1+\overline{m}_b)^2 - G\overline{m}_b\left(2 - \tfrac{1}{2}\overline{n}_{ob}^2\right) - \tfrac{1}{2}\overline{m}_b^2 \approx -G\overline{m}_b\left(2 - \tfrac{1}{2}\overline{n}_{ob}^2\right), \\ a_3 = \tfrac{1}{2}G\overline{m}_b^2 + G\overline{m}_b(2 - \tfrac{1}{2}\overline{n}_{0b}). \end{gathered} \tag{3.73}$$

The approximate value of a_2 is valid because the parameter G is very large for solid-state lasers.

The roots of the cubic equation can be found strictly. However, the procedure becomes much more simple if a priori known that there is a pair of complex-conjugate roots among them and, moreover, that $|\mathrm{Re}\,\lambda / \mathrm{Im}\,\lambda| \ll 1$. In this case we come to

$$\begin{gathered} \mathrm{Im}\,\lambda = \Omega_1 \approx \sqrt{a_2}, \\ \mathrm{Re}\,\lambda = \theta_1 \approx -\frac{a_1 a_2 - a_3}{2a_2}. \end{gathered} \tag{3.74}$$

Since we neglected the higher spatial harmonics of inversion, the obtained results are valid only when the saturation of the laser medium is weak, i.e., close to the laser threshold. Limiting the pumping parameter by condition $A-1 \ll 1$, we have

$$\overline{m}_b = \frac{2}{3}(A-1), \tag{3.75}$$

instead of Eq. (3.71). In this limit expressions (3.74) take the form

$$\Omega_1 = \sqrt{G(A-1)}, \quad \theta_1 = -1/2 . \tag{3.76}$$

Formulas (3.75) and (3.76) coincide with (3.66)–(3.67) in the limit of small excess of the laser threshold.

Compare Eqs. (3.65)–(3.67) with similar formulas (3.15), (3.20) and (3.21) obtained earlier in the model with a spatial uniform field. The expressions for Ω_1 are fully coincident. The steady-state intensities $\overline{m}_b$ coincide when the laser is well above threshold. However, near threshold, in the domain $A \approx 1$, Eq. (3.65) leads to somewhere lower values of $\overline{m}$ than these yielded by Eq. (3.15). The reason for this result is physically apparent: the spatially uniform field interacts with all active molecules over the entire volume, whereas for standing wave mode the field does not interact with the molecules at the nodes at all. As for the damping rate, Eqs. (3.67) and (3.21) coincide at $A \to 1$, while at $A \gg 1$ the spatial inhomogeinity of the field increases the damping rate of small oscillations by a factor of 1.5.

Judging by the results we have obtained, the single-mode lasers with uniform and nonuniform field in the active medium volume differ little in their dynamical behaviours. However, one should remember that we have considered only the case of a spatially uniform excitation of the laser rod. Things are different if the unsaturated inversion is nonuniform: the damping rate of relaxation oscillations is especially sensitive to this factor, and if the quantity alternates in sign, then the steady state can lose temporal stability. This is discussed in Section 7.2.

3.3.3. Power Characteristics

In Section 3.2.2 we estimated the spike duration and the spike repetition rate. The agreement with experimental data is rather satisfactory. The energy characteristics of laser emission can be determined as follows. The field energy in the cavity is expressed through the field amplitude and the cavity volume. Making use of the relation of to the dimensionless intensity m, which is found from Eqs. (3.1), we obtain

$$W = \frac{\hbar^2 \gamma_\parallel \gamma_\perp V_c}{8\pi d^2} m . \tag{3.77}$$

Very often it is more convenient to use the formula containing the transition cross-section rather than the matrix element of the dipole moment of the transition d. These quantities are related by Eq. (1.2). Bearing this in mind, we write the field energy as the number of photons

$$M = \frac{W}{\hbar\omega} = \frac{\gamma_\parallel V_c}{2c\sigma_{tr}} m . \tag{3.78}$$

Provided that the main energy losses in laser are due to the energy withdrawal, the output power is found in a very simple way

$$P_{\text{out}}[\text{W}] = \frac{W}{T_c} = \frac{\hbar\omega\gamma_{\parallel}\kappa V_c}{c\sigma_{\text{tr}}} m \cdot 10^7 . \qquad (3.79)$$

To calculate the laser power in steady state we should substitute $m = A-1$ into Eq. (3.79). The maximum power of a spike can be found from the expression (3.33).

Example 3.5

Nd:YAG laser

$\sigma_{\text{tr}} = 10^{-18}$ cm^2; $\omega = 10^{15}$ s^{-1};	$P_{\text{out}}^{\text{cw}} = 0.3$ W;
$\gamma_{\parallel} = 5 \cdot 10^3$ s^{-1}; $\kappa = 10^8$ s^{-1};	$M^{\text{cw}} = 2 \cdot 10^{10}$ photons;
$A = 1.2$; $\ln(\overline{m}_b / m_{\min}) = 25$;	$P_{\text{out}}^{\max} = 8$ W.
$V_c = 1$ cm^3;	

For a ruby laser with $\sigma_{\text{tr}} = 10^{-20}$ cm^2, $\gamma_{\parallel} = 10^3$ s^{-1}, $A = 5$ the power increases up to $P_{\text{out}}^{\max} = 3$ kW.

These examples show that the simplest theory adequately reflects the energy characteristics of the free-running mode of laser operation.

Single-mode models hold a prominent place in the laser theory. In spite of their simplicity they offer reliable qualitative and even quantitative information about some practically important features of laser emission. Yet, a much more important conclusion is that the spike repetition rate and the spike duration as well as spike energy and power are not very sensitive to small perturbations of the model. Thus, there is no need to recognize such problems when analyzing more sophisticated problems. Among the more subtle laser characteristics is the dumping rate of the pulsations in laser.

3.4. Instabilities and Chaos in a Travelling-Wave Single-Mode Laser

In this section, we shall discuss the dynamics of class C lasers, as governed only by the relationship between the radiation field and the laser medium. Instabilities and complex pulsations in these lasers do not require external forcing of the system or the presence of additional nonlinear elements nor even, generally speaking, mode-mode coupling. Each of these additional factors can influence the behaviour and enrich the pattern but the primary cause of the laser instability is more fundamental.

3.4.1. Some history

Many authors have dwelt upon the interesting history of this problem (see, e.g., [200,247]) but each author has the right to propose his own version. So it seems reasonable to have a glimpse at the milestones.

Considerations of the stability of a quantum oscillator go back to the prelaser epoch [203, 204, 216, 248]. In 1958 Gurtovnik [248] showed that the steady-state solutions of the maser equations can be unstable in principle. At first these considerations were provoked by an interest in the stability of the ammonia beam maser but soon problems occurred with paramagnetic masers, the output pulsations of which were observed with CW pumping [69, 70]. Meanwhile the significance of this problem was fully realized when it was found that spike-like operation was inherent in practically all solid-state lasers [249-251].

An attempt to relate the generation of intensity spikes to dynamical instability was made by Korobkin and Uspensky [252, 253]. They investigated the threshold conditions for laser instability. Simultaneously, in the early 60s there were the first numerical computations of the time-dependent processes in quantum oscillators [254-256]. Grasyuk and Oraevsky, as well as Buley and Cummings, even obtained nonperiodic solutions. Unfortunately, those results were underestimated. The point was that the instability was achieved only with parameter combinations (the cavity passband was more than the homogeneous linewidth and the laser threshold was greatly exceeded) which seemed exotic at that time. And, though this might have been related to the new concepts of nonlinear dynamics, it had just started to be formulated and had not become widespread (the basic paper [177] by Lorenz was published in 1963).

The situation was changed radically by Haken's paper [176] in 1975. Haken found that the two-level laser model, used in papers [252-256], is fully isomorphic to the Lorenz system. Against the background of universal interest in dynamical chaos phenomena in relatively low-dimensional systems this fact put the experimental achievement of the laser version of the Lorenz attractor on the agenda.

Since solid-state lasers had the firm reputation of being unstable systems, the experimentalists' primary attention was paid above all to these devices. Meanwhile Casperson and Yariv [195] added consideration of a laser of quite a different type having detected undamped pulsations in the gas-discharge xenon laser output. Of course, a full concordance with the Lorenz–Haken model was not possible, since the gain line at 3.51 mm in xenon was inhomogeneously broadened. However, earlier Yakubovich [257] had paid attention to the specific features of inhomogeneously broadened media. Thereafter Idiatulin and Uspensky [258] analyzed a single-mode laser model with a two-component gain line and showed, using this simplest example, that inhomogeneous broadening led to a remarkable lowering of the instability threshold. A little bit later Fradkin [215, 259] discussed the possible manifestation of this effect under somewhat different conditions (a smooth profile of inhomogeneous broadening, multimode cavity). The further course of events was greatly influenced by the papers of Casperson [194, 196, 200, 260].

His theoretical analysis focused, unlike previous work, on explaining the particular experiment with a xenon laser.

Achievements in the early 80's changed crucially the experimental situation with laser chaos and they included the results of Abraham and coworkers. They realized and investigated the unstable behaviour first at 3.51 mm in the He-Xe laser [197, 198, 201, 202] and then at 3.39 mm in the He-Ne laser [33].

Inhomogeneous broadening greatly facilitates the task of an experimentalist pursuing the aim of realizing and investigating complex dynamical processes in lasers. Equally, this is a headache for a theoretician dealing with very high-dimensional models dependent on a large number of parameters. Although some results have been obtained analytically [261-270], much more progress has been achieved by numerical methods [270-276]. Nevertheless, "a detailed study of all candidate dynamical evolutions appears to be a Herculean task" (Abraham, Mandel and Narducci [247]).

Thus, it is clear why, being tempted to thoroughly verify the predictability of the modern theory of nonlinear oscillations, the experimentalists could not avoid the realization of Lorenz chaos in lasers. Weiss and Klische [38] soon showed that laser-pumped FIR lasers were most promising in this respect. A little bit later the experiments with ammonia lasers [39, 40, 42, 178–180] confirmed many predictions: the instability was reached once the laser threshold is well exceeded, higher-order bifurcations were observed, and chaos was achieved. Meanwhile, the experiments revealed an inconsistency between these scenarios and the Lorenz–Haken model. It should be most natural to attribute this inconsistency to the coherency of the pumping. The nonlinear interaction between the pumping and the laser field can be taken into account consistently but this is useful only if the three-level model widens appreciably the possibilities for interpretation of the experimental facts [277–287]. However, the experimental realization of the Lorenz attractor in such lasers requires breaking of the coherency of the laser field-monochromatic pumping interaction. Attempts in this direction were made, but the main prize for the experimental realization of the Lorenz attractor in lasers has not been awarded.

3.4.2. The Lorenz–Haken Model

Let us turn to a single-mode two-level model of a traveling wave laser with incoherent pumping, which in its simplest version is isomorphing to the Lorenz system. This laser can be described by Eqs. (3.2). Since the time normalization to $\gamma_\perp^{-1}$ is convenient in this case, these equations take the form

$$\frac{\mathrm{d}f}{\mathrm{d}\tau} - i\tilde{\kappa}\Delta_c f = \tilde{\kappa}(p - f), \tag{3.80a}$$

$$\frac{\mathrm{d}p}{\mathrm{d}\tau} - i\Delta_0 p = nf - p, \tag{3.80b}$$

$$\frac{\mathrm{d}n}{\mathrm{d}\tau} = \tilde{\gamma}\left[A - n - \frac{1}{2}(fp^* + f^* p)\right]. \tag{3.80c}$$

Here we use the following normalization: $\tilde{\kappa} = \kappa/\gamma_\perp, \tilde{\gamma} = \gamma_\parallel/\gamma_\perp$, i.e., $\hat{t} = 1/\gamma_\perp$.

These equations can be written in their real form in different ways. Changing from complex amplitudes to modules and arguments, $f = F\exp(i\varphi_e), p = P\exp(i\varphi_p)$, we transform Eqs. (3.80) to

$$\frac{\mathrm{d}F}{\mathrm{d}\tau} = \tilde{\kappa}(P\cos\Phi - F), \tag{3.81a}$$

$$\frac{\mathrm{d}P}{\mathrm{d}\tau} = nF\cos\Phi - P, \tag{3.81b}$$

$$\frac{\mathrm{d}n}{\mathrm{d}\tau} = \tilde{\gamma}(A - n - FP\cos\Phi), \tag{3.81c}$$

$$\frac{\mathrm{d}\Phi}{\mathrm{d}\tau} = \Delta_{0c} + \left(\frac{nF}{P} - \tilde{\kappa}\frac{P}{F}\right)\sin\Phi, \tag{3.81d}$$

where $\Phi = \varphi_e - \varphi_p$ and $\Delta_{0c} = (\omega_0 - \omega_c)/\gamma_\perp$. The advantage of this form of equations is that there is no unknown laser frequency given explicitly, and the disadvantage is that the right-hand terms of the phase equation (3.81d) can reach infinity. This last fact is unpleasant in the numerical simulation. Thus, the equations are often organized so that the variables are the real and imaginary parts of the complex amplitudes: $f = f' + if'', p = p' + ip''$. We therefore arrive at

$$\frac{\mathrm{d}f'}{\mathrm{d}\tau} + \tilde{\kappa}\Delta_c f'' = \tilde{\kappa}(p' - f'), \tag{3.82a}$$

$$\frac{\mathrm{d}f''}{\mathrm{d}\tau} - \tilde{\kappa}\Delta_c f' = \tilde{\kappa}(p'' - f''), \tag{3.82b}$$

$$\frac{dp'}{d\tau} + \Delta_0 p'' = nf' - p', \tag{3.82c}$$

$$\frac{dp''}{d\tau} - \Delta_0 p' = nf'' - p'', \tag{3.82d}$$

$$\frac{dn}{d\tau} = \tilde{\gamma}(A - n - f'p' - f''p''). \tag{3.82e}$$

There is an apparent paradox in that the Eqs. (3.82) have one equation more than the set (3.81). The contradiction is eliminated by noting that one of the two phases comprising Φ can be chosen arbitrary. This gives grounds to consider the corresponding variable a real one. Assuming $f'' = 0$, from Eqs. (3.82) we find the fourth-order set

$$\frac{df'}{d\tau} = \tilde{\kappa}(p' - f'), \tag{3.83a}$$

$$\frac{dp'}{d\tau} + \Delta_0 p'' = nf' - p', \tag{3.83b}$$

$$\frac{dp''}{d\tau} - \Delta_0 p' = -p'', \tag{3.83c}$$

$$\frac{dn}{d\tau} = \tilde{\gamma}(A - n - f'p'). \tag{3.83d}$$

The fifth equation has become algebraic,

$$\Delta_c f' = -p'', \tag{3.84}$$

actually, it determines the unknown lasing frequency ω, since $\Delta_c = (\omega - \omega_c)/\kappa$.

The case of resonant tuning, $\Delta_0 = \Delta_c = \Delta_{0c} = 0$, is the simplest. If we introduce, following [288], a variable $Z = FP\sin\Phi$, then using Eqs. (3.81) we can see that the equation $dZ/d\tau = -(\tilde{\kappa} + 1)Z$ is valid. Obviously, Z tends to zero during the laser action. There are two options, $\Phi = 0$ and $\Phi = \pi$, for the phase difference when Eqs. (3.81) transform to Lorenz–Haken system

$$\frac{dF}{d\tau} = \tilde{\kappa}(P - F), \tag{3.85a}$$

$$\frac{dP}{d\tau} = nF - P, \tag{3.85b}$$

$$\frac{dn}{d\tau} = \tilde{\gamma}(A - n - FP). \tag{3.85c}$$

It is sufficient to put $\Delta_0 = 0$ to obtain Eqs. (3.85) from Eqs. (3.83).

3.4.3. Bifurcations and Their Sequences

Besides the trivial steady-state solution

$$\overline{F}_a = \overline{P}_a = 0, \quad \overline{n}_a = A, \tag{3.86}$$

Eqs. (3.85) can also have nontrivial steady-state solutions

$$\overline{F}_b = \overline{P}_b = \pm\sqrt{A-1}, \quad \overline{n}_b = 1. \tag{3.87}$$

The steady state (3.86) becomes unstable when the laser self-excitation condition

$$A > 1 \tag{3.88}$$

is met, which, simultaneously, is the condition for physical existence of the equilibrium states (3.87), since the steady state coordinate cannot be a complex number.

Stability of the nonzero steady-state solution is investigated by linearization of Eqs. (3.85) in the vicinity of the fixed point b [27, 204, 237, 246, 252, 253, 255]. Assuming $F = \overline{F}_b + \delta F$, $P = \overline{P}_b + \delta P, n = \overline{n}_b + \delta n$ and retaining the terms linear with respect to these small deviations, we get a linearized set of equations

$$\frac{d}{d}(\delta F) = \tilde{\kappa}(\delta P - \delta F),$$

$$\frac{d}{d}(\delta P) = \overline{n}_b \delta F + \overline{F}_b \delta n - \delta P, \tag{3.89}$$

$$\frac{d}{d}(\delta n) = -\tilde{\gamma}(\delta n + \overline{F}_b \delta P + \overline{P}_b \delta F).$$

Supposing that the solution has the form $\{\delta F, \delta P, \delta n\} = \{\delta F_0, \delta P_0, \delta n_0\} e^{\lambda\tau}$, we reduce this uniform set of differential equations to an set of algebraic equations. Vanishing of the determinant means the existence of nontrivial solutions. We thus arrive at a cubic characteristic equation

$$\sum_{j=1}^{3} a_j \lambda^{3-j} = 0$$

with the coefficients

$$a_0 = 1, \quad a_1 = \tilde{\kappa} + \tilde{\gamma} + 1, \quad a_2 = \tilde{\gamma} + (A + \tilde{\kappa}), \quad a_3 = 2\tilde{\gamma}\tilde{\kappa}(A-1). \tag{3.90}$$

The presence of roots with the positive real part implies that the corresponding steady-state solution is unstable.

To obtain the instability condition in analytical form we can make use of the Routh–Hurwitz criterion. Meanwhile, a better result can be achieved under the assumption of complex roots meeting the inequality $|\mathrm{Re}\lambda/\mathrm{Im}\lambda| << 1$: approximate values of the roots can be found, as in Section 3.3.2, by linearization of the characteristic equation in respect with $\mathrm{Re}\lambda$ [289]. Taking this route, we find:

$$\lambda_{1,2} \approx \frac{a_3 - a_1 a_2}{2a_1^2} \pm i\sqrt{\frac{a_3}{a_1}}, \quad \lambda_3 \approx -a_1. \tag{3.91}$$

The instability condition $a_3 > a_1 a_2$ leads to two inequalities

$$\tilde{\kappa} > \tilde{\gamma} + 1, \tag{3.92}$$

$$A > A_{cr} = \tilde{\kappa}\frac{\tilde{\kappa} + \tilde{\gamma} + 3}{\tilde{\kappa} - \tilde{\gamma} - 1}. \tag{3.93}$$

The first of them requires that the cavity passband exceeds the homogeneous linewidth. The inequality (3.93) defines the height of the so-called second laser threshold (assuming that the laser self-excitation threshold is the first one). The quantity A_{cr} has the minimum at $\tilde{\kappa} = \tilde{\kappa}_m = \tilde{\gamma} + 1 + \sqrt{(\tilde{\gamma}+1)(2\tilde{\gamma}+4)}$. In turn, $\tilde{\kappa}_m$ decreases with $\tilde{\gamma}$ reaching its least value $\tilde{\kappa}_m = 3$ at $\tilde{\gamma} << 1$. According to Eq. (3.93), this least value corresponds to $A_{cr} = 9$.

What are the real possibilities for both instability conditions to be met? Only gases can satisfy the inequality (3.92), since a narrow gain line rather than a wide cavity band is the decisive factor (from this point of view the widespread term *bad cavity condition* is not quite suitable). To estimate the maximum attainable pumping parameter we turn to a relation

$$A = \frac{\pi\omega d}{\hbar\kappa\gamma_\perp} N_S D^{(0)} = A_{max} D^{(0)},$$

which enters (3.1). Assuming parameter values $\omega = 2\cdot10^{13}\ s^{-1}$, $\gamma_\perp = 10^7\ s^{-1}$, $\kappa = 10^7\ s^{-1}$, $d = 1$ Db and $N_S = 10^{14}\ cm^3$, we find $A_{max} \approx 10^4$. Therefore, the excess of laser threshold, required in Eq. (3.93), is practicable in low-pressure lasers.

The curve (3.93) is shown in Fig. 3.4 as the solid line 1. The remaining data of this phase diagram are obtained by numerical integration of Eqs. (3.85) [285, 290, 291]. In the region above dash line 2 only steady state solutions of Eqs. (3.87) are stable. In the region below solid line 1 the steady state solutions are unstable. Between these lines there is a bistability zone within which both steady state and self-modulations are locally stable. Bistability is accompanied by the hysteresis phenomenon with a typical

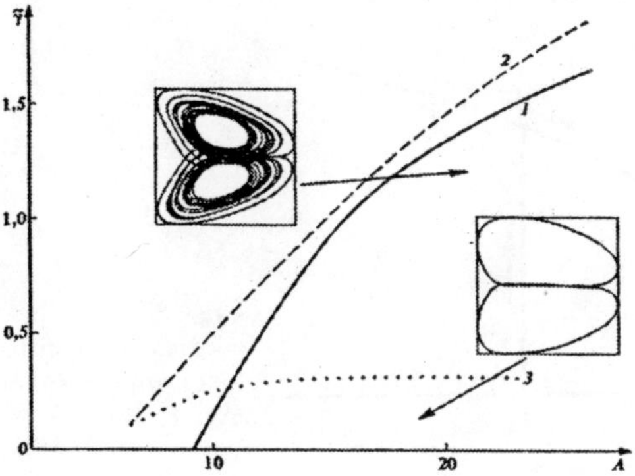

Fig. 3.4. Phase diagram of a single-mode two-level laser model in the control parameter plane $(A, \tilde{\gamma})$ at $\tilde{\kappa} = 4.0$ [285, 290]: *1* – boundary of the domain of instability of nonzero steady-state solutions, *2* – boundary of the domain, in which the nonzero steady-state solutions of Eqs. (3.85) are uniquely stable; *3* – boundary between zones of chaotic and regular behaviour. Types of the attractors corresponding to the different areas in the diagram are shown

jump-like transition from one to another stable branch of solutions at the instability zone boundaries. If the control parameters are varied quasistatically so that operating point on the phase diagram moves along a trajectory *s*, which crosses both boundaries, the observed sequence of states corresponds to the loop in Fig. 3.5. The diameter of the attractor in the phase space is shown on the ordinate. The steady state conforms to a point attractor with $D = 0$. The finite value of D means the presence of a limit cycle or a strange attractor[1].

The existence of an attractor with a finite diameter, immediately after the instability boundary $A = A_{cr}$ is crossed, indicates that a subcritical Hopf bifurcation occurs. The transient process from an unstable steady state to developed pulsations is shown in Fig. 3.6. At first there is amplitude modulation with gradually increasing depth and neither variable reverses sign at this stage. At some time, however, this process is replaced by a qualitatively different mode of oscillations with sign reversal of the field and the

[1]In numerical simulations the quasistatic parameter variation is imitated as follows. The first realization is calculated with arbitrary chosen initial values. Then the control parameter is varied be a chosen discrete step and the new realization is calculated from the set of variable values at which the preceding solution was interrupted. These initial conditions make it possible to avoid the transient process and exclude the solution from entering the attractor basin of the alternative attractor in the bistability zone. The initial conditions for the field and polarization, which are close to zero, we assigned only if the aim of the numerical experiment is to define the boundaries of hard excitation of pulsations) dash boundary 3 in Fig. 3.4.

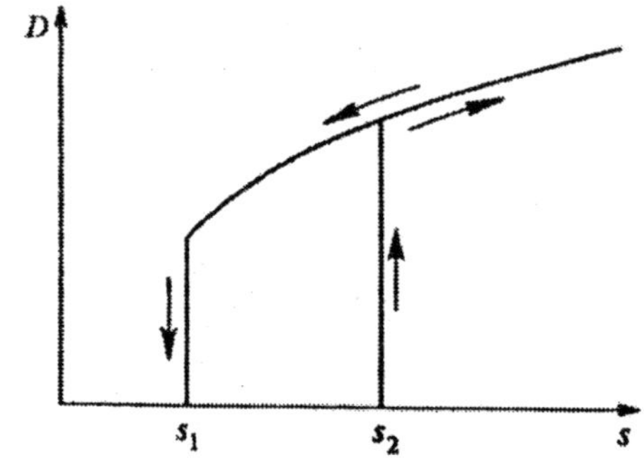

Fig. 3.5. Diagram illustrates the coexistence of different attractors in the laser phase space (bistability and hysteresis).

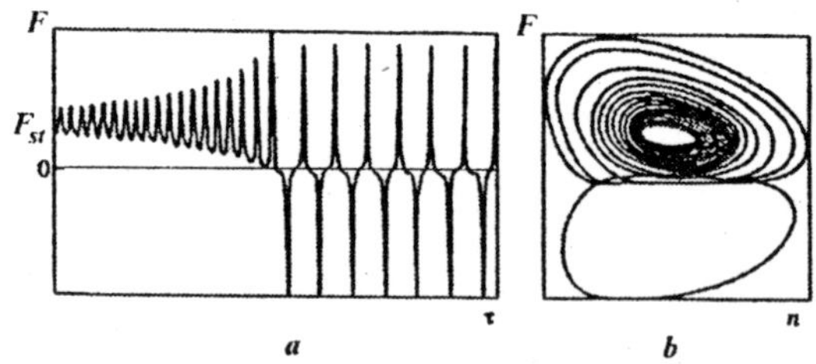

Fig. 3.6. (*a*) Transition from unstable equilibrium state to regular undamped pulsations in the Lorenz–Haken model and (*b*) phase space trajectory projection onto the plane n, F [291]. $\tilde{\kappa}=4.0$; $\tilde{\gamma}=0.10$; $A=12.0$.

polarization amplitudes F and P. Excess of odd harmonics is a distinctive feature of this radiation field envelope spectrum (Fig. 3.7). It is reasonable to call the symmetrical pulsations 'beats', since the corresponding spectrum of the laser field does not have a central mode [290]. In the simplest version, the beats are represented by two spectral components. It is well known that the presence of a central mode (carrier frequency) is necessary in the case of amplitude modulation, moreover, this component dominates.

The frequency of small oscillations near the steady state is given by

$$\Omega_1 = \sqrt{\frac{a_3}{a_1}} = \Omega_R \sqrt{\frac{2\tilde{\kappa}}{\tilde{\kappa}+\tilde{\gamma}+1}}, \tag{3.94}$$

which is derived from Eq. (3.91) using Eq. (3.90). Here $\Omega_R = \sqrt{\tilde{\gamma}(A-1)}$ is the dimensionless Rabi frequency in the CW radiation field (3.87) with intensity $A-1$. For class C the coefficient at Ω_R in Eq. (3.94) only slightly differs from unity; in the limiting case $\tilde{\kappa} \gg 1+\tilde{\gamma}$ Eq. (3.94) reduces to $\Omega_1 = \sqrt{2}\Omega_R$.

The frequency of settled beat-like pulsations can be estimated by assuming solutions of Eqs. (3.85) as $\{F, P, n\} = \{F_1, P_1, n_1\}\exp(i\Omega\tau) + \text{c.c.}$ disregarding all higher beat harmonics. Omitting calculations we give the

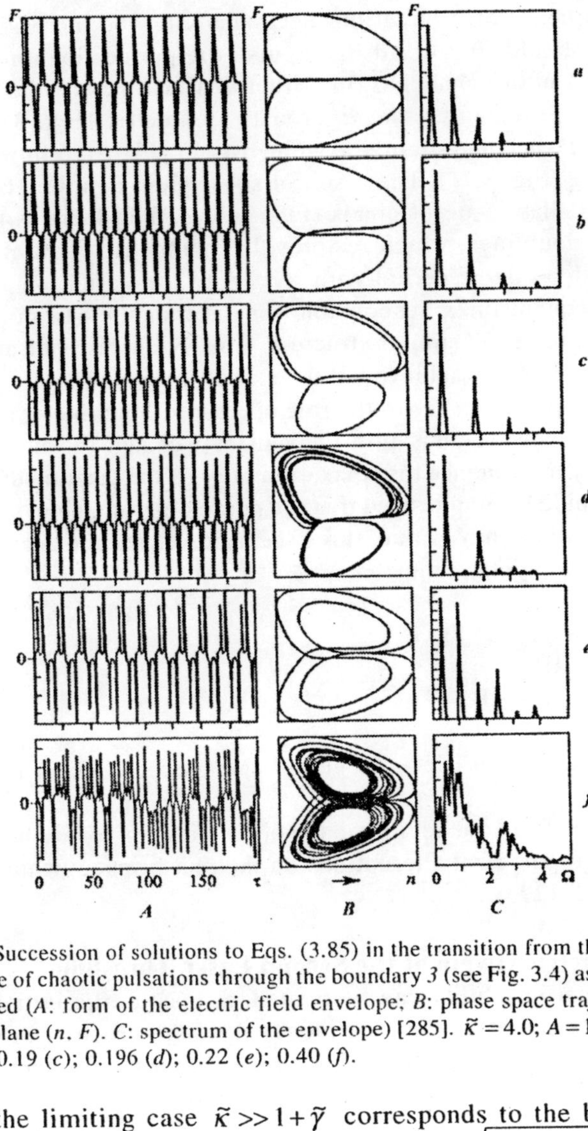

Fig. 3.7. Succession of solutions to Eqs. (3.85) in the transition from the zone of regular to the zone of chaotic pulsations through the boundary *3* (see Fig. 3.4) as the parameter $\tilde{\gamma}$ is increased (*A*: form of the electric field envelope; *B*: phase space trajectory projection onto the plane (*n*, *F*). *C*: spectrum of the envelope) [285]. $\tilde{\kappa} = 4.0$; $A = 12.0$; $\tilde{\gamma} = 0.10$ (*a*); 0.17 (*b*); 0.19 (*c*); 0.196 (*d*); 0.22 (*e*); 0.40 (*f*).

result: the limiting case $\tilde{\kappa} >> 1 + \tilde{\gamma}$ corresponds to the beat frequency $2\Omega = \Omega_R$. In the natural dimension we have $\omega_R = \sqrt{\gamma_\parallel \gamma_\perp (A-1)}$ which yields $\omega_R / 2\pi \approx 2$ MHz at $\gamma_\perp = 10^7\ s^{-1}$, $\gamma_\parallel = 10^6\ s^{-1}$ and A=11. Experiments with molecular FIR lasers confirm that the dynamical pulsations have megahertz frequencies [247]. Undamped pulsations above the second laser threshold can be both regular and chaotic. In Fig. 3.4 the domains of regular and chaotic behaviour are separated by a dash line. Switching from one to another is not abrupt and includes a complex hierarchic sequence of bifurcations sketched in Fig. 3.7.

In the sequence of bifurcations accompanying the increase in parameter $\tilde{\gamma}$ one should first of all distinguish the period doubling chain. In turn, each link of this chain has fine structure illustrated by the first link as example. As $\tilde{\gamma}$ increases, the rigorously symmetric (Fig. 3.7*a*) regular pulsations (RP) become slightly asymmetric (Fig. 3.7*b*), followed by periodic doubling (Fig. 3.7*c*). The result of subsequent doubling bifurcations is the intermediate state of chaos (Fig. 3.7*d*). Then there is the inverse sequence of doublings, which restores the symmetric form of the pulsations but with a double period with respect to the initial one (Fig. 3.7e). All the subsequent links, responsible for the transition from period 2 to period 4, etc. have the similar structure. Finally chaotic pulsations (CP), shown in Fig. 3.7*f*, are achieved at the end of this chain.

In the example given above the role of a control parameter is played by $\tilde{\gamma}$. In practice, however, it is more convenient to change the pumping power rather than try to influence the relaxation rates of the laser medium. In this respect, it should be emphasized that the phase diagram, shown in Fig. 3.7, admits, depending on $\tilde{\gamma}$ value, three scenarios of regime change with increase of the pumping parameter A:

1) CW—CP,
2) CW—RP,
3) CW—CP—RP.

Note, however, that isles of regular behaviour may occur in the sea of chaos [292].

The passage from one region on the phase diagram to another is indicative of some topological alternations on the phase space, which were described in Ref. [293].

3.4.4. Parametric Origin of the Second Laser Threshold

This dynamical instability can be treated, at least at the initial stage, as parametric excitation of the side modes, i.e., the process, in which the oscillating mode plays the role of pumping. In terms of quantum theory, we are dealing with a four-photon interaction, which satisfies the frequency matching condition $2\omega_0 = \omega_+ + \omega_-$. Using these premises it is easy to find the instability criteria [289]. Assume that the laser generates a three-harmonic field

$$F = \overline{F}_b + F_1 e^{i\Omega\tau} + F_{-1} e^{-i\Omega\tau}. \tag{3.95}$$

Until the side mode have small amplitudes, $|F_{\pm 1}/\overline{F}_b| \ll 1$, their harmonics will not matter so that the material variables can be presented in a similar form:

$$P=\overline{P}_b+P_1e^{i\Omega\tau}+P_{-1}e^{-i\Omega\tau}, \quad n=\overline{n}_b+n_1e^{i\Omega\tau}+n_{-1}e^{-i\Omega\tau}. \tag{3.96}$$

The assumed smallness of all sideband components permits one to neglect the reaction of $\overline{F}_b, \overline{P}_b$ and $\overline{n}_b$ to a perturbation and make use of Eqs. (3.87).

Substituting Eqs. (3.95) and (3.96) into Eqs. (3.85), separating the groups of terms with similar frequency dependence and neglecting the products of small quantities, we obtain a set of homogeneous algebraic equations

$$(\widetilde{\kappa}+i\Omega)F_1-\widetilde{\kappa}P_1=0, \tag{3.97a}$$

$$\overline{n}_bF_1-(1+i\Omega)P_1+\overline{F}_bn_1=0, \tag{3.97b}$$

$$\widetilde{\gamma}\overline{F}_bF_1+\widetilde{\gamma}\overline{F}_bP_1+(\widetilde{\gamma}+i\Omega)n_1=0. \tag{3.97c}$$

The complex characteristic equation, which corresponds to the vanishing determinant, falls into a pair of real equations:

$$\Omega^2=2\widetilde{\gamma}\widetilde{\kappa}\overline{F}_b^2/(\widetilde{\kappa}+\widetilde{\gamma}+1), \tag{3.98}$$

$$\Omega^2=\widetilde{\gamma}(\overline{F}_b^2+1+\widetilde{\kappa}). \tag{3.99}$$

The Eq. (3.99) coincides with Eq. (3.94) and become consistent with Eq. (3.98) when $\overline{F}_b^2=A_{cr}-1$, where A_{cr} is given by Eq. (3.93).

A different variety of the weak-sideband approach, which is widely practiced [194, 260, 267-269, 296], goes back to the early papers [294, 295]. The main point is to determine the susceptibility of the laser medium in the presence of a strong oscillating mode. Expression

$$\frac{P_1}{F_1}=\frac{\widetilde{\gamma}(A-2)-i\Omega}{\Omega^2-\widetilde{\gamma}A-i\Omega(1+\widetilde{\gamma})}=\widetilde{\chi}(\Omega). \tag{3.100}$$

can be derived by solving a set of two last equations in (3.97). The profiles of side components gain and dispersion, specified by the relation (3.100), are shown in solid lines in Fig. 3.8.

Self-modulation regime of operation exists if Eq. (3.100) is compatible with Eq. (3.97a). It is convenient to represent the latter equation as

$$\frac{P_1}{F_1}=1+i\frac{\Omega}{\widetilde{\kappa}}. \tag{3.101}$$

The plots $\mathrm{Re}\,\widetilde{\chi}_1=1$ and $\mathrm{Im}\,\widetilde{\chi}=\Omega/\widetilde{\kappa}$ are given in Fig. 3.8 by dashed lines. The intersection points in Fig. (3.8) determine the frequencies of modes degenerate with respect to wavenumber. The presence of several frequencies like these gives grounds to talk of 'mode splitting' following the terminology of the work [297]. The physical interpretation of the instability conditions is that the points of intersection enter the domain of sufficiently strong gain.

The graphic method is not the most convenient in application to a two-level homogeneously broadened laser. Several ways to analytical solution of the laser stability problem have been shown above. We add that comparison of Eqs. (3.100) and (3.101) leads to Eqs. (3.98) and (3.99) as well. Meanwhile, this method can be preferable in more complicated situations, in an inhomogeneously broadened laser, for example.

3.4.5. Effect of Detuning on the Laser Dynamic Properties

Detuning between the cavity mode frequency and the center frequency of the gain line increases the dimension of the laser model. One might think that this leads to increased complexity of laser behaviour but the actual tendency is opposite: the instability threshold increases and chaotic pulsations in the domain of instability are replaced by regular ones.

There is a specific complication of the problem when a detuning is introduced since besides the variables f, p, n we have one more unknown quantity – the laser frequency, hidden in Δ_0 and Δ_c. This causes problems even in the linear stability analysis of the zero intensity branch of the steady-state solution. A way out is to turn from Eqs. (3.2) to Eqs. (3.4), which do not contain the laser frequency. Linearization of Eqs. (3.4) near the zero singularity leads to a characteristic equation $\sum a_j \lambda^{4-j} = 0$ with the coefficients

$$a_0 = 1, \quad a_1 = 4(\tilde{\kappa}+1), \quad a_2 = 5(\tilde{\kappa}+1)^2 - 4\tilde{\kappa}(A-1) + \Delta_{0c}^2, \tag{3.102}$$

$$a_3 = 2(\tilde{\kappa}+1)^3 - 8\tilde{\kappa}(\tilde{\kappa}+1)(A-1) + 2(\tilde{\kappa}+1)\Delta_{0c}^2, \quad a_4 = -4\tilde{\kappa}(\tilde{\kappa}+1)(A-1) + 4\tilde{\kappa}\Delta_{0c}^2.$$

Using the Routh–Hurwitz criterion we find the laser self-excitation condition $a_4 < 0$, or

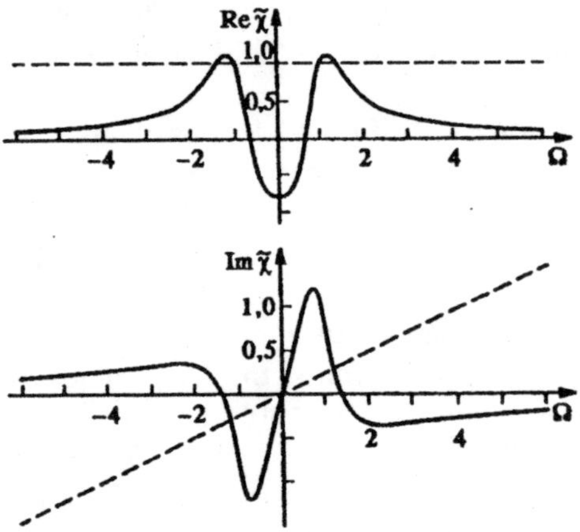

Fig. 3.8. Plots illustrate the induced mode splitting effect: *a* – gain, *b* – dispersion.

$$A > 1 + \frac{\Delta_{0c}^2}{(1+\tilde{\kappa})^2}. \quad (3.103)$$

The nontrivial steady-state solution of Eqs. (3.4) is given by

$$\overline{n}_b = 1 + \frac{\Delta_{0c}^2}{(1+\tilde{\kappa})^2}, \quad \overline{m}_b = A - 1 - \frac{\Delta_{0c}^2}{(1+\tilde{\kappa})^2}. \quad (3.104)$$

Meanwhile, it is more convenient to complete a linear stability analysis by turning to Eqs. (3.83) and (3.84) as in papers [247, 292, 298]. If we find the steady state directly from Eqs. (3.83) and (3.84) we have

$$\overline{n}_b = 1 + \Delta_{0c}^2, \quad \overline{m}_b = A - 1 - \Delta_{0c}^2, \quad \Delta_c = -\Delta_0, \quad p' = f, \quad p'' = -\Delta_c f. \quad (3.105)$$

The relations (3.105) agree with (3.104) provided $\Delta_c^2 = \Delta_{0c}^2/(\tilde{\kappa}+1)^2$. This should be readily apparent if we make use of the equality $\Delta_0 = -\Delta_c$. Linearization of Eqs. (3.83)-(3.84) near the fixed point (3.105) leads to a quartic characteristic equation with the coefficients

$$a_0 = 1, \quad a_1 = 2\tilde{\kappa} + 2 + \tilde{\gamma}, \quad a_2 = 2\tilde{\gamma}(\tilde{\kappa}+1) + (\tilde{\kappa}+1)^2 + \tilde{\gamma}\overline{m}_b + \Delta_c^2(\tilde{\kappa}-1)^2 \quad (3.106)$$

$$a_3 = \tilde{\gamma}(\tilde{\kappa}+1)^2 + \tilde{\gamma}(3\tilde{\kappa}+1)^2\overline{m}_b + \Delta_c^2\tilde{\gamma}(\tilde{\kappa}-1)^2, \quad a_4 = 2\tilde{\gamma}\tilde{\kappa}(\tilde{\kappa}+1)\overline{m}_b.$$

The assumption of complex roots $\lambda = i\Omega + \theta$ with a small real part permits one to write down, instead of the characteristic equation, the equalities

$$(a_1 + 4\theta)\Omega^2 = a_3 + 2a_2\theta, \quad \Omega^4 - a_1\Omega^2 + a_4 = (3a_1 - a_3)\theta.$$

The first one determines the frequency of small oscillations

$$\Omega \approx \sqrt{a_3/a_1}, \quad (3.107)$$

and the second governs the damping factor

$$\theta = \frac{a_3^2 + a_1a_4 - a_1a_2a_3}{a_1^2(3a_1 - a_3)}. \quad (3.108)$$

Generally, finding the bifurcation value of m_{cr} from $\theta = 0$ requires cumbersome calculations, but in the limit $\tilde{\kappa} \gg 1$ (the necessary instability condition is retained with detuning) we are to solve the quadratic equation

$$3\tilde{\gamma}m_{cr}^2 + 2\tilde{\kappa}(1 - 3\Delta_c^2)m_{cr} - 2\tilde{\kappa}^3(1+\Delta_c^2)^2 = 0. \quad (3.109)$$

At $\Delta_c^2 = 1/3$ one coefficient of this equation turns to zero. The laser behaviour is different on either side of this detuning [298].

If the dependence $m_{cr}(\Delta_c)$ and Eq. (3.105) are known, it is rather easy to obtain $A_{cr}(\Delta_c)$. The form of the latter is illustrated in Fig. 3.9 where

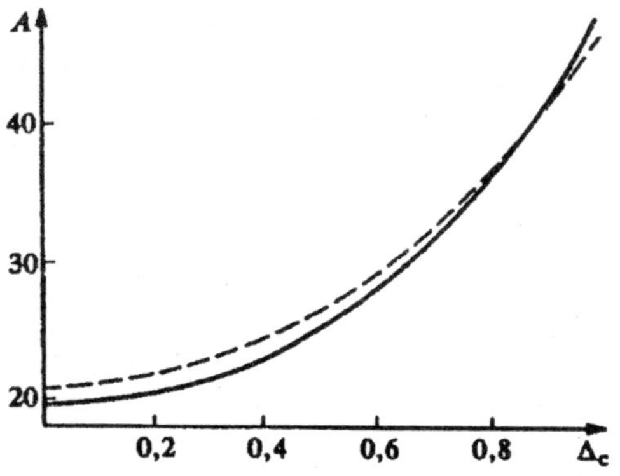

Fig. 3.9. Phase diagram of the single-mode two-level la]ser model in the control parameter plane (Δ_c, A). The sense of the boundaries is the same as that in Fig. 3.4 [292].

this dependence is shown by a solid line. Above this line the steady-state solutions are unstable. We deal with a bistable system when the operating point enters the band between the solid and the dash lines. Note that these lines are intersected. Below the intersection point the CW laser action coexists of finite amplitude pulsations (a stable fixed point and a large-diameter limit cycle). Above the intersection point a small-diameter limit cycle is an alternative to a large-diameter one. Correspondingly, the bifurcation changes at the second laser threshold: from subcritical with small detuning, the bifurcation becomes supercritical with large detuning. The position of above mentioned bifurcation point in the control parameter plane A, Δ_c depends on the coefficients. In the limit $\tilde{\kappa} >> 1$ this point is near $\Delta_c^2 = 1/3$ and shifts toward the larger Δ_c with decreasing $\tilde{\kappa}$.

Possible scenarios of laser behaviour depending on the control parameter variation are illustrated in Fig. 3.10. The additional information on Lorenz–Haken's system, as compared to Fig. 3.4, refers to laser operation far above threshold (A>60). Windows of regular behaviour occur exactly in the chaotic domain. An increase in detuning restricts this domain: an upper limit is seen for $\Delta_c = 0.1$ and an upper and a lower limit are seen for $\Delta_c = 0.2$. Chaotic pulsations are cancelled beginning from $\Delta_c = 0.3$ and all higher bifurcations disappear beginning from $\Delta_c = 0.5$. Figure 3.10 shows three routes to enter the chaotic self-modulation domain: an increase in pumping, a decrease in pumping and a decrease in detuning. It is interesting to note that a transition to chaos through a succession of period doubling (Feigenbaum's scenario) occurs in all these cases.

The increased dimension of the system from 3 to 5 raises the question: Is an introduction of detuning accompanied by jump-like restructuring of the phase space of the laser? A negative answer was given in [299]. A special investigation was required because the phase portrait plotted by numerical integration of Eqs. (3.80) is appreciably dependent on the cho-

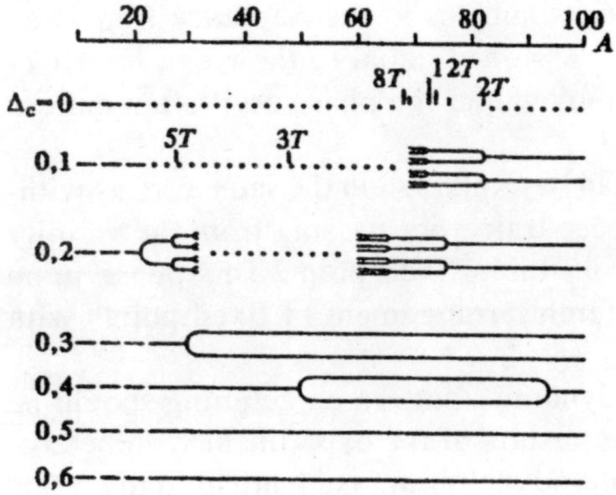

Fig. 3.10. Scenarios of regime change in a two-level laser model at different detuning [292]. The domains of steady-state solutions are shown by dashed lines; the domains of periodic pulsations by solid lines and the domains of irregular pulsations by dotted lines ($\tilde{\kappa} = 3, \tilde{\gamma} = 1$).

sen reference frequency ω, which remains indefinite in the abbreviated equations. Only for a certain reference frequency, which can be determined by the method proposed in [299], the attractor is bilateral, like that shown in Figs. 3.4 and 3.7. Varying ω it is easy to lose this, as well as any other symmetry of the attractor.

3.4.6. Phase Dynamics of a Single-Mode Laser

So far, speaking about the time-dependent processes we meant exactly the time variations in amplitude of all variables. At the end of the Section 3.4.2 we made the sole assertion concerning the phase difference between the field and the medium polarization that Φ can assume only two values: 0 and π, under the exact resonance conditions. The value reversal is abrupt; physically it is switching from the field emission to absorption by the laser medium and back [288]. Most of the time the laser is in the state of emission, and it is the state of absorption only for a short period of time.

The phase difference between the field and the polarization can easily be determined by numerical simulation but definitely not in a real experiment. One can only retrieve the field phase using the optical heterodyning method [175]. Numerical integration of Eqs. (3.85) indicates that the field phase changed by π at the time the field amplitude passes zero. In the phase space of this system the phase discontinuity is associated with the trajectory passing from the vicinity of the fixed point $P = F = \sqrt{A-1}$ to $P = F = -\sqrt{A-1}$ and back.

All these phase dynamics mechanisms are inherent, besides periodic pulsations (Fig. 3.11), in chaotic pulsations (Fig. 3.12) as well.

Detuning changes the phase dynamic suggested by the numerical investigation. We mean that together with the step-like variations, the field phase undergoes a monotonic drift [41, 300]. This drift has a pure kinematic nature and the drift velocity is subject to the reference frequency variation. Specifically, when ω is assigned equal to the mean laser frequency, the drift is completely removed and the phase oscillation pattern is like that shown in Fig. 3.13*b*.

The phase jump phenomenon can be explained in the same way as without detuning, i.e., by the phase space trajectory passing from the vicinity of one to the vicinity of another unstable fixed point. This phase jump differs from π because of the different arrangement of fixed points with and without detuning.

This dependence of the phase dynamics pattern on detuning should be borne in mind when discussing the results of the experiment with heterodyne detection of laser output. Heterodyne tuning is identical to the reference frequency change during the numerical integration.

3.5. Dynamics of Three-Level Lasers with Coherent Pumping

If the laser is pumped by the monochromatic radiation of another laser and the auxiliary and the operation transitions have a common upper level (Fig. 1.2) then a two-level model for the laser medium appears to be insufficient-

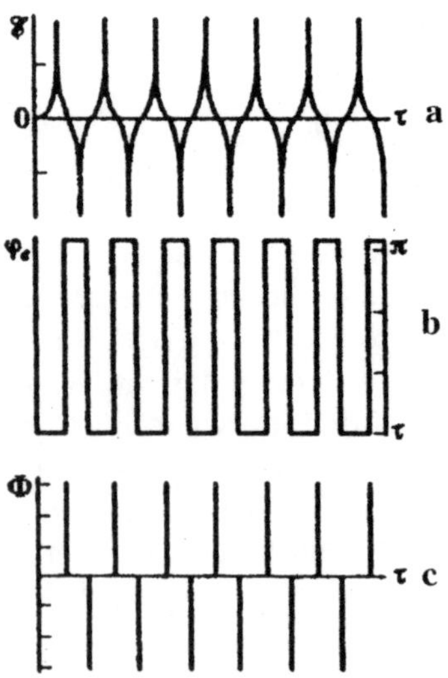

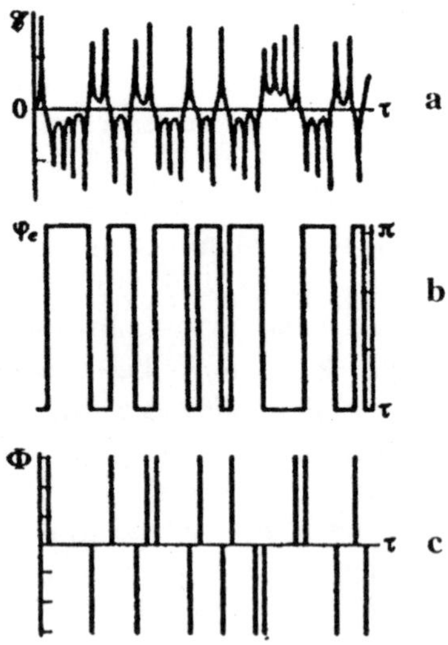

Fig. 3.11. Example of regular solution to Eqs. (3.80) for amplitudes and phases: (*a*) the electric field amplitude envelope; (*b*) the electric field phase φ_e; (*c*) the phase difference of field and polarization Φ.

Fig. 3.12 (right) Example of chaotic solution to Eqs. (3.80). The notation and the parameter values are the same as those given in Fig. 3.11 except for $\tilde{\gamma} = 0.4$.

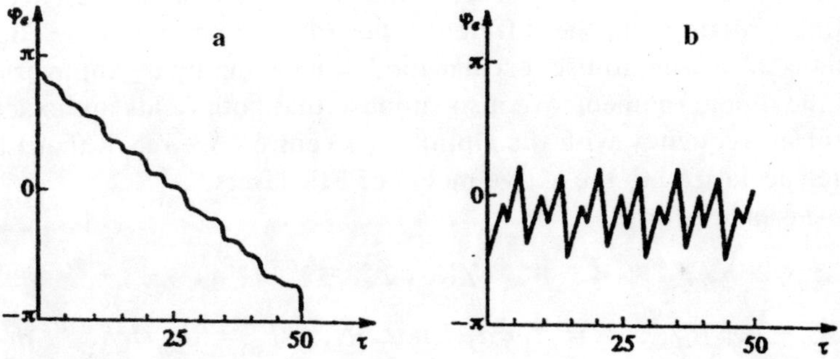

Fig. 3.13. Dynamics of the field phase of a two-level single-mode laser revealed by numerically solving Egs. (3.80) at (*a*) occasional and (*b*) specially chosen value of the reference frequency ω (detuning Δ_c) [300].

cient. Such a system can exhibit the effects caused by the coherent interaction between the pump field and the laser field as well as be the gain line splitting due to the Rabi oscillations in the pumping field [6, 227, 301-303].

We should differentiate between two cases that correspond to opposite limiting cases of detuning. In the first case, considered below, both the pumping and the laser field are exactly resonant with the corresponding molecular transitions. In the second case, the difference between each frequency and the corresponding transition frequency is much greater than the molecular linewidth. This second case corresponds to the Raman laser since the laser action is due to the two-photon interaction of fields with an inversionless transition between level 1 and 2. The dynamics of the Raman laser will not be considered here but it is discussed in [304–306]. In the resonance case the one-photon transitions between level 3 and 2 are of primary importance although the two-photon interactions mentioned above also contribute to the lasing and should not be ignored until the conditions for their confident suppression are provided. Essentially, we are dealing with a resonance Raman laser, but this term has not become widespread.

Mathematically, the transition from a two-level to a three-level laser model markedly increases the number of degrees of freedom and parameters of the system. The new features of laser dynamics include the emergence of the second instability domain on the trivial branch of steady-state solutions. The collection of time-dependent laser regimes and the scenarios of regime switching due to the control parameter variation are much wider.

3.5.1. Self-Consistent Model and Ways to Its Simplification

The model consists of the material Eqs. (2.49) and the laser field Eqs. (2.75a). The pumping field is assumed to be given. We disregard the effects due to the orientation disorder of the laser medium, the presence of a hyperfine structure and the different types of field polarization. These assumptions allow one to use a scalar model, meaning by d_{mn} an average value of the dipole moment. We also suppose that both fields are exactly coincident in frequency with the atomic line centres. We recall that Eqs. (2.49) are specialized to the active media of FIR lasers.

In dimensionless form

$$\tau = t\gamma_{32}, \quad \tilde{\kappa} = \kappa / \gamma_{32}, \quad \tilde{w} = w_{mn} / \gamma_{32}, \quad \tilde{\gamma}_{ml} = \gamma_{ml} / \gamma_{32},$$

$$\tilde{\gamma} = (w_{31} + 2w_{32})/\gamma_{32}, \quad R = (\pi\omega_{32} d_{32}^2 / \hbar\kappa\gamma_{32}) N_S, \quad F_{mn} = (d_{mn} / \hbar\gamma_{32}\tilde{\gamma}^{1/2}) \tilde{F}_{mn}, \tag{3.110}$$

$$P_{3n} = 2iR\tilde{\gamma}^{-1/2}\sigma_{3n}, \quad P_{21} = 2R\sigma_{21}, \quad n_m = R\sigma_{mm}, \quad n = n_3 - n_2,$$

the set of equations of a tree-level laser can be written as

$$\frac{dF_{32}}{d\tau} = \tilde{\kappa}(P_{32} - F_{32}), \tag{3.111a}$$

$$\frac{dP_{32}}{d\tau} = nF_{32} - P_{32} - \frac{1}{2} F_{31} P_{31}, \tag{3.111b}$$

$$\frac{dP_{21}}{d\tau} = -\tilde{\gamma}_{21} P_{21} + \frac{1}{2}\tilde{\gamma}(P_{31} F_{32} + F_{31} P_{32}), \tag{3.111c}$$

$$\frac{dP_{31}}{d\tau} = -\tilde{\gamma}_{31} P_{31} - F_{31}(R - 2n - 3n_2) - \frac{1}{2} P_{21} F_{32}, \tag{3.111d}$$

$$\frac{dn}{d\tau} = -\tilde{\gamma}\left(n + F_{32} P_{32} + \frac{1}{2} F_{31} P_{31}\right) + (\tilde{w}_{21} - \tilde{w}_{31}) n_2, \tag{3.111e}$$

$$\frac{dn_2}{d\tau} = \tilde{w}_{32} n - \tilde{w}_{21} n_2 + \frac{1}{2}\tilde{\gamma} F_{32} P_{32}. \tag{3.111f}$$

Since conditions

$$\{\tilde{\gamma}_{31}, \tilde{\gamma}_{32}\} >> \{1, \tilde{\gamma}, \tilde{\kappa}, F_{mn}^2\} \tag{3.112}$$

are satisfied, we can eliminate adiabatically the variables P_{31}, P_{32} and use, instead of six equations (3.111), four equations

$$\frac{dF_{32}}{d\tau} = \tilde{\kappa}(P_{32} - F_{32}), \tag{3.113a}$$

$$\frac{dP_{32}}{d\tau} = nF_{32} - P_{32}, \tag{3.113b}$$

$$\frac{dn}{d\tau} = \tilde{\gamma}(A - n - F_{32}P_{32}) + (\tilde{w}_{21} - \tilde{w}_{31} - 3\tilde{\gamma}A/R)n_2, \tag{3.113c}$$

$$\frac{dn_2}{d\tau} = \tilde{w}_{32}n - \tilde{w}_{21}n_2 + \frac{1}{2}\tilde{\gamma}F_{32}P_{32}. \tag{3.113d}$$

Adding the conditions

$$\tilde{w}_{31} = \tilde{w}_{21}, \tag{3.114}$$

$$\tilde{\gamma}A/R << \tilde{w}_{21} \tag{3.115}$$

we can separate from Eqs. (3.113) the set of three equations. In order to ensure full coincidence with the Lorenz system (3.85), we have introduced a pumping parameter $A = RF_{31}^2/(2\tilde{\gamma}_{31})$. The number of parameters in Eqs. (3.111) can be reduced without depleting the physical pattern. In particular, we can put $\tilde{\gamma}_{31} = \tilde{\gamma}_{21} = \tilde{\gamma}_1$, $\tilde{w}_{31} = \tilde{w}_{21} = \tilde{w}_1$.

3.5.2. Self-Excitation Conditions

In the absence of laser action, the steady state is described by the following solutions of Eqs. (3.111)

$$F_{32}^{(a)} = P_{32}^{(a)} = P_{21}^{(a)} = 0,\ n_2^{(a)} = (\tilde{w}_{32}/\tilde{w}_1)n^{(a)},$$

$$n^{(a)} = \frac{A}{1+2A/R},\quad P_{31}^{(a)} = \frac{R/\tilde{\gamma}_1}{1+2\tilde{b}A/R}F_{31} \tag{3.116}$$

To make it compact, we have introduced parameter $\tilde{b} = 1 + 3\tilde{w}_{32}/2\tilde{w}_1$.

We take the deviations from steady-state solutions (3.116) as new variables

$$\delta F_{32} = F_{32},\quad \delta P_{32} = P_{32},\quad \delta P_{21} = P_{21}$$

$$\delta P_{31} = P_{31} - P_{31}^{(a)},\quad \delta n = n - n^{(a)},\quad \delta n_2 = n_2 - n_2^{(a)}.$$

The set of equations linearized by these variables is decomposed into two isolated units:

$$\frac{d(\delta F_{32})}{d\tau} = \tilde{\kappa}(\delta P_{32} - \delta F_{32}),$$

$$\frac{d(\delta P_{32})}{d\tau} = n^{(a)}\delta F_{32} - \delta P_{32} - \frac{1}{2}F_{31}\delta P_{21}, \tag{3.117}$$

$$\frac{d(\delta P_{21})}{d\tau} = -\tilde{\gamma}_1 \delta P_{21} + \frac{1}{2}\tilde{\gamma}(P_{31}^{(a)}\delta F_{32} + F_{31}\delta P_{32}),$$

and
$$\frac{d(\delta P_{31})}{d\tau} = -\tilde{\gamma}_1 \delta P_{31} + F_{31}(2\delta n - 3\delta n_2),$$

$$\frac{d(\delta n)}{d\tau} = -\tilde{\gamma}\left(\delta n + \frac{1}{2}F_{31}\delta P_{31}\right), \tag{3.118}$$

$$\frac{d(\delta n_2)}{d\tau} = \tilde{w}_{32}\delta n - \tilde{w}_1 \delta n_2 .$$

Equations (3.118) have no rising solutions. Eqs. (3.117) correspond to a cubic characteristic equation with the coefficients

$$a_0 = 1, \quad a_1 = \tilde{\gamma}_1 + \tilde{\kappa} + 1,$$

$$a_2 = \tilde{\kappa} + \tilde{\gamma}_1(\tilde{\kappa}+1) + \frac{1}{4}\tilde{\gamma}F_{31}^2 - \tilde{\kappa}n^{(a)}, \tag{3.119}$$

$$a_3 = \tilde{\kappa}\tilde{\gamma}(1 - \tilde{c}n^{(a)}) + \frac{1}{4}\tilde{\gamma}\tilde{\kappa}F_{31}^2,$$

where $\tilde{c} = 1 + \tilde{\gamma}/2\tilde{\gamma}_1$.

The Routh–Hurwitz criterion exhibits two possible ways for the roots of the characteristic equation to pass over to the right-hand half-plane: $a_3 < 0$ and $a_1 a_2 - a_3 < 0$. Bifurcation on the boundary $a_3 = 0$ means, as in a two-level laser, the transformation of a stable node to a saddle, accompanied by two nontrivial solutions of the type (3.87), arising simultaneously. At the boundary $a_1 a_2 = a_3$ the Hopf-bifurcation arises, owing to which the damped small oscillations at the frequency $\Omega = \sqrt{a_3 / a_1}$ become undamped. Since such oscillations cannot occur for $a_3 < 0$, the overlapping of two lasing areas is excluded. Satisfying the inequalities $a_1 a_2 - a_3 < 0$ and $a_3 > 0$ means that the laser self-excitation occurs as a time-dependent laser action.

The condition $a_3 = 0$ leads to a quadratic equation

$$\tilde{b}\tilde{\gamma}(A/R)^2 - M(A/R) + 1 = 0 \tag{3.120}$$

$(M = \tilde{c}R - 2\tilde{b}^2 - \tilde{\gamma}/2)$, which defines the first lasing area

$$\frac{A_1}{R} = \frac{M}{2\tilde{b}\tilde{\gamma}}\left[1 \pm \sqrt{1 - \frac{4\tilde{b}\tilde{\gamma}}{M^2}}\right]. \tag{3.121}$$

In a similar fashion, the equation

$$\tilde{b}K_1(A/R)^2-(K_2-2\tilde{b}K_3-K_1/2)(A/R)+K_3=0 \quad (3.122)$$

with the roots coefficients:

$$\frac{A_2}{R}=\frac{K_2-2\tilde{b}K_3-K_1/2}{2\tilde{b}K_1}\left\{1\pm\left[1-\sqrt{\frac{4\tilde{b}K_1K_3}{(K_2-2\tilde{b}K_3-K_1/2)^2}}\right]\right\} \quad (3.123)$$

are obtained from the condition $a_1a_2=a_3$. By K_i we denote the following combination of coefficients:

$K_1=\tilde{\gamma}\tilde{\gamma}_1(\tilde{\gamma}_1+1)$, $K_2=\tilde{\kappa}R(\tilde{\kappa}+1-\tilde{\gamma}/2)$, $K_3=(\tilde{\gamma}_1+\tilde{\kappa})(\tilde{\kappa}+1)+\tilde{\gamma}_1(\tilde{\kappa}+1)^2$.

The pumping parameter A is real and positive. As seen from Eq. (3.121), this fact makes the existence of the first lasing area dependent on satisfying two inequalities: $M^2>4\tilde{b}\tilde{\gamma}$ and $M>0$. Hence, a lower limit is imposed on the molecular concentration (limiting gain) in the laser medium

$$R>R_1\approx 2\tilde{b}/\tilde{c}<2\tilde{b}. \quad (3.124)$$

Similar considerations in application to Eq. (3.123) lead to the condition for existence of the second laser area,

$$\left(K_2-2\tilde{b}K_3-\frac{1}{2}K_1\right)^2>4\tilde{b}K_1K_3, \quad K_2>2\tilde{b}K_3+\frac{1}{2}K_1,$$

and then to

$$R>R_2\approx\tilde{b}[\tilde{\gamma}_1^2+\tilde{\kappa}+\tilde{\gamma}_1(\tilde{\kappa}+1)]/\tilde{\kappa}>2\tilde{b}. \quad (3.125)$$

It is seen that $R_2=2\tilde{b}(\tilde{\kappa}+1)/\tilde{\kappa}$ at $\tilde{\gamma}_1=1$ and quickly increases with increasing $\tilde{\gamma}_1$. In the limit $R>>A$ the asymptotic form is valid:

$$A_1^{(-)}\approx\frac{R}{M}\approx\frac{1}{\tilde{c}}, \quad A_1^{(+)}\approx\frac{RM}{\tilde{b}\tilde{\gamma}}\approx\frac{R^2\tilde{c}}{\tilde{b}\tilde{\gamma}},$$

$$A_2^{(-)}\approx\frac{2RK_3}{K_2}, \quad A_2^{(+)}\approx\frac{RK_2}{\tilde{b}K_1}.$$

It should be readily apparent that $A_1^{(-)}<A_2^{(-)}$, which means that the second lasing area is completely covered by the first area and disappears by virtue of $a_3<0$.

A more thorough consideration of the inequalities $(K_2-2\tilde{b}K_3-K_1/2)^2>4\tilde{b}K_1K_3$ $M^2>4\tilde{b}\tilde{\gamma}$ and $M^2>4\tilde{b}\tilde{\gamma}$ leads to the conclusion that there are the ranges of the parameter values without the real solutions of Eqs. (3.120) and (3.122), respectively. Numerical estimates for $R=5, \tilde{\gamma}_1=1$ lead to the ranges of suppression of the first lasing area $1.8<\tilde{b}<5.2\,(\tilde{\gamma}=0.8)$ and $2.01<\tilde{b}<3.49\,(\tilde{\gamma}=0.2)$ and of the second lasing area $1.01<\tilde{b}<1.2\,(\tilde{\gamma}=0.8)$ and $1.1<\tilde{b}<1.6\,(\tilde{\gamma}=0.2)$. Recall that $\tilde{b}=1$ corresponds to $\tilde{w}_{32} << \tilde{w}_1$.

Figure 3.14 shows the phase diagram illustrating the influence of the parameter $\tilde{\gamma}$ on the laser threshold. As the second control parameter, we have taken the normalized Rabi frequency in the pumping field

$$\Omega_{R31} = \sqrt{\frac{\tilde{\gamma}F_{31}^2}{4}} = \sqrt{\frac{A\tilde{\gamma}\tilde{\gamma}_1}{2R}} = \frac{d_{31}E_{31}}{2\hbar\gamma_{32}}.$$

This choice is reasonable, since Ω_R remains finite for $\tilde{\gamma}\to 0$, where $A_1^{(+)}$ and $A_2^{(+)}$ shift toward the infinity. It is interesting to note that the second lasing area is absent for large $\tilde{\gamma}$ and that the bifurcation boundaries intersect. Clearly, only the outside part of the second area, which is not covered by the first area, is physically meaningful. However, numerical integration of Eqs. (3.111) indicates [281] that the influence of the second area is exhibited as instability of the steady-state solution in the first area (involving the nonzero branch of steady-state solutions, as stated in [281]). For $R \gg A$ this influence is not essential.

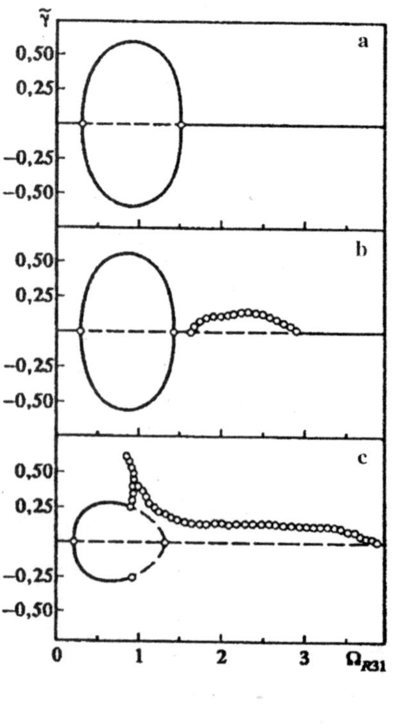

Fig. 3.14. Diagram illustrates the influence of control parameters $\tilde{\gamma}$ and Ω_{R31} on the threshold conditions of a coherently pumped three-level laser [303]. The Rabi frequency in the pump field is laid off on the abscissas axis; the dash line stands for the unstable steady-state solutions branch. $\tilde{\kappa}=10$; $R=5$; $\tilde{\gamma}=1.0$ (*a*); 0.8 (*b*); 0.7 (*c*); 0.6 (*d*); 0.2 (*e*)

3.5.3. Lasing Modes

The lasing branch of steady-state solutions (3.111) is described by a set of equations

$$\overline{P}_{32} = \overline{F}_{32}, \quad \overline{P}_{21} = 2(\overline{F}_{32} / F_{31})(\overline{n} - 1), \quad \overline{P}_{31} = -2(\overline{n} + \overline{F}_{32}^2) / F_{31},$$

$$\widetilde{w}_1 \overline{n}_2 = \widetilde{w}_{32} \overline{n} + \frac{1}{2} \widetilde{\gamma} \overline{F}_{32}^2, \quad 2\widetilde{\gamma}_1 \overline{P}_{21} = \widetilde{\gamma} \overline{F}_{32} (F_{31} + \overline{P}_{31}),$$

$$\widetilde{\gamma}_1 \overline{P}_{21} + F_{31}(R - 2\overline{n} - 3\overline{n}_2) + \frac{1}{2} \overline{P}_{21} \overline{F}_{32} = 0,$$

which are reduced to a biquadratic equation

$$\overline{F}_{32}^4 + \left[F_{31}^2 \left(\frac{3}{2} \frac{\widetilde{\gamma}}{\widetilde{w}_1} \widetilde{r} - 2\widetilde{b} - \frac{1}{2} \right) + 2\widetilde{\gamma}_1 (\widetilde{r} - 1) + 1 \right] \overline{F}_{32}^2 + \widetilde{b} F_{31}^4 + [R\widetilde{r} - 2\widetilde{b}(\widetilde{r} - 1) - \widetilde{\gamma}_1] F_{31}^2 + 2\widetilde{\gamma}_1 (\widetilde{r} - 1) = 0 \tag{3.126}$$

where $\widetilde{r} = 1 + 2\widetilde{\gamma}_1 / \widetilde{\gamma}$. Since the coefficient of $\overline{F}_{32}^2$ is more than zero, the condition for existence of the positive root of Eq. (3.126) is reduced to negative value of the free coefficient of this equation. The task reduces to a search for the roots of the quadratic equation

$$\widetilde{b} F_{31}^4 - [R\widetilde{r} - 2\widetilde{b}(\widetilde{r} - 1) - \widetilde{\gamma}_1] F_{31}^2 + 2\widetilde{\gamma}_1 (\widetilde{r} - 1) = 0, \tag{3.127}$$

between which the area of stationary lasing is situated. Note, however, that Eqs. (3.127) is identical to Eq. (3.120). As was to be expected, the region of existence of the nonzero solutions of the initial system coincides with the region of unstable trivial solutions. It should be mentioned that if criteria (3.112) and (3.115) are satisfied, then Eq. (3.126) is reduced to the same expression for the steady-state output, $\overline{F}_{32}^2 = A - 1$, which was found for the Lorenz model [see Eq. (3.87)].

Information on the time-dependent solutions is obtained by numerical integration of Eqs. (3.111) in two limiting cases. In the first, the inequalities (3.115) are not satisfied and the limits of smallness of parameter *A*/*R* are not observed [281, 303]. This means that both the Rabi splitting of the gain line in the pumping field and the nonlinear interaction of the pumping with the laser field are noticeable. The effect is most pronounced when the two laser domains merge as shown in Fig. 3.14. Irregular solutions are observed here among the numerical solutions of Eqs. (3.111).

In the other limiting case, $A \ll R$, the pumping transition is far from saturation, so that the gain line is not deformed initially. Fig. 3.15 shows a phase diagram of such a laser in the control parameter plane $A, \widetilde{\gamma}_1$, which gives an idea of the influence of the rate of coherence decay in transitions $1 \to 3$ and $1 \to 2$ on the laser dynamics. As in Fig. 3.4 the solid line shows the instability threshold of steady-state solutions and the dash line shows the boundary of hard-excitation of pulsations. The dotted and dash-dotted lines denote the transition bands between the domains of regular and chaotic behaviour.

The case with high $\tilde{\gamma}_1$ is qualitatively the same as the Lorenz case, as was to be expected. However, things differ as $\tilde{\gamma}_1$ decreases. Here, unlike the Lorenz–Haken model, there is a domain of asymmetric pulsations of the amplitude modulation type. The scenario of transition from RP(B) to RP(A) through the chaotic self-modulation band is apparent from Fig. 3.16, the frame of which follow the $\tilde{\gamma}_1$ profile for a fixed parameter A. The sequence of bifurcations between Fig. 3.16*a* and Fig. 3.16*b* imitates the one given in Figs. 3.7*a-e* and omitted. Then, through a chain of successive complications of the symmetric pulsations, we get deterministic chaos (Fig. 3.16*c*) like the one shown in Fig. 3.7*f*. All these bifurcations occur in the band confined by dotted lines 3′ and 3″ in Fig. 3.15.

A further decrease in $\tilde{\gamma}_1$ results in qualitative restructuring of the chaotic process. A double-sided attractor with irregular transition of the phase-space trajectory from the vicinity of a saddle-focus point to the vicinity of an alternative one (the phase of the field envelope undergoes a jump of π) transforms to a single-sided one. The field amplitude oscillations that correspond to a single-sided attractor do not show any phase jump of π, and the odd and even harmonics are equally represented in the spectrum. The attractor escapes from the domain of single-sided chaos into the domain of regular nonsymmetrical pulsations through an inverse period doubling se-

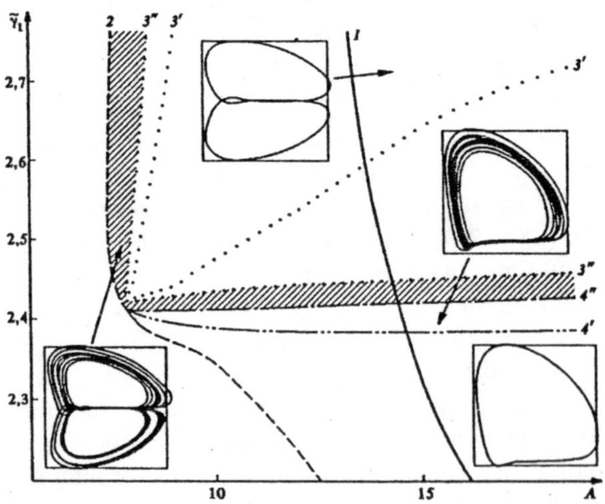

Fig. 3.15. Phase diagram of a single-mode three-level laser model in the control parameter plane ($A, \tilde{\gamma}_1$) [284]. $\tilde{\gamma} = 0.15$; $\tilde{\kappa} = 4.0$; $\tilde{w}_1 = 0.015$; $R = 1000$.

quence (Fig. 3.16*d* shows pulsations with period 8 and Fig. 3.16*e* shows pulsations with period 2) and finally arrives at a simple periodic regime of the amplitude modulation type (Fig. 3.16*f*).

Numerical simulations have shown that the pattern given in Fig. 3.15 is not sensitive to the parameters $\tilde{w}_1$ and R until the inequality (3.115) is satisfied. Quantitatively, the whole process is very dependent on the parameter $\tilde{\gamma}$: as $\tilde{\gamma}$ grows, boundaries 1 and 2 are shifted to the right and the domain of chaotic pulsations is broadened.

3.5.4. Experimental Investigations of Optically Pumped Ammonia Lasers

The scheme of the experimental setup is presented in Fig. 1.22. Its creators did their utmost to satisfy the requirements of the Lorenz model. The

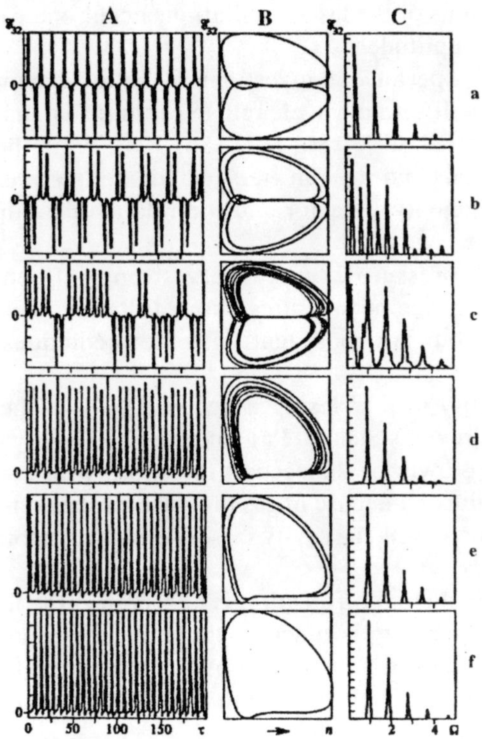

Fig. 3.16. Succession of solutions to Eqs (3.111) that occur in transition from the zone of regular single-sided pulsations to the zone of regular double-sided pulsations through boundaries *3* and *4*, Fig. 3.15 [284].
$\tilde{\gamma}=0.15$; $\tilde{\kappa}=4.0$; $\tilde{w}_1=0.015$; $A=12$; $R=100$; $\tilde{\gamma}_1=2.60$ (*a*); 2.45 (*b*); 2.44 (*c*); 2.42 (*d*); 2.40 (*e*); 2.30 (*f*). *A:* form of the electric field envelope; *B*: phase space trajectory projection onto the plane (n, F_{32}) ; *C*: spectrum of the envelope.

homogeneity of the broadening is favored by the monochromatic pumping. The pumping frequency is specially shifted from the center of the absorption Doppler line with the aim of selectively exciting a group of molecules with a definite nonzero velocity component in the direction of the pumping beam. This is to provide the conditions for a travelling wave generation in the direction opposite to the pumping wave. The linear distortion of the gain line shape by the coherent pump field is a minimum in this case [305].

Heterodyne detection of the laser field is of fundamental importance in this experiment. This method makes it possible to gain information on the field amplitude and phase dynamics while the homodyne detection reveals only the intensity characteristics. However, if the laser intensity alone is recorded, then it is unclear whether or not the field envelope reverses sign during the pulsations or the laser oscillations occur one way from the zero line of the field amplitude.

The results of experimental investigations of the dynamics of ammonia lasers operating at different wavelength (81 mm, 153 mm and 376 mm) are given in Refs. [39–41, 178–180, 182–186, 308]. From these references it is seen that essential control parameters include: the gas pressure, the pumping intensity, the cavity Q-factor and the relative detuning of the cavity from the gain line centre.

The ammonia pressure exerts a very strong influence on the laser behaviour. In the range of pressures above 6 Pa the main features of the dynamical behaviour are consistent with the predictions of the Lorenz-Haken model:

– The instability of CW lasing is reached only in the region of relatively low pressure (less than 10 Pa) since the bad cavity condition is fulfilled here. The growth of the second threshold with respect to the first laser threshold with an increase in gas pressure can be compensated, within some limits, by increasing the cavity losses (using a variable iris diaphragm, for example).

– The subcritical Hopf bifurcation is observed at the instability threshold when the laser cavity is tuned to the gain line centre. The resulting large-amplitude pulsations can be both regular ($^{15}NH_3$ laser with λ = 376 μm) and chaotic ($^{15}NH_3$ laser with λ = 153 μm and $^{14}NH_3$ laser with λ = 81 μm).

– For central tuning the chaotic process looks like a sequence of ascending pulses terminating for random times (the ‘spiral chaos’). The transition from one pulse sequence to another is accompanied by a jump of the field phase by π (Fig. 3.17). This indicates that the attractor has a double-sided spiral form characteristic of the Lorenz model. Within each monotonically growing pulse sequence the pulse shows a linear evolution. The

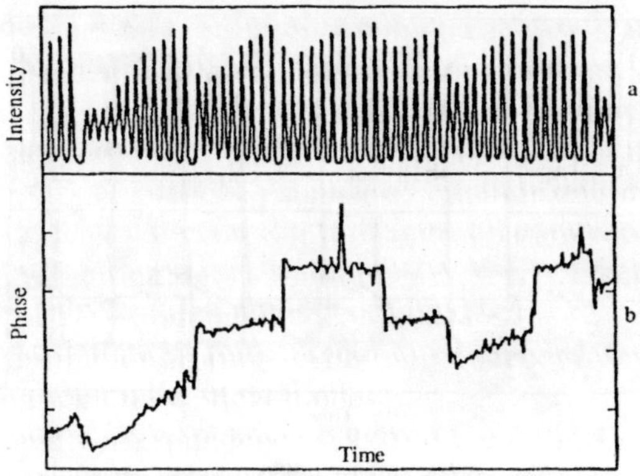

Fig. 3.17. Double-sided chaos observed in the output of an ammonia FIR laser at a wavelength 153 μm at a pressure 9 Pa [41]; (*a*) intensity; (*b*) phase of the laser field (the vertical scale reads to π).

average slope of the phase during a spiral depends on the reference (heterodyne) frequency and on the cavity detuning. However, the linear evolution is affected by the phase modulation synchronous with the intensity pulsing. The modulation index is anomalously large from the viewpoint of the Lorenz model.

– In the case of detuning of the laser cavity from resonance the laser behaviour is similar to that predicted by the complex Lorenz model at least in the range of relatively high gas pressure from 6 to 10 Pa. As the laser is tuned towards line centre its dynamics starts with CW emission, which is then followed by periodic pulsing. This is followed by a periodic-doubling cascade ending in an intermediate chaotic range. The latter has high periodic windows and then it followed by the period-three attractor. The final point of these transformations is the Lorenz-like spiral chaos approximately at the line center[1].

– The correlation dimension d_2, which was calculated using the experimental data, was found slightly above two. About the same values are offered for spiral chaos by the Lorenz model calculations with laser parameters.

The difference between the laser behaviour and the Lorenz model increases with a decrease in gas pressure and an increase in pumping power since these two factors make the nonlinear phenomena due to the coherence of pumping more important. Thus, even for a pressure below 8 Pa the threshold of instability of steady-state lasing is lower than the allowed for

[2]These results have been obtained in the experiments with a laser operating at wavelength 81 mm.

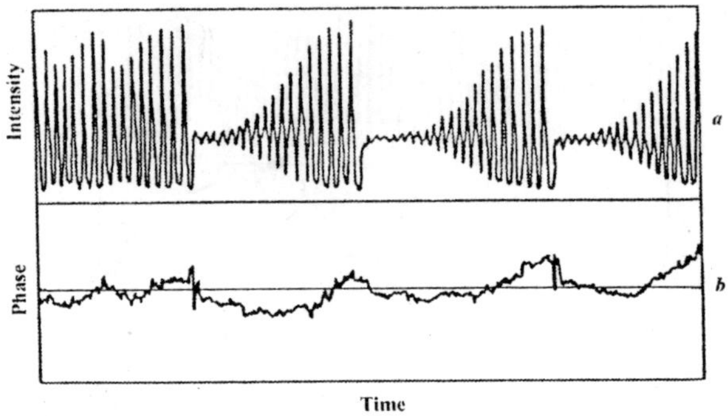

Fig. 3.18. Single-sided chaos observed in the output of an ammonia FIR laser at a wavelength 153 μm at a pressure 5 Pa [41].

the Lorenz model. At low pressures, instead of the double-sided (Lorenz) attractor, a single-sided (Roessler) attractor can be observed, when the phase space trajectory is localized near one of the unstable fixed points without jumping to the vicinity of another fixed point (Fig. 3.18).

At lower gas pressures and higher pump power the laser can exhibit routes to chaos that differ from the Feigenbaum scenario. For example, various types of intermittency were observed at $\lambda = 81$ μm, if the chosen gas pressure lies in the range 3.5–6 Pa and the pumping power exceeds 3 W/cm^2.

Which model of laser best agrees with the experimental results: a three-level, the Lorenz model or some other model? After comparing the experimental data with the numerical calculations the authors of Refs. [183, 186] were inclined to think that in the optimal region of operating gas pressures the Lorenz model is preferable, while a three-level model (for laser parameters adopted in the calculations) is not so good. This is because the solutions are very sensitive to the choice of a control parameter such as $\tilde{\gamma}_1 = \gamma_{m1} / \gamma_{32}$, and this parameter is known rather approximately. Therefore, it seems to be more correct to say that for parameters used in the numerical calculations (they have been given previously), the best agreement with the experiments is yielded by the Lorenz model. It can be paraphrased as follows: since the Lorenz model is a particular case of a three-level model, it is not excluded that the difference of the parameter, which was used in the calculations, from the true value is beyond the admissible limits.

Discussing the problem of finding an adequate model of a laser with coherent optical pumping it should be kept in mind that there are at least two more factors that attenuate the coherent interaction of the fields. One is due to the hyperfine structure (degeneration) of the laser levels. In Ref.

[309] it is shown that taking into account the hyperfine structure of the upper laser level, common to the pumping and laser transitions, makes the laser behaviour dependent on the field polarization. The laser behaviour is most similar to that predicted by the Lorenz model when the fields polarization are linear and orthogonal.

The second factor is the inhomogeneous (Doppler) line broadening, a trace of which remains in spite of the monochromatic pumping. The print is that the common upper level of the molecule can split under the action of both the pumping field and laser field (AC Stark effect). Accordingly the absorption line is distorted, and that strengthens the interaction of the pumping with different groups of molecules. In Ref. [310] the authors calculated numerically three-level laser model represented by 217 equations, which means the partition of the active medium molecules into 24 monovelocity groups. By its dynamical properties, this model is more similar to the Lorenz model than a three-level homogeneously broadened laser model.

3.6. Effect of Inhomogeneous Broadening on the Laser Dynamic Characteristic

As mentioned above, the dynamic instability phenomenon in lasers can be attributed to the nonlinear deformation of the gain line. In a homogeneously broadened medium, the unique mechanism to ensure favourable conditions for the growth of sideband components in the radiation spectrum is due to Rabi oscillations. The second mechanism is due to the hole burning by selective saturation of part of the line by a monochromatic field and is characteristic of inhomogeneously broadened media. The instability threshold is lowered in the second case since the saturating power is much less than required for Rabi oscillations.

3.6.1. Lowering of the Instability Threshold

The generalization of Eqs. (2.75) to the case of inhomogeneously broadened laser medium leads to the following traveling wave laser equations:

$$\frac{\partial F}{\partial t}+c\frac{\partial F}{\partial z}+[\kappa-i(\omega-\omega_{c}\}]F=2\pi i\omega N_{S}d\int\sigma(t,\omega_{0})\mathrm{d}\omega_{0}\,,$$

$$\frac{\partial\sigma}{\partial t}+[\gamma_{\perp}-i(\omega-\omega_{0})]\sigma=-\frac{i}{2\hbar}dFD\,,\qquad(3.128)$$

$$\frac{\partial D}{\partial t}+\gamma_{\parallel}(D-D^{(0)})=-\frac{i\beta_{a}}{2\hbar}d(F^{*}\sigma-F\sigma^{*})\,.$$

Confining ourselves to the case of fine adjustment of one mode to the inhomogeneous line centre ω_{00} we keep to the equality $\omega = \omega_c = \omega_{00}$, which, of course, is valid for a symmetrical gain line. We transform Eqs. (3.148) to a dimensionless form in accordance with Eqs. (3.1) assuming $\bar{t} = \gamma_\perp^{-1}$ and adding $u = (\omega_0 - \omega_{00})/\gamma_\perp, \zeta = z\gamma_\perp/c$. If the inhomogeneous broadening has a Doppler nature, then $u = kU/\gamma_\perp$. In this way, Eqs. (3.128) can be transformed to

$$\left(\frac{\partial}{\partial\tau} + \frac{\partial}{\partial\zeta}\right)f = \tilde{\kappa}\left(\int_{-\infty}^{\infty} p\mathrm{d}u - f\right),$$

$$\frac{\partial p}{\partial\tau} - iup = nf - p, \tag{3.129}$$

$$\frac{\partial n}{\partial\tau} = \tilde{\gamma}\left[\Lambda(u) - n - \frac{1}{2}(f^*p + p^*f)\right].$$

First of all, we will find the steady-state solutions of Eqs. (3.129). If the gain and loss are uniformly distributed over the cavity perimeter, then $\bar{f}$ is independent of the space coordinate. Assuming that all the derivatives in Eqs. (3.129) vanish we find

$$\bar{p} = \Lambda(u)\bar{f}\frac{1+iu}{1+\bar{m}+u^2}, \quad \bar{n} = \Lambda(u)\frac{1+u^2}{1+\bar{m}+u^2}, \tag{3.130}$$

where $\bar{m} = |\bar{f}|^2$ is determined from

$$\int_{-\infty}^{\infty}\frac{\Lambda(u)\mathrm{d}u}{1+\bar{m}+u^2} = 1. \tag{3.131}$$

Based on relation (3.131), we define the pumping parameter. We note that when the laser is at threshold, Eq. (3.131) transforms to

$$\int_{-\infty}^{\infty}\frac{\Lambda\mathrm{d}u}{1+u^2} = 1. \tag{3.132}$$

Putting $\Lambda(u) = \Lambda(0)h(u)$ and dividing Eq. (3.131) by Eq. (3.132), we arrive at

$$A = \frac{\Lambda(0)}{\Lambda_{\text{thr}}} = \int_{-\infty}^{\infty}\frac{h(u)\mathrm{d}u}{1+u^2}\left(\int_{-\infty}^{\infty}\frac{h(u)\mathrm{d}u}{1+\bar{m}+u^2}\right)^{-1}. \tag{3.133}$$

In the limiting case of homogeneous broadening, where $h(u) = \delta(0)$, Eq. (3.133) is reduced to $A = 1 + \bar{m}$, which is identical to Eq. (3.87).

Of the distribution functions capable of approximating a real situation in a homogeneous broadened laser medium, the Lorentzian function

$$h(u)=\frac{1}{\pi}\frac{u_0}{u^2+u_0^2}$$

seems to be preferable mathematically, since it permits integration on closed form. Thus, Eq. (3.131) reduced to

$$\sqrt{1+\overline{m}}(\sqrt{1+\overline{m}}+u_0)=A\,. \tag{3.134}$$

We now investigate the steady-state resonant lasing mode for stability with respect to its own perturbation following references [262–264]. Generally, this analysis uses the scheme represented in Section 3.4. Since all other modes, except for the resonant mode, are beyond the scope of our consideration, there is no need to retain the derivative $\partial/\partial\zeta$. Therefore, assuming that f is a real quantity and $p=p'+ip''$ Eqs. (3.129) can be rewritten as

$$\begin{aligned}
&\frac{\mathrm{d}f}{\mathrm{d}\tau}+\tilde{\kappa}f=\tilde{\kappa}\int p'\mathrm{d}u\,,\\
&\frac{\mathrm{d}p'}{\mathrm{d}\tau}+p'=-up''+nf\,,\\
&\frac{\mathrm{d}p''}{\mathrm{d}\tau}+p''=up'\,,\\
&\frac{\mathrm{d}n}{\mathrm{d}\tau}=\tilde{\gamma}(A-n-fp')\,.
\end{aligned} \tag{3.135}$$

We then linearize these equations, substituting $\{f,p,n\}=\{\bar{f},\bar{p},\bar{n}\}+\{\delta f,\delta p,\delta n\}e^{\lambda\tau}$ into Eqs. (3.135) and discarding the nonlinear terms with respect to small deviations in $\delta f,\delta p,\delta n$. We thus arrive at

$$(\lambda+\tilde{\kappa})\delta f=\tilde{\kappa}\int\delta p'\mathrm{d}u\,, \tag{3.136a}$$

$$(\lambda+1)\delta p'=-u\delta p''+\bar{n}\delta f+\bar{f}\delta n\,, \tag{3.136b}$$

$$(\lambda+1)\delta p''=u\delta p'\,, \tag{3.136c}$$

$$(\lambda+\tilde{\gamma})\delta n=-\tilde{\gamma}(\bar{f}\delta p'+\bar{p}'\delta f)\,. \tag{3.136d}$$

The last three equations of (3.136) can be reduced to the relation

$$\delta p'=(\lambda+1)\frac{(\lambda+\tilde{\gamma})\bar{n}-\tilde{\gamma}\bar{f}\bar{p}'}{(\lambda+1)^2(\lambda+\tilde{\gamma})+\tilde{\gamma}\overline{m}(\lambda+1)+u^2(\lambda+\tilde{\gamma})}\delta f$$

which can de inserted into Eq. (3.135a). Taking into account Eq. (3.130) leads to a characteristic equation

$$(\lambda+\tilde{\kappa})\frac{\lambda+\tilde{\gamma}}{\lambda+1}=\tilde{\kappa}\int_{-\infty}^{\infty}\Lambda\frac{\lambda+\tilde{\gamma}-\overline{m}\tilde{\gamma}+u^2(\lambda+\tilde{\gamma})}{(B+u^2)(1+\overline{m}+u^2)}du, \tag{3.137}$$

in which

$$B=(\lambda+1)^2+\tilde{\gamma}\overline{m}\frac{\lambda+1}{\lambda+\tilde{\gamma}}.$$

Transforming the integrand we rewrite Eq. (3.137) as

$$(\lambda+\tilde{\gamma})\frac{\lambda+\tilde{\kappa}}{\lambda+1}=\tilde{\kappa}\frac{(\lambda+\tilde{\gamma})(1-B)-\tilde{\gamma}\overline{m}}{1+\overline{m}-B}\int_{-\infty}^{\infty}\Lambda\left(\frac{1}{B+u^2}-\frac{1}{1+\overline{m}+u^2}\right)du+\tilde{\kappa}(\lambda+\tilde{\gamma})\int_{-\infty}^{\infty}\Lambda\frac{du}{1+\overline{m}+u^2},$$

and, in view of Eq. (3.131), as

$$\left((\lambda+\tilde{\gamma})\frac{\lambda+\tilde{\kappa}}{\lambda+1}-\tilde{\kappa}\right)+\tilde{\kappa}\frac{(\lambda+\tilde{\gamma})(1-B)-\tilde{\gamma}\overline{m}}{1+\overline{m}-B}\left(1-\int_{-\infty}^{\infty}\Lambda\frac{du}{B+u^2}\right)=0. \tag{3.138}$$

By integration for the Lorentz distribution function and taking Eq. (3.134) into account we find from Eq. (3.138)

$$\frac{\lambda(\lambda+\tilde{\gamma})}{\lambda+1}(1-\tilde{\kappa})=\tilde{\kappa}[(\lambda+\tilde{\gamma})(1-B)-\tilde{\gamma}\overline{m}]\frac{\sqrt{B}+u_0+\sqrt{1+\overline{m}}}{\sqrt{B}(u_0+\sqrt{B})(\sqrt{1+\overline{m}}+\sqrt{B})}. \tag{3.139}$$

The last difficulty on the way to an analytical solution of the laser stability problem is due the irrational form of Eq. (3.139). We can remove this headache by multiplying Eq. (3.139) by

$$\frac{\lambda(\lambda+\tilde{\gamma})}{\lambda+1}(1-\tilde{\kappa})=\tilde{\kappa}[(\lambda+\tilde{\gamma})(1-B)-\tilde{\gamma}\overline{m}]\frac{\sqrt{B}-u_0-\sqrt{1+\overline{m}}}{\sqrt{B}(u_0-\sqrt{B})(\sqrt{1+\overline{m}}-\sqrt{B})},$$

thus creating extra roots. In general, such a method involves cumbersome algebraic transformations. The procedure is much easier in two limiting cases:

$u_0=0$ for the pure homogeneous broadening,

$u_0\to\infty$ for the limiting inhomogeneous broadening.

The first case was considered above. In the second case Eq. (3.139) transforms to

$$\frac{\lambda(\lambda+\tilde{\gamma})}{\lambda+1}(1-\tilde{\kappa})\sqrt{1+\overline{m}}\sqrt{B}=\tilde{\kappa}[(\lambda+\tilde{\gamma})(1-B)-\tilde{\gamma}\overline{m}]-B\frac{(\lambda+\tilde{\gamma})(1-\tilde{\kappa})\lambda}{\lambda+1}. \tag{3.140}$$

Squaring both parts of Eq. (3.140) yields an eight-power equation. This is reduced to a six-power equation in the particular case $\tilde{\gamma}=1$, in which Eq. (3.140) becomes

$$\lambda(\tilde{\kappa}-1)\sqrt{1+\overline{m}}\sqrt{B}=\tilde{\kappa}[(\lambda+1)(B-1)+\overline{m}]-B\lambda(\tilde{\kappa}-1), \tag{3.141}$$

where $B=(\lambda+1)^2+\overline{m}$. It can be easily seen that among the roots of the characteristic polynomial obtained by squaring Eq. (3.141), there is $\lambda=-2$. Thus, we should only analyze the roots of the equation

$$\sum_{n=0}^{5}a_n\lambda^{n-5}=0 \tag{3.142}$$

with the coefficients

$$a_0=1,\quad a_1=2(\tilde{\kappa}+1),\quad a_2=-\overline{m}\tilde{\kappa}^2+2(2+\overline{m})\tilde{\kappa}+1+\overline{m}, \tag{3.143}$$

$$a_3=2(1+3\overline{m})\tilde{\kappa},\quad a_4=\overline{m}(2-m)\tilde{\kappa}^2+2\overline{m}(1+\overline{m})\tilde{\kappa},\quad a_5=2\overline{m}^2\tilde{\kappa}.$$

Assuming the Hopf bifurcation and substituting the bifurcation value of the root $\lambda=\pm i\Omega$ into Eq. (3.142) we obtain a pair of real equations

$$\Omega^4-a_2\Omega^2+a_4=0, \tag{3.144a}$$

$$a_1\Omega^4-a_3\Omega^2+a_5=0, \tag{3.144b}$$

which rule out the pulsation frequency and the critical intensity m_{cr}. It is easy to switch from Eqs. (3.144) to an equality

$$\Omega^2=\tilde{\kappa}m_{cr}\frac{m_{cr}(\tilde{\kappa}^2-2)-2(\tilde{\kappa}+1)}{m_{cr}(\tilde{\kappa}^3-\tilde{\kappa}^2-1)-(2\tilde{\kappa}+1)^2}, \tag{3.145}$$

a substitution of which into Eq. (3.144a) leads to a cubic equation

$$\sum_{n=0}^{3}b_n m_{cr}^{3-n}=0$$

with the coefficients

$$b_0=(\tilde{\kappa}-1)^5,\quad b_1=-7\tilde{\kappa}^4+14\tilde{\kappa}^3-4\tilde{\kappa}^2-4, \tag{3.146}$$

$$b_2=6\tilde{\kappa}^4+12\tilde{\kappa}^3-10\tilde{\kappa}^2-15\tilde{\kappa}-5,\quad b_3=-(2\tilde{\kappa}+1)^2(\tilde{\kappa}+1).$$

It is seen that in the limit $\tilde{\kappa} \to 1$ all b_n except b_0 are negative. Hence, a positive definite solution of the cubic equation can exists only in the case $\tilde{\kappa} > 1$. Recall that for a homogeneously broadened laser the corresponding inequality would look like $\tilde{\kappa} > 1 + \tilde{\gamma}$ or $\tilde{\kappa} > 2$ for $\tilde{\gamma} = 1$. From Eqs. (3.146) it is apparent that the asymptotic form $m_{cr} \to (\tilde{\kappa} - 1)^{-5}$ is valid in the limit $\tilde{\kappa} \to 1$. In the limit $\tilde{\kappa} >> 1$ the bifurcation value

$$m_{cr} = 4/(3\tilde{\kappa}) \tag{3.147}$$

can be found by applying the Routh–Hurwitz criterion to Eq. (3.142) with coefficients (3.143). The small oscillation frequency at the instability threshold is given by

$$\Omega = \begin{cases} 1, & \text{for } \tilde{\kappa} >> 1, \\ \tilde{\kappa} m_{cr}^{1/2}, & \text{for } \tilde{\kappa} \to 1. \end{cases} \tag{3.148}$$

3.6.2. Time-Dependent Laser behaviour

In the numerical simulation of time-dependent processes in gas lasers there is no need to give up the Maxwellian molecule velocity distribution. Such investigations confirm that in the case of inhomogeneous broadening there are the parameter ranges where the laser exhibits the bistable properties [311–313]. The phase diagrams presented in Fig. 3.19 show only quantitative differences in the position of the time-independent and time-dependent laser mode stability boundaries for different values of the broadening ratio of the laser medium. The tendency is that as the broadening ratio u_0 increases, the instability threshold is lowered and the instability band is narrowed.

In the numerical investigation of the steady-state processes in inhomogeneously broadened lasers there is always the question of what is an adequate laser model. Rigorously speaking, one should take into ac-

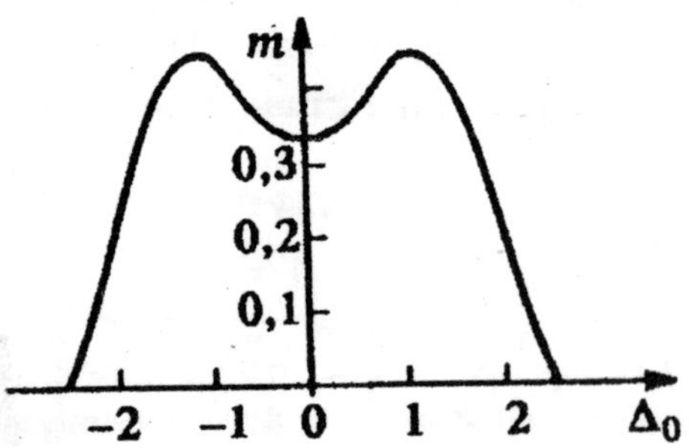

Fig. 3.19. Phase diagrams of a single-mode laser model in the control parameter plane $1/\tilde{\kappa}, A$ at $\tilde{\gamma} = 0.1$ (*a*); 0.2 (*b*) for three values of inhomogeneous broadening parameter $u_0 = 0.5$; 1.0; 2.0. The boundaries mean the same as in Fig 3.4 [318].

count the features inherent in real gas lasers such as the spectral cross-relaxation or the finite lifetime of the lower laser level. The influence of each of these factors was investigated in [31, 274]. Numerical integration of Eqs. (2.72) and their simple versions were performed for the set of parameters corresponding to a xenon gas laser. It should not be a surprise that a more general model ensures a more detailed agreement with the experiments. Stepwise elimination of cross-relaxation terms and reducing to a two-level system influences the quantitative estimates without leading to qualitative changes. Even adiabatic elimination of the field amplitude ($\mathrm{d}f/\mathrm{d}\tau = 0$) does not have dramatic influence on the laser processes. Only in the rate equation form ($\mathrm{d}p/\mathrm{d}\tau = 0$) does the steady-state solution remain stable for all the parameter values assigned in [274]. This should be natural, since among the conditions for adiabatic elimination of the polarization (3.12) there is $\tilde{\kappa} \ll 1$ in contradiction to the necessary instability condition $\tilde{\kappa} > 1$. These results prove the qualitative efficiency of a two-level model (2.70), which is generally adopted in theoretical papers.

Numerical simulations indicate that there are two types of time-dependent solutions for laser intensity. They differ in form of structure elements: there is either a pulse of smooth shape or a burst with oscillatory damping [31, 199, 271, 274, 275]. Both of them are observed in the xenon laser. The oscillating bursts are contributed by a decrease in the parameter $\tilde{\gamma}$ and an increase in pumping. At $\tilde{\gamma} = 2$ (the limiting admissible value), immediately above the instability threshold, the field envelope oscillations are nearly harmonic, but the process ceases to be regular as the pumping increases [271]. The Feigenbaum period-doubling scenario is followed in the route to chaos in this case. In general, it is found that the different regions of periodic pulsations are separated by zones of chaotic behaviour in the parameter space. This is illustrated in Fig. 3.20 where calculations are compared to experiment.

As in the solutions to the model of the homogeneously broadened laser, the field amplitude can be symmetric and asymmetric with respect to the zero value [199]. The first type of solutions corresponds to the field spectra without power at the carrier frequency and the second type corresponds to the spectra with a pronounced central component. The domains of regular symmetric and regular asymmetric pulsations are separated by zones of chaotic behaviour in the parameter space.

The influence of cavity detuning on laser dynamics needs further investigation. Clearly, a determining role is played by the decrease in gain as the frequency recedes from the line centre. Therefore, the laser mode change due to shifting toward the inhomogeneous gain line wing is defined by the regime of the tuned laser. The general tendency is toward a simplification of the process as detuning increases [199]. However, we

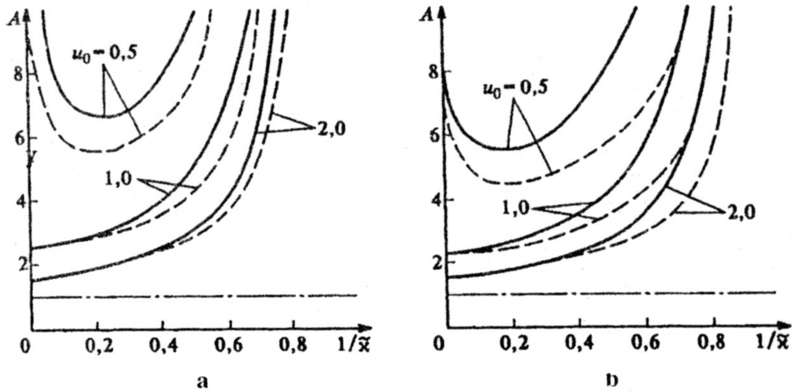

Fig. 3.20. Theoretical (lines) and experimental (circles) pulsation frequencies of a xenon gas discharge laser as the function of pumping parameter. Dash lines indicate the regions of chaotic output [31].

should not attribute the restructuring entirely to the change in effective gain, since the numerical calculation shows a variety of transitions to chaos. Chaotization following a quasiperiodic sequence of Ruelle-Takens, where incommensurate frequencies arise in the pulsation spectrum, is mentioned in [276]. In this paper it is also found that the steady state can turn to chaos in a hard manner as the cavity Q-factor decreases.

Since the dynamic theory mainly describes the travelling wave regime, the experimental investigations of unidirectional ring lasers are of primary interest. The experimental results were partly mentioned at the end of Section 1.2.3. The data obtained by way of heterodyne signal measurement revealed the existence of the regime of sharply asymmetric pulsations, in the spectrum of which the central component dominates, and the regime of symmetric pulsations with the sideband components prevailing [199]. The experimental data are in good agreement with the numerical investigations of a two-level model.

Also, the behaviour of the field spectrum components generated by a single-mode standing wave laser in time-dependent regular regimes (a He-Xe laser with λ = 3.51 μm and a He–Ne laser with λ = 3.39 μm) was thoroughly investigated by optical heterodyning [314]. The results can be interpreted as an exhibition of effects of the Lamb dip on each spectral component when it returns through the line centre region. The presence of several components, each interacting with two groups of atoms, shed new light upon the nature of complicated behaviour of standing wave lasers.

Chapter 4

Multimode Lasers with Frequency-Nondegenerate Modes

In the previous chapter, the fundamental mechanism of laser dynamics was considered, which is the coherent interaction of laser field with the active medium inside the cavity. This mechanism works also in multimode lasers where it leads to Risken-Nummedal-Graham-Haken instability of stationary oscillations, which is analogous to the Lorenz instability in single-mode lasers. However, in multimode lasers the dynamic behaviour is more often determined by nonlinear mode interaction.

4.1. Rate Equations Model with Spatial Mode Competition and Its Time Independent

Solutions

There are several varieties of the rate-equation laser models. In this section we consider the simplest of these models, which take into account only additive saturation of the laser medium by lasing modes. The nonstationary processes in such lasers are limited by relaxation oscillations that form the set of low-frequency eigenoscillations of the model

4.1.1. Combination Tone Mode-Mode Coupling. The Rate Equations of the Multimode Lasers

The rate equations can be generalized to describe a multimode laser following the procedure proposed in [316]. Unfortunately, the conditions of their validity remain unknown. Inequalities (3.12) now become insufficient, since there are intermode beat frequencies in the envelope field spectrum if many nondegenerate modes are excited simultaneously. In a nonlinear system, such as the laser, the beats lead to a combination coupling between modes, which is absent in the rate-equation model. Therefore it is advisable to move carefully along the whole route from the general nonadiabatic system to the rate equations [289, 317, 318].

First of all, adding to Eq. (3.1) some new dimensionless quantities,

$$v=V/V_c,\quad \psi_k=E_k(\mathbf{r}),\quad \Delta_k=(\omega_k-\omega_0)/\gamma_\perp,\quad \Delta_{ck}=(\omega_{ck}-\omega_0)/\kappa, \tag{4.1}$$

and denoting the mode amplitude as ϕ_k, let us write equations (2.76) in a dimensionless form

$$\frac{d\phi_k}{d\tau}+i\tilde{\kappa}\Delta_{ck}\phi_k=\tilde{\kappa}\left(\int p\psi_k dv-\phi_k\right), \tag{4.2a}$$

$$\frac{\partial p}{\partial\tau}=\tilde{\gamma}_\perp\left(n\sum\phi_k\psi_k-p\right), \tag{4.2b}$$

$$\frac{\partial n}{\partial\tau}=\tilde{\gamma}_\parallel\left[A-n-\frac{1}{2}\sum\psi_k(\phi_k^*p+\phi_kp^*)\right]. \tag{4.2c}$$

It is convenient to choose the reference frequency ω equal to ω_0 because in this case $\Delta_0=0$ and this quantity is absent in Eq. (4.2b).

Let us present solutions of the set (4.2) in the form

$$\phi_k=f_k\exp(-i\tilde{\gamma}_\perp\Delta_k\tau),\quad p=\sum p_k\exp(-i\tilde{\gamma}_\perp\Delta_k\tau),\quad n=n_b+\sum_{\mu,\nu}n_{\mu\nu}\exp[-i\tilde{\gamma}_\perp(\Delta_\mu-\Delta_\nu)\tau] \tag{4.3}$$

We assume that the time dependence of variables $f_k, p_k, n_b, n_{\mu\nu}$ is slow on the scale of the intermode beats. Substituting Eq. (4.3) into Eq. (4.2b) and using the harmonic balance principle, let us pick out terms with the same frequency dependence in the resulting relation. Assuming that conditions (3.12) are met and disregarding $dp_k/d\tau$, we come to

$$p_k=\tilde{F}_k\left(n_b\psi_kf_k+\sum_{\mu,\nu,\rho}n_{\mu\nu}\psi_\rho f_\rho\delta_{k\rho\mu\nu}\right). \tag{4.4}$$

Here $\tilde{F}_k=(1-i\Delta_k)^{-1}$. The designation

$$\delta_{k\rho\mu\nu}=\exp[i\tilde{\gamma}_\perp(\Delta_k-\Delta_\rho-\Delta_\mu+\Delta_\nu)\tau]$$

means that we keep in Eq. (4.4) only the terms with $\Delta_\rho+\Delta_\mu-\Delta_\nu\cong\Delta_k$.

In the next step we transform Eq. (4.2c) using Eqs. (4.3) and (4.4). A rather simple result such as

$$n_{\mu\nu} = \frac{n_b \psi_\mu \psi_\nu}{2[1 + i(\Delta_\nu - \Delta_\mu)/\widetilde{\gamma}} (F_\mu + F_\nu^*) f_\mu f_\nu^* , \tag{4.5}$$

can be obtained if the combination sum in Eq. (4.4) is small and the inequality

$$\left| \frac{1}{n_{\mu\nu}} \frac{\mathrm{d}n_{\mu\nu}}{\mathrm{d}\tau} \right| << \widetilde{\gamma}_\perp \, | \Delta_\nu - \Delta_\mu |$$

is fulfilled.

We then turn Eq. (4.4) replacing $n_{\mu\nu}$ by Eq. (4.5):

$$p_k = \widetilde{F}_k \left\{ n_b \psi_k f_k - \frac{1}{2} \sum_{\mu,\nu,\rho} \frac{n_b \psi_\mu \psi_\nu \psi_\rho}{1 + i(\Delta_\nu - \Delta_\mu)/\widetilde{\gamma}} (F_\mu + F_\nu^*) f_\mu f_\nu^* f_\rho \delta_{k\rho\mu\nu} \right\} . \tag{4.6}$$

Using Eqs. (4.3) and (4.6) we can transform Eq. (4.2a) to

$$\frac{\mathrm{d}f_k}{\mathrm{d}\tau} - i(\widetilde{\gamma}_\perp - \widetilde{\kappa}\Delta_{ck}) f_k = \widetilde{\kappa} f_k \left\{ \widetilde{F}_k \left[n_{kk} - \frac{1}{2} \sum_{\mu,\nu,\rho} \frac{(F_\mu + F_\nu^*) n_{k\mu\nu\rho}}{1 + i(\Delta_\nu - \Delta_\mu)/\widetilde{\gamma}} \frac{f_\mu f_\nu^* f_\rho}{f_k} \delta_{k\rho\mu\nu} \right] - 1 \right\} \tag{4.7}$$

Here

$$n_{kk} = \int n_b \psi_k^2 \mathrm{d}v, \quad n_{k\mu\nu\rho} = \int n_b \psi_k \psi_\mu \psi_\nu \psi_\rho \mathrm{d}v . \tag{4.8}$$

The first term in the curly brackets describes the saturated gain coefficient of the k-th mode. The second term originates from the combination scattering of modes on the inversion gratings vibrations: each pair of modes induces oscillations of inversion with the beats frequency. These oscillations scatter the third mode thus producing a combination tone near the fourth mode. In particular, the combination sum contains terms with $\nu = \rho$, $\mu = k$, which describe the situation when the scattering component coincides with that of interacting modes. These terms differ from others because they do not contain phases. Adding equation (4.7), which governs the dynamics of the slow component of the inversion, we obtain a closed set of equations

$$\frac{\mathrm{d}f_k}{\mathrm{d}\tau} - i(\widetilde{\gamma}_\perp - \widetilde{\kappa}\Delta_{ck}) f_k$$

$$= \widehat{\kappa} f_k \left\{ \widetilde{F}_k \left[n_{kk} - \frac{1}{2} \sum_{\rho} \frac{(F_k + F_\rho^*) n_{kk\rho\rho}}{1 + i(\Delta_\rho - \Delta_k)/\widetilde{\gamma}} | f_\rho |^2 - \frac{1}{2} \sum_{\mu,\nu,\rho} \frac{(F_\mu + F_\nu^*) n_{k\mu\nu\rho}}{1 + i(\Delta_\nu - \Delta_\mu)/\widetilde{\gamma}} \frac{f_\mu f_\nu^* f_\rho}{f_k} \delta_{k\rho\mu\nu} \right] - 1 \right\}, \tag{4.9a}$$

$$\frac{\partial n_b}{\partial \tau} = \widetilde{\gamma}_\| \left[A - n_b \left(1 + \sum \widetilde{L}_k \psi_k | f_k |^2 \right) \right], \tag{4.9b}$$

where $\widetilde{L}_k = \operatorname{Re} \widetilde{F}_k = (1 + \Delta_k^2)^{-1}$ is the Lorentzian function of the line shape.

Neglecting terms responsible for the four-wave mixing we reduce this set to the rate equations

$$\frac{dm_k}{d\tau} = 2\widehat{\kappa} m_k \left(\widetilde{L}_k \int n_b \psi_k^2 dv - 1 \right), \tag{4.10a}$$

$$\frac{\partial n_b}{\partial \tau} = \widetilde{\gamma}_\| \left[A - n_b \left(1 + \widetilde{L}_k \psi_k^2 m_k \right) \right]. \tag{4.10b}$$

The assumptions we made in order to pass from Eq. (4.2) to (4.10) need comments.

1. The harmonic balance principle is valid if the intermode beats have frequencies higher then the other oscillatory processes (the relaxation oscillations, for instance).

2. By use of Eq. (4.5) the condition $| n_{\mu\nu} / n_b | << 1$ can be rewritten as

$$\left| \frac{\psi_\mu \psi_\nu (F_\mu + F_\nu^*)}{2[1 + i(\Delta_\nu - \Delta_\mu)/\widetilde{\gamma}]} f_\mu f_\nu^* \right| << 1 \tag{4.11}$$

It does not necessarily mean that the combination sum in Eq. (4.6) is small. If the laser spectrum is nonequidistant and the beats are noncoherent, the sum can be small even when the condition (4.11) is not satisfied. If the spectrum is equidistant ($\Delta_\nu - \Delta_\mu = l\Delta$ where Δ is the intermode frequency spacing, and l is an integer), the smallness of the combination sum can be guaranteed by a more rigid condition

$$N | f_1 |^2 << | 1 + i\Delta/\widetilde{\gamma} | = \begin{cases} \Delta/\widetilde{\gamma} & \text{for } \Delta >> \widetilde{\gamma}, \\ 1 & \text{for } \Delta << \widetilde{\gamma}. \end{cases} \tag{4.12}$$

which has the sense of a higher limit of Eq. (4.11). Here N is the number of lasing modes and f_1 is the strongest one.

3. Expression (4.6) resembles the series expansion of the polarization in terms of field, which is widely used in nonlinear optics [3, 6, 319]. In both cases the higher terms of the series correspond to multiphoton processes. The difference is due to the choice of an unperturbed state. In the

approach developed here the unperturbed inversion is taken as the saturated inversion n_0. In the traditional expansion of the polarization, which is used in nonlinear optics, the unperturbed state is assumed to be the state of the medium in the absence of a radiation field, and saturation is not distinguished from other nonlinear effects[1]. In this context it should be remembered that the temporary local relation between the polarization and the field is in the form of Eq. (4.6) (the ε-description according to the terms adopted in [320]) is due to the specific features of media with $\gamma_\parallel << \gamma_\perp$. For such media the saturating field $F_{\text{sat}} \sim (\gamma_\parallel \gamma_\perp)^{1/2}$ is much less than $F_{\text{coh}} \sim \gamma_\perp$, where the field interaction with the medium becomes coherent and the susceptibility ceases to be an adequate characterization of the medium in an unsteady state.

The consideration is restricted to the third-order terms of the polarization expansion, which correspond to the four-wave mixing in a laser medium. Obviously, the rate Eqs. (4.10) take into account only the one-photon processes while Eqs. (4.9) include the four-photon processes that satisfy the frequency matching condition $\omega_k + \omega_\mu = \omega_p + \omega_\nu$.

Unfortunately, the derivation of the equations does not give a clear answer to the question when the rate-equation approach, which dominates in the laser dynamics, is fully justified and when neglecting of the combination terms in Eqs. (4.9) is incorrect. A necessary condition for using the rate equations is the smallness of the combination sum in Eqs. (4.9). If $\Delta / \tilde{\gamma} >> 1$, then this sum enters the equations with a small parameter $\tilde{\gamma} / \Delta$. These are satisfied in solid-state and other class *B* lasers. We do not know, however, which degree of smallness of the parameter is sufficient for confident discarding of the combination sum. For example, if we turn to dye lasers, then $\tilde{\gamma} / \Delta \geq 1$ and the rate-equation approach becomes problematic. The discussion of these problems will be continued later in this Chapter. Here we only make the obvious assertion that intrinsically the rate-equation approximation cannot be applied where the phase relations between modes are important.

4.1.2. Stationary Oscillation Spectrum of the Fabry-Perot Laser

First of all, we consider the simple case of a solid-state laser with plane-parallel mirrors [321–323]. If d/dτ =0, Eqs. (4.10) become the following set of equations:

$$\overline{n} = A\left(1 + \sum \tilde{L}_l \psi_l^2 \overline{m}_l\right)^{-1}, \tag{4.13a}$$

$$\tilde{L}_k \int \overline{n} \psi_k^2 \mathrm{d}\nu = 1 + \beta_k, \tag{4.13b}$$

which define the time-independent solution. Here β_k are extra losses of

[1]Lamb's theory is built on the same basis [315].

k-th cavity mode normalized to the losses of the reference mode. Equations (4.13) can be solved without specifying the form of cavity eigenfunctions only near the laser threshold. The right-hand side of Eq. (4.13a) can be expanded in a series of $\sum \tilde{L}_l \psi_l^2 \overline{m}_l$ and, by retaining of the linear term of expansion, Eq. (4.13b) is reduced to

$$A\tilde{L}_k\left(1-\sum_l \tilde{L}\overline{m}_l S_{kl}\right)=1+\beta_k\,, \tag{4.14}$$

where $S_{kl}=\int \psi_k^2\psi_l^2 \mathrm{d}v$.

More general results can be obtained for the axial modes. Substitution of the longitudinal mode eigenfunctions $\psi_l=\sqrt{2}\sin(\pi q_l\zeta)$ into Eq. (4.13) and taking the series expansion of

$$\sum \tilde{L}_l\overline{m}_l\cos(2\pi q_l\zeta)/\left(1+\sum \tilde{L}_l\overline{m}_l\right)$$

as proposed in Ref. [321], yield

$$\overline{n}=\frac{1}{1+\sum \tilde{L}_l\overline{m}_l[1-\cos(2\pi q_l\zeta)]}\approx\frac{1}{1+\sum \tilde{L}_l\overline{m}_l}+\sum\frac{\tilde{L}_l\overline{m}_l\cos(2\pi q_l\zeta)}{\left(1+\sum \tilde{L}_l\overline{m}_l\right)^2}\,. \tag{4.15}$$

This expression is not limited only to small-amplitude modes and it has higher precision since a larger number of modes are involved. Introducing Eq. (4.15) into (4.13), integrating over the longitudinal limits the medium (from ζ_1 to ζ_2) and taking into account that the number $q=2L/\lambda$ is very large, we obtain

$$A\tilde{L}_k\left\{\frac{\zeta_2-\zeta_1}{1+\sum \tilde{L}_l\overline{m}_l}-\frac{1}{2(1+\sum \tilde{L}_l\overline{m}_l)^2}\left[\tilde{L}_k\overline{m}_k(\zeta_2-\zeta_1)+\sum_{l\neq k}\tilde{L}_l\overline{m}_l\frac{\sin[2\pi(q_l-q_k)\zeta_2]-\sin[2\pi(q_l-q_k)\zeta_1]}{2\pi(q_l-q_k)}\right]\right\}$$
$$=1+\beta_k\,. \tag{4.16}$$

This equation indicated that the form of the laser spectrum is also dependent on the length of the laser rod and its location inside the optical cavity. Nonuniform filling of the cavity with the laser medium results in discrimination against some modes even if the laser rod ends are antireflection coated.

The spectrum is most easily calculated if the cavity is completely and uniformly filled with the laser medium. In this case $\zeta_1=0, \zeta_2=1$, $\sin[2\pi(q_l-q_k)\zeta_{1,2}]=0$ and Eq. (4.16) is transformed to

$$\frac{A\tilde{L}_k}{1+\sum \tilde{L}_l\overline{m}_l}\left[1-\frac{\tilde{L}_l\overline{m}_l}{2(1+\sum \tilde{L}_l\overline{m}_l)}\right]=1+\beta_k\,. \tag{4.17}$$

For convenience, the last relation can be rewritten in the form of qualratic equation with respect to $1+\sum \tilde{L}_l \overline{m}_l$:

$$2\frac{1+\beta_k}{A\tilde{L}_k}\left(1+\sum \tilde{L}_l \overline{m}_l\right)^2 - 2\left(1+\sum \tilde{L}_l \overline{m}_l\right) + \tilde{L}_k \overline{m}_k = 0 . \tag{4.18}$$

Assuming that all the cavity modes have the same losses, i.e., $\beta_k = 0$ and the frequency of one mode coincides with the line centre, we sum both sides of Eq. (4.18) over all modes. Since $\tilde{L}_l^{-1} = 1+(q_l - q_0)^2 \Delta^2$, and, therefore

$$\sum_{l=-j}^{l=j} \tilde{L}_l^{-1} = 2j+1+\frac{1}{3}\Delta^2 j(j+1)(2j+1),$$

the summation transforms Eq. (4.18) to

$$\frac{2}{A}(2j+1)\left[1+\frac{\Delta^2}{3}j(j+1)\right]\left(1+\sum \tilde{L}_l \overline{m}_l\right)^2 - (4j+1)\left(1+\sum \tilde{L}_l \overline{m}_l\right) - 1 = 0 .$$

The positive solution of the latter is

$$1+\sum \tilde{L}_l \overline{m}_l = A\frac{4j+1+\sqrt{(4j+1)^2 + 8A^{-1}(2j+1)[1+\Delta^2 j(j+1)/3]}}{4(2j+1)[1+\Delta^2 j(j+1)/3]} . \tag{4.19}$$

In order to find $2j+1$, the total number of modes excited in the steady state at a given pumping level, one should demand in Eq. (4.17) that the amplitudes of the modes with the indices $l = \pm(j+1)$ vanish:

$$1+\sum_{l=-j}^{l=j} \tilde{L}_l \overline{m}_l = A\tilde{L}_{\pm(j+1)} = \frac{A}{1+\Delta_{\text{las}}^2} . \tag{4.20}$$

First of all, it is useful to define the upper limit on the number of lasing modes. For this we should substitute the expression (4.19) into (4.20) and use the limit $A \to \infty$. The resultant equation

$$\frac{1+j(j+1)\Delta^2/3}{1+(j+1)^2\Delta^2} = \frac{4j+1}{4j+2},$$

is satisfied by the value $j_{\max}$ bounded by $1 \ll j_{\max} \ll \Delta^{-1}$, which is approximately

$$j_{\max} = \left(\frac{3}{8\Delta^2}\right)^{1/3} . \tag{4.21}$$

The maximal admissible spectral width of the laser emission

$$2\Delta_{las}^{max} \approx 2j_{max}\Delta = (3\Delta)^{1/3} \qquad (4.22)$$

reduces as the intermode spacing decreases. Meanwhile, the total number of modes grows. Thus, the relative radiation spectral width of a typical laser with $\Delta = 10^{-3}$ does not exceed $2\Delta_{las}^{max} = 1.5\times10^{-1}$.

Knowing that $\Delta_{las}^{max} << 1$, it is easy to find from Eq. (4.20) the dependence of the number of lasing modes on parameter A:

$$j = \left[\frac{3(A-1)}{8\Delta^2 A}\right]^{1/3}, \quad \Delta_{las} = \left[\frac{3\Delta(A-1)}{8A}\right]^{1/3}. \qquad (4.23)$$

Strictly speaking, in deriving Eqs. (4.23) it was assumed that the magnitude of j is large. However, the dependence $j(A)$ is such that these formulas are useful at almost any A. Actually, if the number of modes is $j_{max}/2$, then A=8/7, i.e., the pumping is close to threshold value.

Using Eq. (4.17) we also find the shape of radiation spectrum. If $j >> 1$ and $j^2\Delta^2 << 1$, then the radiation intensity falls off quadratically with detuning from the line center; thus using relation (4.23) we get a simple formula

$$\overline{m}_l = 2A\Delta^2[j^2 - (q_l - q_0)^2] = 2A(\Delta_{las}^2 - \Delta_l^2). \qquad (4.24)$$

The physical reasons for multimode emission are following: there is nonumiform saturation of the active medium by each individual mode field and the spatial structures of the intensities of the modes do not coincide. The greater the number of modes involved in laser action, the more uniform the resultant inversion. Nevertheless, as the uniformity of the saturated inversion grows, the conditions for excitation of each individual mode are impaired, so that they are no longer satisfied for modes sufficiently detuned from the line centre. This is the mechanism that limits the number of lasing modes. As the intermode spacing decreases, the discrimination between modes weakens, thus increasing the number of modes and, simultaneously, narrowing the mode spectrum.

By increasing the density of the cavity eigenfrequency spectrum it is possible to narrow the laser radiation line. One way is to increase the cavity length, but for practical purposes it seems more reasonable to use cavities with spherical mirrors since their transverse mode spectrum is nearly degenerate provided the curvature of the mirrors is chosen appropriately [98, 324].

All this refers to ideal cavities, the modes of which are assumed to be purely standing waves. However, due to the inevitable losses on the resonator boundaries the modes differ from standing waves (amplitudes of the counter-running waves in a Fabry–Perot cavity are not necessarily equal in real conditions). As a consequence, the induced inversion gratings are slightly smoothed out. This leads to the further narrowing of the laser emission spectrum [325]. Recall that the travelling wave saturates the laser medium uniformly over the whole length. Therefore, multimode CW generation is not possible in the travelling wave laser. Similar conditions are provided in a standing mode cavity if the laser rod is rapidly moved along its axis, or if the positions of the nodes and crests of standing wave are modulated with respect to a fixed rod, as mentioned in Chapter 1.

Another mechanism of smoothing out the longitudinal spatial inhomogeneity of the inversion is the diffusion of molecular excitation. In solid-state lasers excitation can be transmitted from one active centre to another close to it. In luminescent crystals and glasses this diffusion is usually too slow to have noticeable effect on laser action [316]. However, the diffusion of excitations can have a significant impact on laser action in modern materials with higher density of dopant, and the mentioned in Section 1.2.3 LNP crystal is among them. The diffusion of carriers plays a significant role in semiconductor lasers.

As mentioned above, the form of the radiation spectrum depends on the size of the laser rod and its position inside the cavity. This can be explained in a simple way. The boundary conditions on the mirrors are the same for all modes. As one moves away from the mirror, the spacing between the nodes of each pair of modes increases reaching a maximum equal to $\lambda/4$ at some point and then decreases to zero etc. the number of points along the length of the laser cavity with the maximum distance between the nodes is equal to the difference of the axial mode indices.

This is illustrated in Fig. 4.1, which gives an idea about what form of radiation spectrum can be realized in each particular case. A short laser rod ($L_a/L \ll 1/\Delta q_{max}$) near one mirror can only support single-mode laser operation since the structure of all modes is nearly identical within the rod volume and, therefore, the mode competition is the maximum. Evidently, this result is easily obtained from (4.16). If a thin noninverted layer remains on the centre of the cavity while the laser medium fills the whole

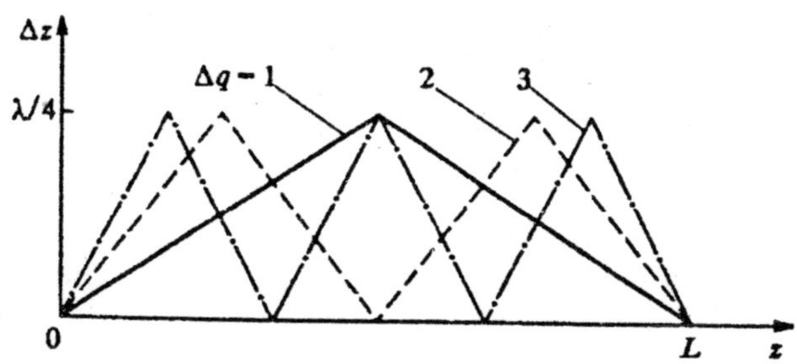

Fig. 4.1. Dependence of the distance between the nodes of longitudinal modes (standing waves) on the coordinate along the cavity axis.

cavity, then the modes with even index difference dominate [80].

We now consider the transverse mode spectrum of a laser with plane-parallel cavity. Assume that a uniformly pumped medium fills the cavity over the entire cross-section. We will use Eq. (4.10) and consider, for simplicity's sake, only the two-dimensional problem assuming the mirrors to be infinitely long in one direction. Then the transverse part of the eigenfunctions is adequately approximate by the sinusoids $\psi_l = \sqrt{2}\sin[\pi(q+1)\zeta_\perp]$ [326] one can use for the calculation the method developed above [327, 328]. In the presented formula $\zeta_\perp = x/(2b)$ is a dimensionless coordinate in the cavity cross-section and b is the transverse dimension of the mirror.

The condition under which the $(q+1)$-th transverse mode is involved in the laser process is given by

$$A > \frac{(1+\beta_q)^2}{1-\beta_q(4q^2-1)/6}. \tag{4.25}$$

The quantity β_q describes the additional part of the losses of the q-th mode, which is conditioned by different values of the diffraction losses of this mode and the referent one. For a plane-parallel cavity the additional (discrimination) part of the losses obeys the dependence $\beta_q = \beta q^2$ and

$$\beta_0 = \frac{\pi^2}{16b^3\,|\ln R_1R_2\,|}\left(\frac{L\lambda}{2\pi}\right)^{3/2}, \tag{4.26}$$

where $R_{1,2}$ are the energy reflectivities of the mirrors. From (4.25) it follows that the transverse pattern on the laser output is formed by modes with indexes between q=0 and $q_{\max}$ where

$$q_{\max} = 0.9\beta_0^{-1/3}. \tag{4.27}$$

Example 4.1

$\lambda = 7\cdot10^{-5}$ cm; $L = 100$ cm;	$\beta_0 = 10^{-3}$;
$R_1R_2 = 0.9$; $b = 0.5$ cm;	$q_{\max} = 9$.

The directions of the maximum radiated intensity for the q-th mode form the angles $\vartheta_q = \pm q\lambda/(4b)$ with the cavity axis. Hence, the angle of divergence of the output laser beam is given by

$$\Delta\vartheta = q\lambda/(2b), \tag{4.28}$$

which is $\Delta\vartheta_{\max} = 2.5$ for the parameter values chosen in the previous numerical example. Relations (4.27), (4.28) imply that the limiting angle of divergence does not depend on the cavity aperture since $\beta \sim b^{-3}$. However the divergence does depend on the cavity length as

$$\Delta\vartheta \sim L^{-1/2}.$$

Using Eq. (4.27), we can exactly specify the conditions for stable single-mode operation. One should put $q_{max} = 1$ in (4.27) and, substituting the computed value of β_0 in (4.26), resolve the last equation with respect to b. Making use of the same parameter values as these given above in the numerical example, we find the aperture required for higher mode discrimination: $2b \leq 1.5$ mm.

4.1.3. Uniqueness of the Stable Steady-State Solution

Let us turn to the problem of the uniqueness of the steady-state solution of the multimode rate equations. There is an assertion belonging to the Novosibirsk [115, 133, 329] that the set of rate equations

$$\frac{dm_k}{d\tau} = Gm_k\left(\tilde{L}_k \int n\psi_k^2 dv - 1\right), \tag{4.29a}$$

$$\frac{\partial n}{\partial \tau} = A - n\left(1 + \sum \psi_q^2 \tilde{L}_q m_q\right), \tag{4.29b}$$

has a globally stable steady-state solution with $\overline{m}_k \neq 0$, and this solution is unique regardless of the number of lasing modes at least in the case when the population difference does not change the sign inside the cavity.

This theorem is proved in the following way. Denoting the pure gain

$$G_k^{net} = \tilde{L}_k \int n\psi_k^2 dv - 1,$$

we rewrite Eq. (4.29a) in the form

$$\frac{dm_k}{d\tau} = Gm_k G_k^{net}.$$

Correspondingly, for a steady state we have

$$\overline{m}_k \overline{G}_k^{net} = 0,$$
$$\overline{n}\left(1 + \sum \psi_q^2 \tilde{L}_q \overline{m}_q\right) = A.$$

For any deviation from the given stationary solution ($\delta n = n - \overline{n}$) the laser equations can be written as follows

$$\frac{dm_k}{d\tau} = Gm_k\left(\tilde{L}_k \int \delta n\,\psi_k^2 dv + \overline{G}_k^{net}\right), \tag{4.30a}$$

$$\frac{\partial(\delta n)}{\partial \tau} = A - (\bar{n} + \delta n)\left(1 + \sum \psi_q^2 \tilde{L}_q m_q\right). \tag{4.30b}$$

We can transform Eq. (4.30b) to a more convenient form

$$\frac{\partial(\delta n)}{\partial \tau} = -\delta\bar{n}\left(1 + \sum \psi_q^2 \tilde{L}_q m_q\right) + A - \bar{n}\left[1 + \sum \psi_q^2 \tilde{L}_q m_q + \sum \psi_q^2 \tilde{L}_q (m_q - \bar{m}_q)\right]$$

$$= -\delta\bar{n}\left(1 + \sum \psi_q^2 \tilde{L}_q m_q\right) - \bar{n} \sum \psi_q^2 \tilde{L}_q (m_q - \bar{m}_q).$$

Multiplying both sides of the last equality by $\delta n / \bar{n}$ and integrating over the whole cavity volume, we obtain:

$$\frac{1}{2}\frac{\mathrm{d}}{\mathrm{d}\tau}\int \frac{(\delta n)^2}{\bar{n}} \mathrm{d}v + \sum \tilde{L}_q (m_q - \bar{m}_q) \int \delta n \psi_q^2 \mathrm{d}v = -\int \frac{(\delta n)^2}{\bar{n}} \left(1 + \sum \tilde{L}_q \psi_q^2 m_q\right) \mathrm{d}v$$

We now rewrite Eq. (4.30a) in a different form:

$$\tilde{L}_q m_q \int \delta n \psi_q^2 \mathrm{d}v = \frac{1}{G}\frac{\mathrm{d}m_q}{\mathrm{d}\tau} - \bar{G}_q^{net} m_q\,,$$

and then, after multiplying by $\bar{m}_q / m_q$, we get

$$\tilde{L}_q \bar{m}_q \int \delta n \psi_q^2 \mathrm{d}v = \frac{1}{G}\bar{m}_q \frac{\mathrm{d}(\ln m_q)}{\mathrm{d}\tau} - \bar{G}_q^{net} m_q\,.$$

This makes it possible to modify the previous equation:

$$\frac{\mathrm{d}}{\mathrm{d}\tau}\left[\frac{1}{2}\int \frac{(\delta n)^2}{\bar{n}} \mathrm{d}v + \frac{1}{G}\sum (m_q - \bar{m}_q \ln m_q)\right] = -\int \frac{(\delta n)^2}{\bar{n}} \left(1 + \sum \tilde{L}_q \psi_q^2 m_q\right) \mathrm{d}v + \bar{G}_q^{net} m_q. \tag{4.31}$$

We are free to add any arbitrary constant to the function under the derivative sign and, therefore, replace $m_q - \bar{m}_q \ln m_q$ by $m_q - \bar{m}_q - \bar{m}_q \ln(m_q / \bar{m}_q)$. Now it can be asserted that an arbitrary deviation from the steady-state value of the variables will decrease to zero in time. This statement is based on the fact that the right-hand side of equality (4.31) is negative ($\bar{G}_q^{net} = 0$), and, consequently, the function under the derivative sign must decrease monotonically until the vanishing of the left-hand side stops the process. This will occur only when $\delta n = 0$ and $m_q = \bar{m}_q$.

The fact that the result holds for an arbitrary deviation from the steady-state solution is indicative of the global stability of the solution and the unique way to choose it.

The proven theorem of global stability of the steady-state solution of the rate equations puts an end to the myth that multimode operation is a universal reason for the undamped spiking of solid-state lasers. The absence of an instability shows that even this model is inadequate and makes one seek for causes of undamped spiking of class *B* lasers outside the nature of the interaction between the radiation field and laser medium.

It should be noted that there is generalization of the presented theorem on the inhomogeneously broadened lasers. The inconstant laser medium saturation does not change the situation, leading, generally speaking, to only the dynamical deformation of the laser modes. And only the existence of domains with the opposite sign of population difference, which act as a saturable absorber, can lead to the laser instability. Only by going beyond the rate equations can we rapidly change the character of these models. Such is the case for class *C* lasers considered in Chapter 11. Meanwhile, a weaker departure from the rate scheme may be sufficient sometimes. In particular, the undamped oscillations can be produced by the effect known as combination tone mode-mode coupling through secondary beats.

4.2. Relaxation Oscillations as Low Frequency Normal Laser Modes

There are many steady states for a multimode laser. It is shown in what follows that only one of these states is stable, and it shows global stability. Normal oscillations of the system, the number of which coincides with the number of modes, are all damped i.e., they are exactly relaxation oscillations. The spectrum and some features of these relaxation oscillations were first investigated in the early works [330–334]. The modern conception of the low-frequency laser dynamics was formulated in Ref. [335–351].

4.2.1. The Model with Spatially Extended Laser Medium

As in the theory of a single-mode laser, we do not start with expanding the inversion in a series of spatial harmonics. In order to avoid using too much algebra, we assume that the number of lasing modes is very large and the modes are plane standing waves [330]. There are many dangerous points in such an approach, but it allows to make clear in a simple and direct way some very interesting features of relaxation oscillations. Supposing that the steady states are known (no other information on these states is required) we linearize the set of Eqs. (4.29). The linearized equations

$$\frac{\mathrm{d}(\delta m_k)}{\mathrm{d}\tau} = G\overline{m}_k\widetilde{L}_k\int\delta n\psi_k^2\mathrm{d}\zeta\ , \tag{4.32a}$$

$$\frac{\partial(\delta n)}{\partial\tau} = \frac{A}{\overline{n}}\delta n - \overline{n}\sum\psi_l^2\widetilde{L}_l(\delta m_l) \tag{4.32b}$$

are reduced by substitution $\{\delta m_k, \delta n\} = \{\delta m'_k, \delta n'\}\exp(\lambda\tau)$ to algebraic ones, yielding a characteristic equation

$$\lambda\delta m'_k + G\overline{m}_k \overline{L}_k \sum \widetilde{L}_l \delta m'_l \int \frac{\overline{n}\,\psi_k^2\psi_l^2}{\lambda + A/\overline{n}} \mathrm{d}\zeta = 0. \tag{4.33}$$

This equation can be simplified by bearing in mind that in the case of large number of modes:

(a) the laser medium is spatially uniform, such that we can assume $\overline{n} = 1$;

(b) the laser spectrum is rather narrow, such that $\widetilde{L}_k \approx 1$;

(c) the modes are orthogonal and, consequently

$$\int_0^1 \psi_k^2\psi_l^2 \mathrm{d}\zeta = 1 + \frac{1}{2}\delta_{l,k}. \tag{4.34}$$

Thus, Eqs. (4.33) transform to a homogeneous linear set of equations

$$\delta m'_k[\lambda(\lambda + A)] + \frac{1}{2}G\overline{m}_k \sum \delta m'_l = 0. \tag{4.35}$$

Equating to zero the determinant

$$\begin{vmatrix} P_1(\lambda) & G\overline{m}_2 & \cdots & G\overline{m}_N \\ G\overline{m}_1 & P_2(\lambda) & \cdots & G\overline{m}_N \\ \cdots & \cdots & \cdots & \cdots \\ G\overline{m}_1 & G\overline{m}_2 & \cdots & P_N(\lambda) \end{vmatrix} = 0,$$

where $P_l(\lambda) = \lambda(\lambda + A) + \frac{1}{2}G\overline{m}_l$ $(l = 1,2,\ldots,N)$, we arrive at the characteristic equation

$$\left[1 + \sum_{l=1}^{N} \frac{G\overline{m}_l}{\lambda(\lambda + A) + \frac{1}{2}G\overline{m}_l}\right] \prod_{l=1}^{N} \left[\lambda(\lambda + A) + \tfrac{1}{2}G\overline{m}_l\right] = 0. \tag{4.36}$$

Since the left-hand side is factored, the Eq. (4.36) can be decomposed into

$$1 + \sum_{l=1}^{N} \frac{G\overline{m}_l}{-p + \frac{1}{2}G\overline{m}_l} = 0, \tag{4.37}$$

and

$$-p + \frac{1}{2}G\overline{m}_k = 0, \tag{4.38}$$

where $p = -\lambda(\lambda + A)$. From the structure of these equations it should be readily apparent that p is a real positive quantity.

Consider first Eq. (4.37) assuming that the mode intensities are different and decrease with an increase in mode index ($\overline{m}_1 \geq \overline{m}_2 \geq \overline{m}_3 ...$). The boundaries of the roots can be defined graphically. Rewrite the Eq. (4.37) in the form $\sum f_j(p) = 1$, where $f_j(p) = G\overline{m}_j(p - G\overline{m}_j/2)^{-1}$, and represent the family of functions $f_j(p)$ in Fig. 4.2. One can see that the largest root has a lower limit imposed by the condition $p_1 > G\overline{m}_1/2$ and each of the remaining N–1 roots is bounded by $G\overline{m}_j/2 > p_j > G\overline{m}_{j+1}/2$. The characteristic roots satisfy the equation $\lambda_j(\lambda_j + A) + p_j = 0$ and, consequently, have the form

$$\lambda_j = -\frac{A}{2} \pm \sqrt{\left(\frac{A}{2}\right)^2 - p_j}.$$

Hence, under the condition $G \gg 1$, the relaxation frequencies $\Omega_j \approx \sqrt{p_j}$. The largest one stands apart from the group. The damping rates are the same for all modes, and they are equal to $A/2$.

If the intensities of any two modes converge, i.e., $\overline{m}_j \to \overline{m}_{j+1}$, then $p_j \to G\overline{m}_j/2$. The limiting cases $p_j = G\overline{m}_j/2$ are beyond the scope of Eq. (4.37), but these are contained in Eq. (4.38). The power of Eq. (4.37) is reduced in accordance with the multiplicity of the degeneracy with respect to the mode intensities. If the system is completely degenerate, then Eq. (4.37) has an unique root $p_1 = GN\overline{m}$. The main difference between the relaxation oscillations is that those, which are characterized by the roots of Eq. (4.38) are compensated and are not manifested in the total radiation intensity. This is apparent from Eq. (4.35) which transforms to

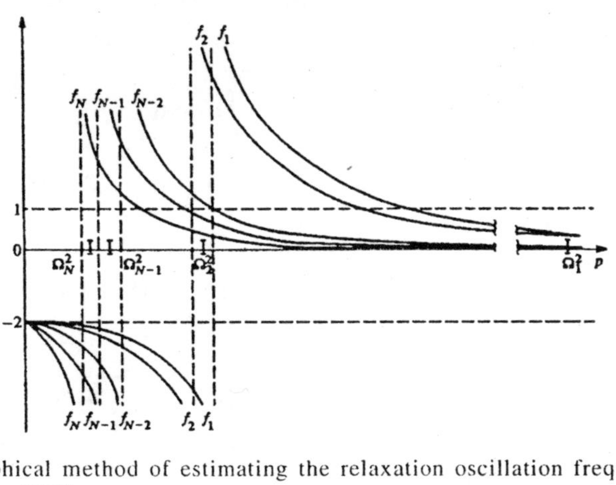

Fig. 4.2. Graphical method of estimating the relaxation oscillation frequencies of a multimode laser [330].

Eq. (4.38) if $\sum \delta m_l' = 0$. The roots of Eq. (4.37) correspond to uncompensated oscillations.

The highest frequency of relaxation oscillations is sometimes called the main relaxation frequency, since it coincides with the sole frequency present in the relaxation oscillations of a single-mode laser. This frequency is characteristic of the in-phase amplitude oscillations of all modes. It is impossible to obtain information about the structure of other relaxation oscillations from the simplified approach. For this purpose, it is necessary to find the eigenvectors of the system and this will be done in the next section.

4.2.2. Approximation of the Spatial Inversion Gratings

The analysis of the model with spatial extended inversion has a very limited generality. Nevertheless, it leads to three important conclusions. First, the absence of characteristic roots with positive real parts confirms the stability of the steady-state solutions. Then, the number of types of relaxation oscillations cannot be more than the number of lasing modes because the order of the characteristic polynomial is equal to 2N. Finally, it is found that all totality of relaxation oscillations is divided into two groups: compensated oscillations and uncompensated ones. This corresponds to experimental facts given in Section 1.2. But many questions remain open and we need to return to the problem of relaxation oscillations using a less rough model.

Let as hold the same scenario as in Section 3.2 and present the inversion as a set of periodic gratings the number of which coincides with the number of lasing modes. Limiting the consideration by longitudinal modes, for the amplitudes of the lowest order inversion gratings we have the following expressions:

$$n_0 = \int_0^1 n \mathrm{d}\zeta, \quad n_k = -\int_0^1 n \cos(2\pi q_k \zeta) \mathrm{d}\zeta, \tag{4.39}$$

and neglecting the higher spatial harmonics Eqs. (4.29) transform into a set of ordinary differential equations

$$\frac{\mathrm{d}m_k}{\mathrm{d}\tau} = G m_k [\tilde{L}_k (n_0 + n_k) - 1 - \beta_k], \tag{4.40a}$$

$$\frac{\mathrm{d}n_0}{\mathrm{d}\tau} = A_0 - n_0 \left(1 + \sum_{j=1}^{N} \tilde{L}_j m_j \right) - \sum_{j=1}^{N} \tilde{L}_j m_j n_j, \tag{4.40b}$$

$$\frac{\mathrm{d}n_k}{\mathrm{d}\tau} = -n_k\left(1+\sum_{j=1}^{N}\tilde{L}_j m_j\right) - \frac{1}{2}\tilde{L}_k m_k n_0 . \qquad (4.40c)$$

This set is valid under the condition that the laser medium uniformly fills the whole cavity volume. The order of the system is equal to $2N+1$, where N is the number of lasing modes. The same is the order of the set of equations linearized in the vicinity of nontrivial steady state. This means that the number of pairs of complex-conjugate characteristic roots and, consequently, the number of relaxation oscillations cannot be more than N.

The further investigation of this problem consists in determination of the eigenvalues and eigenvectors. It can be formulated as

$$\| a_{mn} \| \, \delta V_n = \lambda_m \delta V_m ,$$

where $\| a_{mn} \|$ is the matrix of the system coefficients linearized near the steady state, δV_n is the column of the eigenvector components $(\delta m_j, \delta n_0, \delta n_j)$. This problem can be solved only by using the numerical calculations. However, it is possible to find approximate analytical expressions for the stationary values of the variables [167].

The structure of eigenvectors is shown in Fig. 4.3. The frequencies of five lasing modes are situated symmetrically in respect to the gain line centre. The mode losses are assumed to be equal. The number of relaxation oscillations is also equal to five and this is the maximum number. The length of the arrow corresponds to the modulation depth of the mode amplitude, while its orientation gives its relative phase.

The complete set of eigenvectors presented in Fig. 4.3 testifies to the existence of two groups of relaxation oscillations. The unique representative of the group with in-phase oscillations is the relaxation oscillation with the highest frequency, which is inherent in all class *B* laser models, regardless of the type of the cavity and the number of lasing modes. The rest $N-1$ relaxation oscillations belong to the antiphase dynamics.

The most evident but not the most convenient method of investigation of the relaxation oscillations is the direct oscilloscope observation of the transients. More useful is the alternative spectral approach. In 1965 McCumber [352] demonstrated for a single-mode laser, taken as an example, that relaxation oscillations manifested themselves in the spectra of low-frequency intensity fluctuations (power spectra) as the resonance peaks. The origin of this phenomenon is simple enough: spontaneous emission (quantum noise) and the parameter fluctuation systematically disturb the steady state. The spectra of such processes are close to white noise, and the laser response is more intense on the frequency of the normal mode (which is relaxation oscillations).

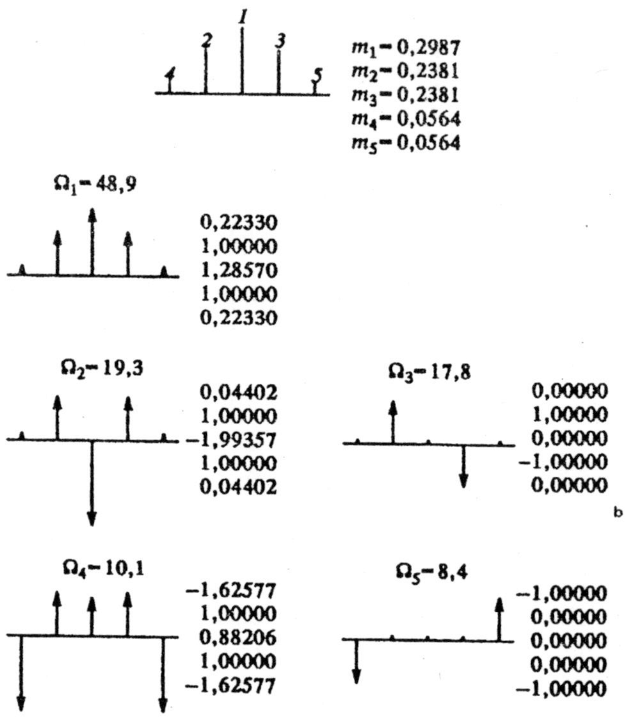

Fig. 4.3. Geometric representation of eigenvectors of the rate equation laser model with five modes [350].

The resonance nature of power spectra is exhibited also in multimode lasers, but the picture here is much more diverse, because one can investigate the behaviour of the total intensity as well as the individual modes (see Fig. 1.20). Note that the origin of in-phase and antiphase relaxation oscillations is different. The first reflect the competition of pumping and stimulated emission processes while the second ones correspond to the mode competition. The antiphase oscillations are manifest mainly in the power spectra of individual modes and they are practically absent in the total intensity where the high-frequency peak corresponding to in-phase oscillations of the mode amplitudes dominates.

A very interesting feature is the correspondence of each antiphase relaxation oscillations to the quite certain cavity mode. If the modes are renumbered in the order of decreasing stationary intensities, and relaxation oscillations are renumbered in the order of decreasing frequencies, then in the power spectrum of the k-th mode a peak on the frequency Ω_k dominates. This is clearly demonstrated in Fig. 1.20.

The linkage of a given relaxation oscillation to a particular cavity mode is confirmed by a laser with selective envelope feedback, which we will discuss in Section 7.1. The optoelectronic feedback circuit serves for the control of power supply of the diode laser, which is the pumping source for the solid-state laser under consideration. The input signal of this circuit could be proportional to intensity of an individual mode or to derivative of this intensity. Experimental findings, which are in good agreement with computational results, indicate that the positive feedback gives rise to a corresponding resonance peak in all power spectra including the total intensity. Changing the cavity mode results in changing the accentuated relaxation oscillation.

Let us focus our attention also on the following circumstances. The number of relaxation oscillations coincides with the number of lasing modes only when the last number is relatively small. Starting with some value of N_{cr} the increase in the number of lasing modes is accompanied by a decrease (not increase!) of the number of relaxation oscillations. The reason is that the frequencies of relaxation oscillations are going up slower than the damping rates. This results in transformation of the complex characteristic roots into the real ones.

4.2.3. Dependence of the Dynamical Features on the Distribution of Unsaturated Gain Over the Perimeter of Laser Cavity

Calculations based on the equations like (4.40) are often not confirmed by the experiment. One of the main reasons is symmetry breaking in real lasers. First of all, this breaking can result from the nonuniform distribution of the unsaturated inversion over the cavity perimeter due to inequality of the laser rod length and the cavity length or from the pumping attenuation in the active element. As a result, in addition to the small-scale gratings with the period $\lambda_l/2$, we have large-scale gratings with the period $L/2r$ ($r =$ 1, 2...) and the inversion is represented by the series

$$n(\tau,\zeta) = n_0(\tau) - 2\sum_{r=1}^{K} D_p(\tau)\cos(2\pi r\zeta) - 2\sum_{l=1}^{N} n_l(\tau)\cos(2\pi q_l\zeta), \quad (4.41)$$

which leads to the set of ordinary differential equations [349]:

$$\frac{dm_k}{d\tau} = Gm_k[\widetilde{L}_k(D_0 + n_k) - 1 - \beta_k], \quad (4.42a)$$

$$\frac{dn_k}{d\tau} = -n_k\left(1 + \sum_{l=1}^{N}\widetilde{L}_l m_l\right) - \frac{1}{2}\sum_{l=1}^{N}\widetilde{L}_l m_l D_{k-l}, \quad (4.42b)$$

$$\frac{dn_0}{d\tau} = A_0 - n_0\left(1+\sum_{l=1}^{N}\tilde{L}_l m_l\right) - \sum_{l=1}^{N}\tilde{L}_l m_l n_l, \tag{4.42c}$$

$$\frac{dD_r}{d\tau} = A_r - D_r\left(1+\sum_{l=1}^{N}\tilde{L}_l m_l\right) - \frac{1}{2}\sum_{l=1}^{N-r}\tilde{L}_l m_l D_{l+r} - \frac{1}{2}\sum_{l=1}^{N}\tilde{L}_l m_l D_{l-r}. \tag{4.42d}$$

Here we used the decomposition of the pumping coefficient

$$A(\zeta) = A_0 + \sum_{r=1}^{R} A_r \cos(2\pi r\zeta), \tag{4.43}$$

in which $R = N$ and coefficients are

$$A_r = \int_0^1 A(\zeta)\cos(2\pi r\zeta)d\zeta. \tag{4.44}$$

Since we neglect all higher harmonics, the set (4.42) consists of N equations (a) for the mode intensities and the same number of equations (b) for the amplitudes of the small-scale inversion gratings n_k plus one equation (c) for the mean spatial value n_0. The number of equations for amplitudes of large-scale gratings is determined both by the gain profile (characterized by the number R) and the number of lasing modes.

The situation here is not simple since it is impossible to assume that N is independent on R. When the small-signal gain profile is relatively smooth, then the main factor of forming the large-scale gratings are the spatial beats of lasing modes and the number of equations is equal to N–1. Thus, the order of the set (4.42) is minimum 3N and this is much more than we have in other models. Nevertheless, this does not affect the number of relaxation oscillations, which is N as in all multimode models considered above. The large-scale gratings increase the order of the model but do not add any new complex characteristic roots.

The method of investigation of the model (4.42) is the same as was used above for equations (4.40). Numerical simulation makes it possible to find the steady-state solutions, eigenvalues and eigenvectors of the system, and to observe its dependence on the control parameters such as pumping and the cavity filling factor [350].

Consider the simplest case of the uniform pumping with $\zeta_1 = 0$ and $L_a < L$ when everything is determined by the filling factor $\xi = L_a / L$

$$A_r = A_0 \frac{\sin(2\pi r\xi)}{2\pi r\xi}.$$

Stationary dependences of the total intensity and intensities of five modes symmetrically situated around the centre of the gain line on the pumping parameter are presented in Fig. 4.4 together with frequencies of relaxation oscillations at $\xi = 0.4$. The mode losses are equal.

This case corresponds to the following distribution of the gain ($g_k = \tilde{L}_k / \tilde{L}^{\max}$) among modes:

$$g_{1,5} < g_{2,4} < g_3 = 1.$$

Note that in the interval $1.3 < A_0 < 1.9$ the intensity of the central mode, which has maximal linear (unsaturated) gain, is less than intensities of closest side modes. Such a result is possible only if $\xi < 1$. This effect was experimentally observed in [167, 351]. The number of lasing modes also depends on the pumping level. Ideal symmetry in the initial gain distribution provides the entrance of new modes in generation by pairs. Each such event gives birth to an additional pair of relaxation oscillations, one of which is compensated while the other is uncompensated. The frequencies

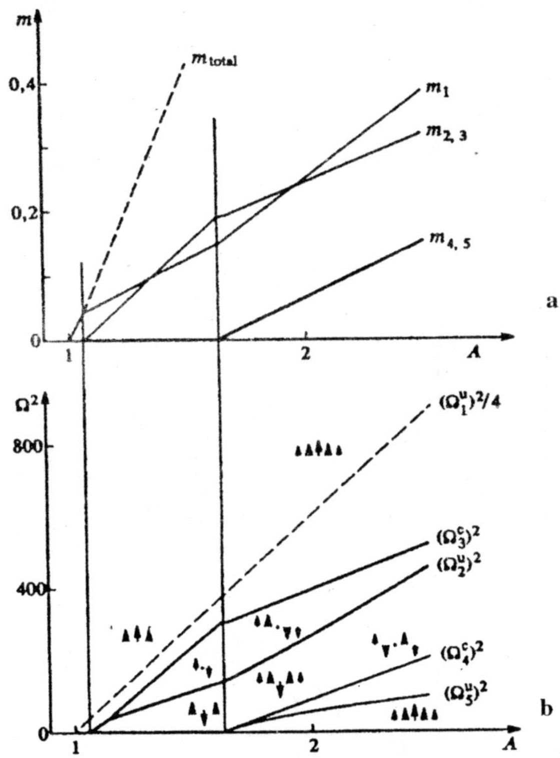

Fig. 4.4. Intensities (*a*) and frequencies (*b*) of relaxation oscillations as functions of the pumping parameter: $G = 2500$; $\Delta = 0.132$; $\xi = 0.4$ [350].

of these oscillations are different.

The influence of the filling factor ξ on the intensities of laser modes and frequencies of relaxation oscillations under the given pumping parameter is demonstrated in Fig. 4.5. In the interval $0.1 < \xi < 0.4$ the central mode is completely suppressed, and the number of lasing modes is equal to four. The suppressed mode is reflected in the number and alternation of relaxation frequencies.

The mentioned peculiarities of the behaviour of multimode lasers must be taken into account when one compares the results of theoretical and experimental investigations. The sensitivity of the dynamical characteristics to the symmetry breaking can be used to reveal the fact of this breaking.

Let us summarise the features of the considered model of multimode class *B* laser.

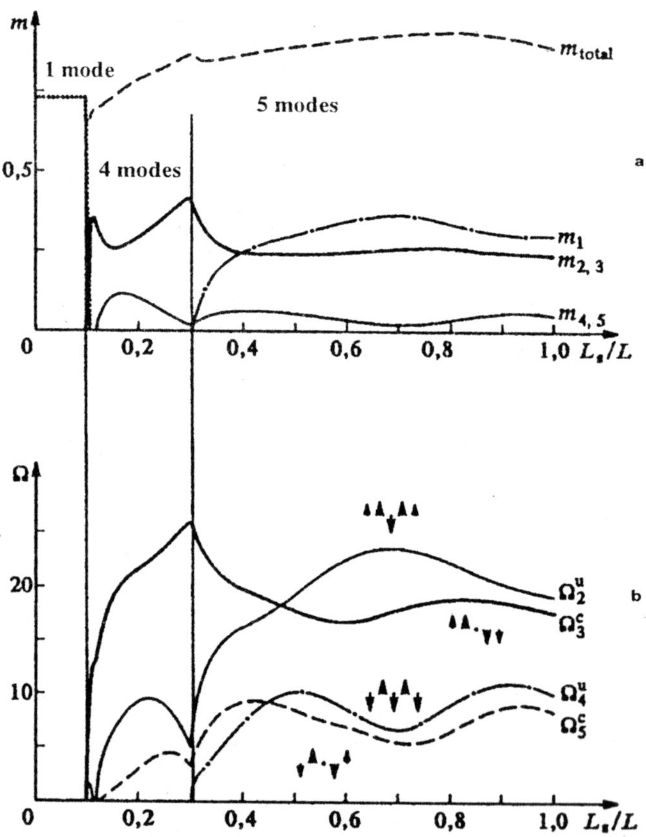

Fig. 4.5. Intensities (*a*) and frequencies (*b*) of relaxation oscillations as functions of the cavity filling factor: $G = 2500$; $\Delta = 0.132$; $\xi = 0.4$ [350].

1. There is only one stable solution among the steady-state solutions.

2. The time-dependent processes after deviation of the system from the equilibrium are realized in the form of relaxation oscillations.

3. The in-phase oscillations of the mode intensities possess the maximal frequency, and the existence of such an oscillation does not depend on the cavity geometry and the values of the control parameters.

4. The number of antiphase oscillations is N–1 if N is relatively small and it decreases for very high numbers of modes.

5. Relaxation oscillations form a set of low-frequency laser eigenmodes, which exists along with the optical modes.

6. There is some correspondence between the optical and relaxation modes.

4.3. Time-Dependent Processes

4.3.1. Features of Spiking in Multimode Lasers

Hence, the spatial competition among modes does not explain the phenomenon of undamped oscillations. Nevertheless, it has a noticeable effect on free running. A single-mode model suggests that the spike emission is preceded by a linear development stage, which starts when the threshold is attained. The linear period is much longer than the spike duration and depends appreciably on mode properties. The same is true for a multimode laser. However, since the modes differ in loss and gain, they reach the threshold at different moments of time. During the linear stage the difference in mode intensities increases [353, 354]. Only those modes, which reach threshold almost simultaneously, retain similar amplitudes before the beginning of the nonlinear stage. This group of modes defines the modal composition of the spike. A few special cases merit consideration.

1. The group consists of a very large number of modes to ensure uniform saturation of the laser medium over the entire volume. This means that the most favourable conditions will be retained for the same modes and the spectral composition of the radiation will not change from spike to spike. The case corresponds to the regular damped or undamped pulsations.

2. The mode discrimination is so strong that for a leading mode the lasing conditions are restored after the spike is ended before the other modes reach the laser threshold. As in the previous case the laser kinetics will be regular, but a single-mode regime will take place.

3. The first single-mode spike has time to be emitted before the other modes achieve a significant intensity. If the mode discrimination is not very great, then a rival mode is ahead in a subsequent period of time, etc.

the emission of each spike modifies the spatial distribution of active molecules and make it irregular. The spikes form a chaotic sequence in time, which is gradually attenuated until the intensity becomes time-independent.

To bring the theory up to the concrete results we should assign the active molecules distribution, i.e., confine ourselves to the time interval preceding the spike. As an illustration, we consider the model with the uniform distribution of pumping in the laser medium. Using the assumption of this model we find the gain $\int n\psi^2 \mathrm{d}v = \xi n$. The filling factor ξ is the same, accurate to λ/L, for all modes and is equal to L_a/L. Since we intend to investigate the formation of the first spike after pumping is switched on, we should introduce a term taking into account the spontaneous emission. Thus, the initial set of equations will become:

$$\frac{\mathrm{d}m_k}{\mathrm{d}\tau} - Gm_k\left[\frac{n}{1+\Delta_k^2} - (1+\beta_k)\right] = \frac{G\varepsilon_{sp}}{1+\Delta_k^2}(n+n_S)\,, \tag{4.45a}$$

$$\frac{\mathrm{d}n}{\mathrm{d}\tau} + n - A = -n\sum \frac{m_q}{1+\Delta_q^2}\,. \tag{4.45b}$$

During the time before the first spike, the induced emission does not influence the inversion. Therefore, the modes do not interact ad can be treated as independent. The linear stage of lasing is started at the time the threshold is reached (n=1) by the 'reference' mode with $\Delta_k = \beta_k = 0$. We take this time as $\tau = 0$. In the stage of linear development the inversion variation is described by Eq. (4.14); substituting this into (4.45a) yields a linear equation for m_k. Its approximate solution, if the small quantities Δ_k^2 and β_k are retained only in the exponential terms, is expressed by

$$m_k(\tau) = m_0(\tau)\exp[-G(A-1)\Delta\tau_k\tau]\,. \tag{4.46}$$

The intensity of the reference mode, $m_0(\tau)$, is given by Eq. (3.52). The quantity

$$\Delta\tau_k = \frac{\Delta_k^2 + \beta_k}{A-1} \tag{4.47}$$

represents the delay of the k-th mode with respect to the reference mode.

A plane-parallel cavity features a very weak dependence of the eigenfrequencies on the transverse mode structure, so that it can be assumed

$$\Delta^2_{kmn} = (\pi c / L\gamma_\perp)^2 (q_k - q_0)^2 . \tag{4.48}$$

By contrast, the difference in losses is determined exclusively by the mode structure and is independent of the axial index. For square mirrors we have

$$\beta_{kmn} = \beta_0 (m^2 + n^2) , \tag{4.49}$$

where the coefficient β_0 is given by Eq. (4.26). The delay between the axial modes is, therefore,

$$\Delta\tau_q = \frac{(\pi c \Delta q)^2}{\gamma_\perp^2 L^2 (A-1)} , \tag{4.50}$$

and the delay of the transverse mode with respect to the axial mode is

$$\Delta\tau_{mn} = \beta_0 \frac{m^2 + n^2}{A-1} . \tag{4.51}$$

In order to estimate the number of effectively excited modes, the delay should be equated to $\tau_p / 2$, half of the spike duration, which is calculated by (3.34). Using this equality we find the boundaries of the axial mode spectrum and the limiting index of the transverse mode:

$$\Delta q_{\text{las}} = \frac{L\gamma_\perp}{\pi c} \sqrt{\frac{1}{2} \tau_p (A-1)} , \tag{4.52}$$

$$(m^2 + n^2)_{\text{las}} = \frac{\tau_p}{2\beta_0} (A-1) .$$

Example 4.2

$L = 10^2$ cm, $\gamma_\perp = 10^{12}$ s^{-1},	$\Delta q_{\text{las}} = 60$,
$A = 5$, $\beta_0 = 2 \cdot 10^{-3}$, $\tau_p = 2 \cdot 10^{-3}$,	$(m^2 + n^2)_{\text{las}} = 2$.

The total number of excited axial modes is of the order of hundred, and that of transverse modes is of the order of unity. This means that the spiking is irregular in such a cavity.

The model of laser with a plane-parallel cavity and a uniformly pumped rod has been chosen owing to the simplicity and clarity of the calculation rather than due to its practical insignificance. In practice these conditions are very seldom provided. In the most commonly used laser rods of cylindrical form achieving uniform inversion is impracticable. These active elements are also heated nonuniformly by the pump and laser fields and acquire lens properties, thus changing the cavity configuration. The problem of the mode composition of the first spike, with all these factors taken into account, was first considered in [355]. Solving this problem is facilitated by the following circumstances:

1. The structure of the light beam can be assumed to be identical in all parallel cross-sections of the active element provided $L_a/L \ll 1$. Thus, in the calculations the real laser rod can be replaced by an infinitely thin sheet with finite gain.

2. Only the difference in gain and loss is important in the linearized problems. Therefore, instead of the initial model we are free to consider an equivalent laser model with uniform gain across the cavity and nonuniform loss.

3. The inversion density in the active element cross-section is described approximately by a quadratic law [356]. Hence, the Gaussian distribution law is valid for the gain and the problem reduces to considering a cavity with the Gaussian diaphragm. The gain (loss) nonuniformity enhanced the discrimination against transverse modes. The significance of this factor depends of how rapidly the gain falls off from the cavity axis. Hence, the effect should be greater in a small-diameter rod. Even the use of spherical mirrors does not lead necessarily to multimode generation within the limits of spike. Only near the concentric configuration, where the focal spot is extremely small and the active element is at the cavity centre, the mode discrimination is weak enough to allow many transverse modes to participate in laser action simultaneously. The estimates of the number of modes, given in Refs. [355], have been confirmed experimentally.

Rigorously speaking, what has been said above refers only to the first spike. After this spike is over, the inversion distribution changes, so that one has to solve a complicated problem to find it. The form of the distribution, and therefore, the problem on the whole, increases in complexity with each new spike. The problem is solvable only by numerical methods [357].

4.3.2 Onset of Laser Spectrum

Besides the period of intermode beats and the period and time of damping of relaxation oscillations, the multimode laser features one more time scale, which characterizes the rate of onset of the radiation spectrum. For many lasers the spectrum dynamics is the slowest process occurring in the laser. This is due to the smallness of the frequency spacing between modes compared to the gain linewidth, which makes the difference in discrimination of the neighbouring axial modes negligible.

Since the laser time constants differ in magnitude, the onset of lasing falls into few stages. After the threshold is exceeded, the total radiation intensity is established first. After this stage is completed, over a time of order T_1 $(1/\gamma_\parallel)$, the inversion can be considered a constant as well. Therefore, at all the subsequent stages we can assume $dn/d\tau = 0$ and use, instead of Eq. (4.29), the equation

$$\frac{dm_k}{d\tau} = Gm_k\left(\tilde{L}_k \int \frac{A\psi_k^2 d\nu}{1+\sum \tilde{L}_q \psi_q^2 m_q} - 1 - \beta_k\right). \tag{4.53}$$

Since we are dealing with longitudinal modes, we take the expansion (4.15) and pass over from Eq. (4.53) to the simplest equation

$$\frac{dm_k}{d\tau} = Gm_k\left\{\frac{A\tilde{L}_k}{1+\sum \tilde{L}_q m_q}\left[1 - \frac{A\tilde{L}_k}{2(1+\sum \tilde{L}_q m_q)}\right] - 1 - \beta_k\right\}. \tag{4.54}$$

At the last stage of the transient and in the steady state the right-hand side of Eq. (4.54) must be fully retained. However, at an intermediate stage where the narrowing of the radiation spectrum mainly occurs, we can omit the second term in the square brackets owing to its smallness and use the equation $(\beta_k = 0)$

$$\frac{dm_k}{d\tau} = Gm_k\left(\frac{A\tilde{L}_k}{1+\sum \tilde{L}_q m_q} - 1\right). \tag{4.55}$$

The main difficulty in using this equation is that it relates the evolution of the k-th mode to all other modes. However, taking the equations for two modes, the k-th and n-th, we can eliminate the sum and obtain the equation relating the intensities of only these two modes:

$$\tilde{L}_n \frac{d(\ln m_k)}{d\tau} - \tilde{L}_k \frac{d(\ln m_n)}{d\tau} = G(\tilde{L}_k - \tilde{L}_n).$$

Integrating this equation, we get

$$\frac{m_k(\tau)}{m_k(0)} = \left[\frac{m_n(\tau)}{m_n(0)}\right]^{\tilde{L}_k/\tilde{L}_n} \exp\left[-\left(1 - \frac{\tilde{L}_k}{\tilde{L}_n}\right)G\tau\right]. \tag{4.56}$$

Based on the assumption of a relatively narrow radiation spectrum, we will neglect the difference of $\tilde{L}_k$ from $\tilde{L}_n$ in the power of the first term. The index k is retained to designate the current mode and the index n is assigned to the reference (resonance) mode. Hence we assume $\tilde{L}_n = 1$ and $1 - \tilde{L}_k \approx \Delta_k^2$. Finally, Eq. (4.56) takes the form

$$\frac{m_k(\tau)}{m_k(0)} = \left[\frac{m_n(\tau)}{m_n(0)}\right]\exp(-G\Delta_k^2\tau)\,. \tag{4.57}$$

Thus, we can express the amplitudes of all modes through the reference mode and write

$$\sum \tilde{L}_q m_q(\tau) = \frac{m_n(\tau)}{m_n(0)}\sum \tilde{L}_q m_q(0)\exp(-G\Delta_k^2\tau)\,, \tag{4.58}$$

reducing the problem to the evolution of the reference mode [358].

We then can make use of the adiabatic slowness of the spectral evolution and take for the reference mode, instead of Eq. (4.55), its quasistationary version

$$1 + \sum \tilde{L}_q m_q(\tau) = A\,,$$

or, in view of Eq. (4.58),

$$\frac{m_n(\tau)}{m_n(0)}\sum \tilde{L}_q m_q(0)\exp(-G\Delta_q^2\tau) = A - 1\,. \tag{4.59}$$

We now replace the summation by integration using, instead of the discrete quantities, the continuous ones

$$m(q,\tau) = m_q(\tau)/\Delta, \quad \tilde{L}(q) = \tilde{L}_q, \quad \Delta q = \Delta\,.$$

Noting the smoothness of the functions $m(q)$ and $\tilde{L}_q$ compared with $\exp(-G\Delta_q^2\tau)$, we make an approximate integration

$$\int_{-\infty}^{\infty} m(q,0)\tilde{L}_q\exp(-G\Delta_q^2\tau)dq \approx m(0,0)\tilde{L}(0)\int_{-\infty}^{\infty}\exp(-G\Delta_q^2\tau)dq = m(0,0)\tilde{L}(0)\sqrt{\frac{\pi}{G\tau}} = \frac{m_q(0)}{\Delta}\sqrt{\frac{\pi}{G\tau}}. \tag{4.60}$$

Using this relation in Eq. (4.59) we obtain

$$m_n(\tau) = (A-1)\Delta\sqrt{G\tau/\pi}\,,$$

and substituting this into Eq. (4.57) we finally arrive at

$$m_k(\tau) = \frac{m_k(0)}{m_n(0)}(A-1)\Delta\sqrt{\frac{G\tau}{\pi}}\exp(-G\Delta_k^2\tau) \tag{4.61}$$

[359, 360]. Generalization to a laser with selective and time-dependent mode losses is given in Ref. [359].

From Eq. (4.61) it is seen that at each moment of time the form of the spectrum is nearly the Gaussian with halfwidth of halfheight:

$$\Delta_{\text{las}}(\tau) = \sqrt{\frac{\ln 2}{G\tau}}\,. \tag{4.62}$$

Eq. (4.61) also yields the time constant of the mode amplitude variation

$$\tau_k = \frac{1}{G\Delta_k^2}. \tag{4.63}$$

According to the initial assumptions $\tau_k >> 1$ these considerations are adequate only where $\Delta_{las}(\tau) << G^{-1/2}$.

At very large times, when form of the radiation spectrum is nearly time-independent, i.e., at $\tau \geq 1/(G\Delta_{las}^2)$ this approach is not appropriate, since one cannot use Eq. (4.55). Thus, the time of onset of the radiation spectrum ranges

$$1 << \tau << 1/(G\Delta_{las}^2). \tag{4.64}$$

For solid-state lasers with typical parameter values $\Delta = 10^{-3}, \Delta_{las} = 10^{-1} - 10^{-2}, G = 10^3 - 10^4$, the condition $G\Delta_{las}^2 << 1$ is barely satisfied or not satisfied at all. Meanwhile, this condition is met for dye lasers with the broader gain lines and the higher rates of population relaxation ($G \leq 1$, $\Delta \approx 10^6, \Delta_{las} \approx 10^{-2} - 10^{-3}$).

4.3.3. Alternative Mechanisms of Laser Multistability

We have considered only one mechanism of multimode lasing – the spatial competition of modes. Remaining within the framework of homogeneous broadened active media we do not consider spectral mode competition. However, we can consider the spontaneous emission as a mechanism of multimode lasing alternative to spatially inhomogeneous saturation of laser medium by separate modes [358, 361].

To analyze the case, we make use of the set of Eqs. (4.29) adding the term, which takes into account the average contribution of spontaneous emission to each mode, by modifying the first equation:

$$\frac{dm_k}{d\tau} - Gm_k\left(\tilde{L}_k \int n\psi_k^2 dv - 1\right) = G\varepsilon_{sp}\tilde{L}_k \int n\psi_k^2 dv, \tag{4.65a}$$

$$\frac{\partial n}{\partial \tau} = A - n\left(1 + \sum \tilde{L}_l \psi_l^2 m_l\right). \tag{4.65b}$$

The right-hand side of Eq. (4.65a) is written in a fashion like that used in Eqs. (3.51) and (4.45), and the parameter ε_{sp} is defined by means of Eq. (3.50). Assume a priori that the number of lasing modes is large and that the inversion is uniformly saturated throughout the laser medium (one-dimensional model in the plane wave approximation). Under these assumptions

$$\int n\psi_k^2 dv = n\int \psi_k^2 dv = n\langle\psi_k^2\rangle v_s = \langle n\rangle,$$

and

$$A\langle\psi_k^2\rangle v_s = \langle A\rangle,$$

where $\langle\psi_l^2\rangle$ is the mean over the active element volume. Transform Eq. (4.65b) by multiplying both parts by $\psi_k^2 v_s$ and integrating over the volume:

$$\frac{\partial\langle n\rangle}{\partial\tau} = \langle A\rangle - \langle n\rangle\left(1 + \sum \tilde{L}_l \psi_l^2 m_l\right).$$

If all $\langle\psi_l^2\rangle$ is equal within λ / L_s, then it should be reasonable to introduce, instead of m, a new variable $\langle m_l\rangle = m\langle\psi_l^2\rangle$, thus transforming Eqs. (4.65) to

$$\frac{\mathrm{d}\langle m_k\rangle}{\mathrm{d}\tau} - G\langle m_k\rangle\left(\tilde{L}_k\langle n\rangle - 1\right) = G\langle\varepsilon_{\mathrm{sp}}\rangle\tilde{L}_k\langle n\rangle, \tag{4.66a}$$

$$\frac{\mathrm{d}\langle n\rangle}{\mathrm{d}\tau} = \langle A\rangle - \langle n\rangle\left(1 + \sum \tilde{L}_l\langle m_l\rangle\right), \tag{4.66b}$$

where $\langle\varepsilon_{\mathrm{sp}}\rangle = \varepsilon_{\mathrm{sp}}\langle\psi_k^2\rangle$

It seems convenient to introduce a new small parameter $\eta = \pi\varepsilon_{\mathrm{sp}} / \Delta$. Then the time-independent version of the equation can be written (with angular brackets omitted)

$$\overline{m}_k(1 - \tilde{L}_k\overline{n}) = (\eta\Delta / \pi)\tilde{L}_k\overline{n}, \tag{4.67a}$$

$$\overline{n} = A\left(1 + \sum \tilde{L}_l\overline{m}_l\right)^{-1}, \tag{4.67b}$$

Since $\overline{n}$ differs little from unity, from (4.67) we find

$$\sum \tilde{L}_l\overline{m}_l = (A / \overline{n}) - 1 = A - 1 + \mathrm{O}(\eta)$$

($\mathrm{O}(\eta)$ is the term of the order of η) and, correspondingly,

$$\frac{1}{\overline{n}} = \frac{1 + \sum \tilde{L}_l\overline{m}_l}{A} = 1 + \frac{\mathrm{O}(\eta)}{A}. \tag{4.68}$$

Putting (4.68) into (4.67a) yields

$$\overline{m}_k = \eta\frac{\tilde{L}_k\overline{n}\Delta}{\pi(1 - \tilde{L}_k\overline{n})} = \eta\frac{\tilde{L}_k\Delta}{\pi(\overline{n}^{-1} - \tilde{L}_k)} \approx \eta\frac{\Delta}{\pi}\left[\Delta_k^2 + \frac{\mathrm{O}(\eta)}{A}\right]^{-1} \tag{4.69}$$

Replacing the summation (4.68) over modes by integration with respect to frequency we find

$$\sum \overline{m}_k = \eta\sqrt{A / \mathrm{O}(\eta)}.$$

Equating $\sum \overline{m}_l$ to $A-1$ we get an equality that implies

$$O(\eta) = \eta^2 A/(A-1)^2, \tag{4.70}$$

and, therefore,

$$\overline{m}_k \approx \frac{\eta}{\pi} \frac{\Delta}{\Delta_k^2 + \eta^2/(A-1)^2}. \tag{4.71}$$

Thus, we have obtained the Lorentzian form of the radiation spectrum with halfwidth at halfheight

$$\Delta_{\text{las}} = \frac{\pi\varepsilon_{\text{sp}}}{\Delta(A-1)}. \tag{4.72}$$

Estimation of the number of excited modes using Eq. (4.72) yields

$$N = \frac{2\Delta_{\text{las}}}{\Delta} = \frac{2\pi\varepsilon_{\text{sp}}}{\Delta^2(A-1)} \approx \frac{1}{100(A-1)},$$

if we assign $\varepsilon_{\text{sp}} = 10^{-13}$ and $\Delta = 10^{-5}$, which corresponds to Rhodamine dye laser. The number of modes will be much more than unity only provided $A-1 \ll 10^{-2}$. The required high-precision proximity to the laser threshold raises doubts that this multimode mechanism applies to such lasers. However, this mechanism is important for semiconductor lasers [362], since the parameter ε_{sp} is many orders higher than that for lasers of other types owing to the small volume of the optical cavity.

The question arises: How can we explain the fact that the spectrum of a CW jet-type laser contains hundreds of longitudinal modes? A possible answer is that in spite of the stability of the total power, its distribution over the spectrum does not reach a steady state. Simple estimation by formula (4.72), which can conveniently be written as

$$N = 2\frac{\Delta_k}{\Delta} \approx \frac{2}{\sqrt{\kappa t_k \Delta^2}} \tag{4.73}$$

shows that at $\Delta = 10^{-5}$, $\kappa = 10^8\ \text{s}^{-1}$, $t_k = 10^{-2}$ s the number of mode $N = 200$. As mentioned in Section 1.2.3, the lifetime of the individual mode of a CW jet-type laser ranges from $10^{-2} - 10^{-3}$ s and the number of modes is a few hundreds.

Thus, in the problem of multimode operation of a dye laser primary emphasis is placed on processes that make the laser spectrum time-dependent.

4.4. Combination Tone Mode-Mode Coupling and Its Influence on Laser Dynamics

When we proceeded to the rate equations in Section 4.1, we have to neglect some terms in the semiclassical description of the system and, therefore, disregard some factors. One of them is the combination tone mode-mode coupling (four-wave mixing). The effects due to the combination interaction are exhibited when the mode frequencies are nearly but not quite equally spaced. Strong interaction of the modes leads to what is termed combination mode locking in which the random spectrum becomes rigorously equidistant, and the influence of intermode beats on the stability of stationary lasing is negligible since the beat frequencies are large with relation to decay rate of inversion. A different situation arises when the dispersion exceeds the nonlinearity and the emitted spectrum remains unequally spaced. If the secondary beats frequencies approach the values of the relaxation frequencies, the influence of the nonlinear mode-mode coupling should be taken into account. The effect predicted in [24,289] was confirmed in subsequent papers [363–366].

In lasers with fast relaxing active media, which do not have relaxation oscillations, the role of combination tone mode-mode coupling can be noticeably. The combination tones do to the action of large number of modes with unequally spaced frequencies ensure, like a complicated external modulation, the complex fluctuations of individual modes [367, 368]. This is a dynamic mechanism causing spectral time dependence, which is commonly present in multimode lasers.

4.4.1. Competition between Dispersion and Nonlinearity of the Laser Medium

Limiting the consideration to only axial laser modes gives us the right to use equations (S.10).

Taking the normalized time $\tau = \gamma_{\|} t$ and introducing the abbreviated notation

$$\widetilde{G}_k^{\text{bal}} = -\frac{1}{2}\widetilde{F}_k(n_0 + n_k), \tag{4.74}$$

$$\widetilde{G}_k^{\text{comb}} = \frac{i}{4}\widetilde{F}_k \sum_{\mu,\nu} \frac{\widetilde{F}_\mu(n_0+n_\mu)+\widetilde{F}_\nu^*(n_0+n_\nu)}{1+i(\Delta_\nu-\Delta_\mu)/\widetilde{\gamma}} \frac{f_\mu f_\nu^* f_{k-\mu+\nu}}{f_k}. \tag{4.75}$$

we rewrite the set of equations (S.10) in a somewhat different form

$$\frac{df_k}{d\tau}+i\left(\frac{\Delta_k}{\tilde{\gamma}}-\frac{G\Delta_{ck}}{2}\right)f_k=\frac{1}{2}Gf_k\left[2i(\tilde{G}_k^{\mathrm{bal}}+\tilde{G}_k^{\mathrm{comb}})-1\right], \tag{4.76a}$$

$$\frac{dn_0}{d\tau}=A_0-n_0\left(1+\sum_{j=1}^{N}\tilde{L}_j m_j\right)-\sum_{j=1}^{N}\tilde{L}_j m_j n_j, \tag{4.76b}$$

$$\frac{dn_k}{d\tau}=-n_k\left(1+\sum_{j=1}^{N}\tilde{L}_j m_j\right)-\frac{1}{2}\tilde{L}_k m_k n_k. \tag{4.76c}$$

The frequency matching condition $\nu=\mu+l, k-\mu+\nu=k+l$ makes it possible to order the modes, and rewrite Eq. (4.75) in a more compact form

$$\tilde{G}_k^{\mathrm{comb}}=\frac{i}{4}\tilde{F}_k\sum_{\mu,\nu}\frac{\tilde{F}_\mu(n_0+n_\mu)+\tilde{F}_{\mu+l}^*(n_0+n_{\mu+l})}{1+i(\Delta_\nu-\Delta_\mu)/\tilde{\gamma}}\frac{f_\mu f_{\mu+l}^* f_{k+l}}{f_k}. \tag{4.77}$$

Substituting $f_k=F_k\exp(i\varphi_k)$ into (4.76a) we isolate the phase equation

$$\frac{d\varphi_k}{d\tau}=\frac{1}{2}G\Delta_{ck}-\frac{1}{\tilde{\gamma}}\Delta_k+G\,\mathrm{Re}(\tilde{G}_k^{\mathrm{bal}}+G_k^{\mathrm{comb}}). \tag{4.78}$$

Since four modes are combined and the expression (4.77) depends on the mode combinations $\Phi_{kl\mu}=\varphi_{k+l}-\varphi_k-\varphi_{\mu+l}+\varphi_\mu$ rather than individual modes it should be reasonable to proceed from (4.78) to equation

$$\frac{d\Phi_{kl\mu}}{d\tau}=G\left(F_{kl\mu}^{\mathrm{nl}}+F_{kl\mu}^{\mathrm{disp}}\right)-\frac{1}{\tilde{\gamma}}\left(\Delta_{k+l}-\Delta_k-\Delta_{\mu+l}+\Delta_\mu\right). \tag{4.79}$$

The function

$$F_{kl\mu}^{\mathrm{nl}}=\mathrm{Re}\left(\tilde{G}_{k+l}^{\mathrm{comb}}-\tilde{G}_k^{\mathrm{comb}}-\tilde{G}_{\mu+l}^{\mathrm{comb}}+\tilde{G}_\mu^{\mathrm{comb}}\right) \tag{4.80}$$

characterizes nonlinearity, and

$$F_{kl\mu}^{\mathrm{disp}}=\frac{1}{2\kappa}(\omega_{ck+l}-\omega_{ck}-\omega_{c\mu+l}+\omega_{c\mu})+\mathrm{Re}\left(\tilde{G}_{k+l}^{\mathrm{bal}}-\tilde{G}_k^{\mathrm{bal}}-\tilde{G}_{\mu+l}^{\mathrm{bal}}+\tilde{G}_\mu^{\mathrm{bal}}\right) \tag{4.81}$$

characterizes the dispersion of the system. Since $\gamma_\perp \gg \kappa$, the second term on the right hand side of (4.81) is rather unimportant, so that the dispersion is conditioned mainly by the cavity spectral features

$$F_{kl\mu}^{\mathrm{disp}}=\frac{1}{2\kappa}(\omega_{ck+l}-\omega_{ck}-\omega_{c\mu+l}+\omega_{c\mu}).$$

The relationship between dispersion and nonlinearity can be different. With strong dispersion one can ignore the nonlinear term; in this case the unequal spacing of the cavity eigenfrequencies (nonequidistancy) leads to about the same nonequidistancy of the output spectrum.

If nonlinearity dominates, a variety of dynamic regimes are possible. In particular, the effect of dispersion can be completely overcome making the radiation spectrum equidistant. This regime is often called mode locking, but it should be born in mind that equal mode spacing itself does not mean the simple phase distribution law inherent in what is usually meant by mode locking.

In order to find the mode-locking criterion in a broad sense we should estimate F^{nl} and compare it to F^{disp}. This estimation is difficult because the modal phases are undefined. Assuming that the phase distribution is uniform, the problem is similar to the summation of N randomly phased oscillations, and for a laser medium with $\Delta/\tilde{\gamma} = \Delta\omega/\gamma_{\|} >> 1, l\Delta\omega/\gamma_{\perp} << 1$

$$F^{nl} \approx \frac{\tilde{\gamma} n F^2 \sqrt{N}}{l\Delta} \approx \frac{\gamma_{\|}}{l\Delta\omega} \frac{A-1}{\sqrt{N}}.$$

The estimates can be simplified assuming that the mode intensities are the same: $F_k^2 = F^2 = (A-1)/N$, and $n_0 + n_k \approx 1$. Supposing that the main contribution to the combination sum is given by terms with $l = 1$ we obtain

$$\frac{F^{nl}}{F^{disp}} \approx \frac{\gamma_{\|}\kappa}{\omega_{kl\mu}\Delta\omega} \frac{A-1}{\sqrt{N}}. \tag{4.82}$$

Going over to normalized nonequidistancy $\Omega'_{kl\mu} = \omega_{kl\mu}/\Delta\omega$ and to repeatedly met value $\Omega_1 = \sqrt{G(A-1)}$ (the latter is relaxation frequency when $G >> 1$) we can formulate (4.82) as

$$\frac{F^{nl}}{F^{disp}} \approx \left(\frac{\Omega_1}{\hat{\Delta}}\right)^2 \frac{1}{\Omega'_{kl\mu}\sqrt{N}}. \tag{4.83}$$

The control parameter

$$\hat{\Delta} = \Delta/\tilde{\gamma} \tag{4.84}$$

defines the inertial properties of the laser medium on the beat period time. According to (4.83) the nonlinearity dominates in the case of small nonequidistancy, which satisfies the inequality

$$\Omega'_{kl\mu} < \Omega_1^2/(\hat{\Delta}^2\sqrt{N}). \tag{4.85}$$

Example 4.3

Nd:YAG laser

$$\gamma_{\parallel} = 5\cdot 10^3\ \mathrm{s}^{-1};\ G = 10^4;$$
$$\kappa = 5\cdot 10^7\ \mathrm{s}^{-1};\Omega_1 = 50;\quad \Omega'_{kl\mu} < 2\cdot 10^{-8};$$
$$\Delta\omega = 10^9\ \mathrm{s}^{-1};\ \widehat{\Delta} = 2\cdot 10^5;\quad \omega_{kl\mu}/2\pi < 3\ \mathrm{Hz}$$
$$N = 9;\ A = 1.5;$$

Example 4.4

Dye laser

$$\gamma_{\parallel} = 2\cdot 10^8\ \mathrm{s}^{-1};\ G = 0.5;$$
$$\kappa = 5\cdot 10^7\ \mathrm{s}^{-1};\Omega_1 = 0.35;\quad \Omega'_{kl\mu} < 5\cdot 10^{-5};$$
$$\Delta\omega = 10^9\ \mathrm{s}^{-1};\ \widehat{\Delta} = 5;\quad \omega_{kl\mu}/2\pi < 8\ \mathrm{kHz}$$
$$N = 10^4;\ A = 1.5;$$

4.4.2. Time-Dependent Regimes Due to Nonlinear Mode Interaction

Let us specify the conditions under which there is noticeable influence of the combination tone mode-mode coupling on the time-dependent processes in lasers. Beginning from the real form of equations (4.76a) we consider the amplitude part of these equations:

$$\frac{\mathrm{d}F_k}{\mathrm{d}\tau} = -\frac{1}{2}GF_k\left[2\,\mathrm{Im}(\widetilde{G}_k^{\mathrm{bal}} + \widetilde{G}_k^{\mathrm{comb}}) + 1\right]. \tag{4.86}$$

The term $\widetilde{G}_k^{\mathrm{comb}}$ plays the role of a perturbation in the rate equation, and we should first estimate its magnitude. To do this we rewrite Eq. (4.77) assuming all $\widetilde{L}_k = 1$ and that the inequalities

$$\widetilde{\gamma} << |\Delta_\nu - \Delta_\mu| << 1 \quad (\gamma_{\parallel} << |\omega_\nu - \omega_\mu| << \gamma_\perp) \tag{4.87}$$

are satisfied. Since we are speaking about condensed media, $\gamma_\perp \geq 10^{12}\ \mathrm{s}^{-1}$, and $\gamma_{\parallel}$ ranges from $5\cdot 10^3\ \mathrm{s}^{-1}$ for Nd:YAG laser to $10^8\ \mathrm{s}^{-1}$ for dye lasers. Frequency spacing between neighbouring longitudinal modes of about $10^9\ \mathrm{s}^{-1}$ corresponds to a cavity length of 100 cm. For solid-state lasers the inequalities (4.87) are met for the transverse mode spectrum as well. Bearing these facts in mind we can pass over (4.77) to the simpler expression

$$\mathrm{Im}\widetilde{G}_k^{\mathrm{comb}} \approx -\frac{\widetilde{\gamma}}{4}\sum_{\mu,l}(2n_0 + n_\mu + n_{\mu+l})\frac{F_\mu F_{\mu+l}F_{k+l}}{F_k}\frac{\sin(\Omega_{kl\mu}\tau + \Phi_{kl\mu})}{\Delta_{\mu+l} - \Delta_\mu}. \tag{4.88}$$

Assuming that $2n_0 + n_\mu + n_{\mu+l} \approx 2,\ F^2 \approx (A-1)/N$, the amplitude of a single term of the sum (4.88) is estimated as

$$\beta_1 = \tilde{\gamma}(A-1)/(N\Delta)\,. \tag{4.89}$$

A noticeable effect on the laser dynamics should be expected, first of all, when the laser modes are not locked, i.e., $\Omega_{kl\mu} \neq 0$. However, this means that time enters Eq. (4.86) explicitly and the problem is beyond the scope of autonomous theory. Therefore, running ahead, we will make use of the results given in Chapter 6. Comparison of Eq. (4.86) with Eq. (6.18a) indicates that the term $2\,\mathrm{Im}\,\tilde{G}_k^{\mathrm{comb}}$ holds the same place in the laser equations as the term responsible for the loss modulation. Thus we can use relations (6.26) and (6.29) to estimate the efficiency of the combination tone mode-mode coupling. There is only one question: What should be put in correspondence with modulation frequency Ω and modulation depth β. The answer to this question is ambiguous. Consider two particular cases.

1. The laser frequency deviations from equidistant spacing are multiples of a quantity Ω_{sb} and the spectrum of secondary beats is composed of its harmonics. Making estimates, we can though cautiously, adopt (4.89) for the lower harmonic amplitude.

2. Deviations from equidistant spacing are random, and the secondary beat spectrum represents a white noise band with its centre at the frequency Ω^*. The combination tone amplitude is estimated following the rules of summation of oscillations with random (drifting in this particular case) phases; thus we have

$$\beta = \tilde{\gamma}\frac{A-1}{\Delta\sqrt{N}}\,. \tag{4.90}$$

The response of a solid-state laser to the presence of oscillating combination terms can be nonlinear if, according to Eqs. (6.26) and (4.90),

$$\frac{G\beta}{A} = \frac{\kappa}{\Delta\omega}\frac{A-1}{A\sqrt{N}} > 1\,. \tag{4.91}$$

From this relation it is seen that the role of the secondajry beats mechanism in changing the mode intensity oscillations is more pronounced when fewer modes participate in the laser action and when the modes are closer to each other in frequency.

Having shown that irregular dynamic behaviour can appear in relatively simple systems with a small (N>3) number of degrees of freedom, many studies of the nonlinear dynamics focused on such systems. These include some of the laser models considered above. However, complex nonperiodic processes are also observed in multimode lasers, as mentioned in Section 1.2.3. The fluctuating intensities often resemble the kind of quantities investigated by the methods of statistical physics, so that one is tempted to attribute their irregular behaviour to random factors. For example, the deep undamped oscillations of intensity of some laser modes in a multimode laser have been ascribed to the presence of fluctuating noise sources resulting from the spontaneous emission from the laser medium [369-371]. Such an explanation seems quite natural if rate equation models are used and if nonlinear effects other than the unavoidable saturation of the laser medium are disregarded. However, keeping in mind the mode coupling due to the nonlinear scattering (including four-wave mixing) dynamical interpretation of the random seeming time dependent processes in a multimode laser could be sought. In contrast to random processes caused by noise, we might attribute the irregularity to deterministic chaos. The true nature of the observed time dependent behaviour of a multimode laser is thus an important matter to investigate, in principle. Theoretical efforts to solve this problem were undertaken in [363-368, 370, 372, 626], and the experimental investigations of the spectral dynamics of dye lasers were performed in [65, 372–374].

Numerical integrations of Eqs. (4.76) have found solutions that feature chaotic behaviour. This is indicated not only by the nonperiodicity of the calculated results in the absence of random forces but also by the inverse dependence of the mode correlation time on the pumping rate [373]. This dependence is expected from the dynamics alone by the fact that the influence of the nonlinear terms, and, therefore, the degree of irregularity they produced, would increase with increasing strength of the average intensity of radiation field.

In what follows we give the results of the numerical investigation of a five-mode model of a travelling-wave dye laser. Although the choice of the model is mainly due to its extreme simplicity, these results are of broader interest than might be supposed.

First, the simplifications of Eqs. (4.76) are due to the fact that the laser modes are co-propagating waves. Therefore, the longitudinal inhomogeneity of inversion is absent. Second, the high rate of relaxation of the inversion allows n to be adiabatically eliminated putting $\mathrm{d}n_0/\mathrm{d}\tau = 0$ and $\mathrm{d}n_k/\mathrm{d}\tau = 0$.

Figure 4.6 is a phase diagram that summarizes results of numerical solution of Eqs. (4.76). The diagram is plotted in the control parameter plane $(\delta\Omega, \Delta/\sqrt{\tilde{\gamma}})$. The mode frequencies deviations from the equal spacing were assumed to be quasiregular, so that $\Omega'_{kl\mu}$ is a multiple of $\delta\Omega = \delta(\Delta\omega)/\Delta\omega$, which was taken as the measure of nonequidistancy.

The solid lines in the diagram confine the control parameter values, which lead to unstable laser action. Upon entering the unstable domain by increasing $\delta\Omega$, the laser dynamics shows a sudden transition to the regime of chaotic compensated oscillations of the amplitudes of individual modes. Close to the inside of the right hand boundary of the unstable region is a band of regular oscillations. With decreasing $\delta\Omega$ the transition from peri-

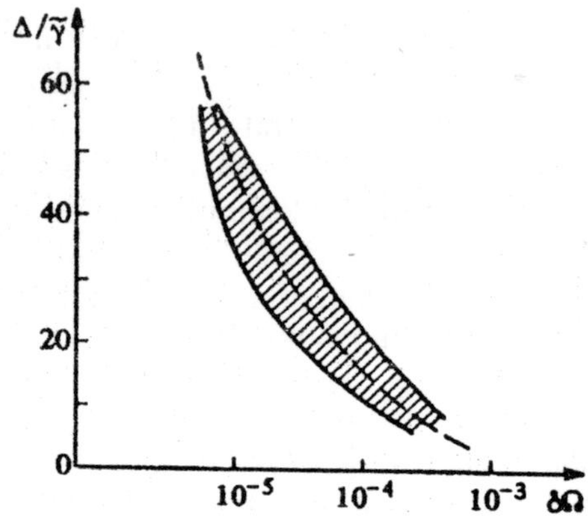

Fig. 4.6. Phase diagram of a five-mode model of travelling wave dye laser in the control parameter plane $(\delta\Omega, \Delta/\tilde{\gamma})$. $A=1.5; \tilde{\gamma}=2\cdot10^{-4}; \kappa/\gamma_{\parallel}=0.1$; The hatching denotes the instability domain [626].

odic oscillations to chaos shows elements both period doubling and quasiperiodicity. Examples of regular and chaotic solutions are given in Fig. 4.7. Outside the instability zone, CW single-mode solutions are established after some transients. The result is quite natural since neither spectral, nor spatial inhomogeinity of the laser medium is assumed in the model. It should be noted, however, that for small deviations of mode positions from the equidistant frequency spacing, when the nonlinearity dominates the dispersion, the transient state is of the form of mode locking, which is unable to sustain itself for a long time. These results support the idea that nonstationary pulsations can be an effective mechanism of multimode lasing.

We should also mention that there are optical bistability and hysteresis in the dynamics observed near the right hand boundary of the instability zone. Depending on the initial conditions, the solution can be either CW single-mode solution or undamped multimode pulsations.

Our attention is drawn to the very steep dependence on the instability band in Fig. 4.6 on the relative intermode beat frequency. The instability zone cannot remain this narrow for a large number of modes, since then a wider range of primary intermode beats between the increasingly spaced modes enter the dynamics. Obviously, this may be the reason for the instability of CW dye laser with hundreds of modes.

The efficiency of combination tone mode-mode coupling as the self-modulation mechanism of the emitted spectrum of a dye laser can be estimated in a fashion similar to that used above. The question arises: Which quantity plays the role of the loss modulation depth? Combining the Eqs. (4.89) and (4.90), we write

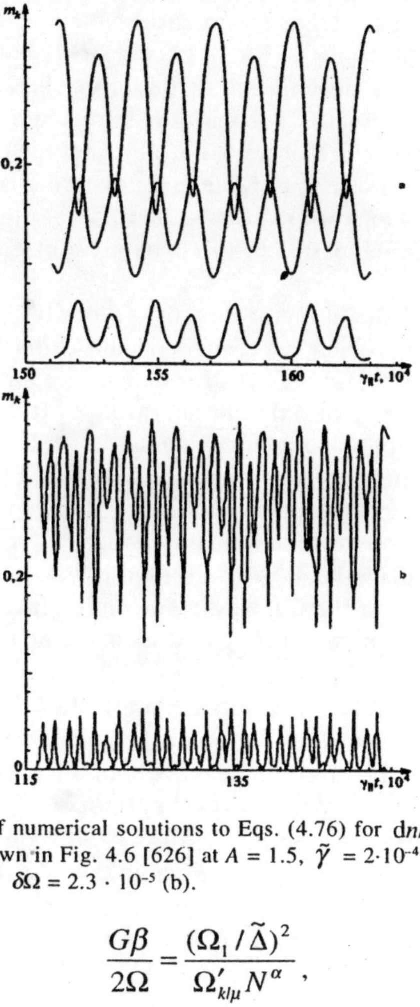

Fig. 4.7. Examples of numerical solutions to Eqs. (4.76) for $dn/d\tau = 0$ and N=5 in the instability region shown in Fig. 4.6 [626] at A = 1.5, $\tilde{\gamma}$ = $2\cdot10^{-4}$; $\kappa/\gamma_\parallel$ = 0.1; $/\tilde{\gamma}$ = 40; $\delta\Omega = 2.5\cdot10^{-5}$ (a) and $\delta\Omega = 2.3\cdot10^{-5}$ (b).

$$\frac{G\beta}{2\Omega} = \frac{(\Omega_1/\tilde{\Delta})^2}{\Omega'_{kl\mu}N^\alpha}\,, \tag{4.92}$$

where $0.5 \le \alpha \le 1$. The modulation depth of the mode will be extremely large if $G\beta/2\Omega > 1$, or, according to (4.92), if

$$\delta\Omega < \frac{\Omega_1^2}{\tilde{\Delta}^2 N^\alpha}\,. \tag{4.93}$$

For $\alpha = 0.5$ this coincides with the condition that the nonlinearity prevails over the dispersion (4.85). The dash line in Fig. 4.6 reproduces the dependence $\delta\Omega = \Omega_1^2/(\tilde{\Delta}^2 N^\alpha)$. The fact that it passes through the narrow instability band indicates that the analogy to the model of laser with selective loss modulation is quite reasonable.

Such modulation behaviour was studied experimentally in [65, 372-

374] for a jet-type dye laser. One fact suggesting the dynamic origin of the time dependence of the lasing (Figs. 1.10 and 1.11) is the existence of bifurcation points for the dependence of the laser characteristics on the spectral power density. From Fig. 4.8 it is seen that sudden variations of the mode correlation time are observed at definite values of the control parameter. The fractal dimension of the attractor also changes abruptly at the same points. A low and not integer-valued fractal dimension between two and six is also the indication of the dynamical origin of the spectral chaos.

In Refs. [373, 374] the attractor dimension as well as Kolmogorov's entropy was determined from experimental data using Grassberger and Procachia's method [375–377]. To realize their procedure, it is enough to know the time dependence of only one of variables (the intensity of one mode). Such an approach is quite understandable while the set of N first-order differential equations can be in principle reduced to one equivalent equation of the -th order. The coordinates in the phase space of such equation are the values of sought function and its derivatives . The results of processing the experimental data are presented in Fig. 4.9. Two jumps are seen in the dependence of the fractal dimension on laser output together with tendency towards increased fractal dimension as the laser power is increased.

These experimental facts merit comments. In Ref. [373], from which Fig. 4.9 is taken, the spectral resolution of the apparatus is insufficient to single out one mode. Therefore, the data, which are processed, represent the time evolution of the total intensify of a group of a few tenth of modes (35 modes in that particular case). Nevertheless, this yields the result predicted by the theory for the intensity of a single mode. Together with the

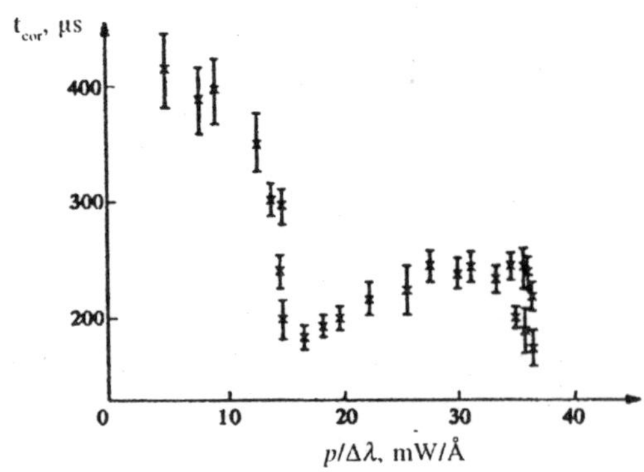

Fig. 4.8. Measured mode correlation time of a jet dye laser as a function of the spectral density of laser output. Discontinued jumps are seen at 14.5 mW/A and 36 mW/A [373].

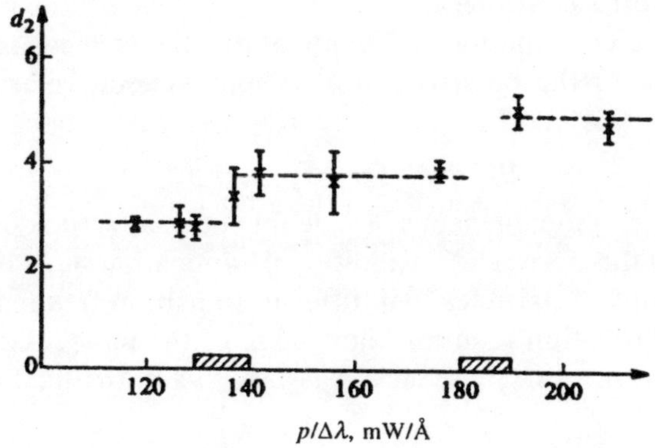

Fig. 4.9. Measured second-order attractor dimension of a multimode jet dye laser as a function of the spectral density. The areas, in which the mode correlation time changed abruptly, are marked on the horizontal axis [373]. Dash lines show the mean values of d_2.

low dimension of the attractor, this fact can be explained by dividing the whole ensemble of the generated modes into a small number of groups, within which the mode correlation is stronger than it is between the modes belonging to different groups. Such a spectral packet, rather than a separate mode, should be evidently identified as a degree of freedom in this case. Selection of a single mode would not change the experimental results as a more recent experiment confirms [65].

The second comment concerns the role of quantum fluctuations in these phenomena. Figures 4.8 and 4.9 do not show the pumping domain near the laser threshold where it has been found that the attractor dimension rises and the mode correlation time falls as pumping decrease (i.e., the tendency opposite to that observed at high pumping level). This contradiction is easily removed if we assume that the time dependence of the lasing near threshold is dominated by quantum fluctuations where the importance of spontaneous emission increases.

4.5. Inhomogeneously Broadened Solid-State Lasers

Although many dynamical properties of a laser are insensitive to the type of the line broadening of the active medium, the spectral characteristics of the laser output are affected by the inhomogeneous broadening. The suggestion that all transition frequencies of all active molecules are exactly the same, i.e., the spectral lines are homogeneously broadened for condensed matter, is not correct in all cases. For example, atoms may occupy different positions in the crystal lattice or may be influenced by nonuniform electric field inside the amorphous hosts.

4.5.1. Mathematical Model

Summing of the contributions of identical molecules leads to the following expression for the polarization of a homogeneously broadened medium:

$$\mathbf{P}_{nm} = N_S(\rho_{mn}\mathbf{d}_{nm} + \rho_{nm}\mathbf{d}_{mn}) . \tag{4.94}$$

In the case of inhomogeneous broadening one can write only part of the polarization in this form and an additional summation must be taken over the whole groups of particles that differ in transition frequency. Calculation of the polarization requires knowledge of the molecular distribution function with respect to frequency, $h(\omega_0)$, with a normalization condition such as

$$\int_0^\infty h(\omega_0)\mathrm{d}\omega_0 = 1 .$$

The form of the expression for polarization depends on the form of density matrix normalization. Assuming

$$\rho_{11}(\omega_0) + \rho_{22}(\omega_0) = h(\omega_0) , \tag{4.95}$$

we have

$$\mathbf{P}_{nm} = N_S\int_0^\infty(\rho_{mn}\mathbf{d}_{nm} + \rho_{nm}\mathbf{d}_{mn})\mathrm{d}\omega_0 , \tag{4.96}$$

while if

$$\rho_{11}(\omega_0) + \rho_{22}(\omega_0) = 1 , \tag{4.97}$$

we get

$$\mathbf{P}_{nm} = N_S\int_0^\infty(\rho_{mn}\mathbf{d}_{nm} + \rho_{nm}\mathbf{d}_{mn})h(\omega_0)\mathrm{d}\omega_0 . \tag{4.98}$$

The inhomogeneous broadening amends twofold corrections in the laser equations.

First, the medium inhomogeneity suggests summation of contributions from all groups of atoms with different transition frequencies.

Second, it is important to take into account the process of spectral cross-relaxation. The form of equations depends on what type of normalization condition is taken: (4.95) or (4.97). Using Eq. (4.95) we get the following set of equations for multimode laser that is generalization of Eq. (2.76):

$$\frac{\mathrm{d}F_\lambda}{\mathrm{d}t} + [\kappa - i(\omega - \omega_{c\lambda})]F_\lambda = 4\pi i\omega d N_S\iint\psi_\lambda\sigma\mathrm{d}\omega_0\mathrm{d}V , \tag{4.99a}$$

$$\frac{\partial\sigma}{\partial t} + [\gamma_\perp - i(\omega - \omega_0)]\sigma = -\frac{id}{2\hbar}D\sum\psi_\nu\tilde{F}_\nu , \tag{4.99b}$$

$$\frac{\partial D}{\partial t}+\gamma_{\|}(D-D^{(0)})+\Gamma\left[D-h(\omega_0)\int_0^{\infty}D\mathrm{d}\omega_0\right]=-\frac{i\beta_a d}{2\hbar}\sum\psi_\nu(F_\nu^*\sigma-F_\nu\sigma^*)$$

(4.99c)

Using the normalization condition (4.97) we replace the integral in right-hand side of (4.99a) by $\iint\psi_\lambda\sigma h(\omega_0)\mathrm{d}\omega_0\mathrm{d}V$ and replace the cross-relaxation term in (4.99b) by $\Gamma[D-\int Dh(\omega_0)\mathrm{d}\omega_0]$.

The formal way to the rate equations is the same as was described in Section 4.1. Using Eq. (4.3) we pass over new variables, then adiabatically eliminate the Fourier components of atomic polarization and neglect of all cross terms in obtained equations. In dimensionless form

$$\frac{\mathrm{d}m_\lambda}{\mathrm{d}\tau}=Gm_\lambda\left[\int_v\int_0^{\infty}\frac{\psi_\lambda^2 n\mathrm{d}\tilde{\omega}_0\mathrm{d}v}{1+(\tilde{\omega}_\lambda-\tilde{\omega}_0)^2}-1\right], \tag{4.100a}$$

$$\frac{\partial n}{\partial\tau}=Ah(\tilde{\omega}_0)-n\left[1+\sum\frac{m_\lambda\psi_\lambda^2}{1+(\tilde{\omega}_\lambda-\tilde{\omega}_0)^2}\right]-\tilde{\Gamma}\left[n-h(\tilde{\omega}_0)\int_0^{\infty}n\mathrm{d}\tilde{\omega}_0\right] \tag{4.100b}$$

the rate equations for class *B* lasers are written when we use the dimensionless variables similar to given by Eqs. (3.1):

$$m_\lambda=\frac{2\pi d^2\beta_a}{\hbar\gamma_{\|}\gamma_{\perp}V_c}|F_\lambda|^2,\quad n=\frac{2\pi\omega d^2N_S}{\hbar\gamma_\perp\kappa}D,\quad A=\frac{2\pi\omega d^2N_S}{\hbar\gamma_\perp\kappa}\int D^{(0)}\mathrm{d}\omega_0, \tag{4.101}$$

$$\tau=\gamma_{\|}t,\quad v=V/V_c,\quad \tilde{\omega}=\omega/\gamma_\perp,\quad G=2\kappa/\gamma_{\|},\quad \tilde{\Gamma}=\Gamma/\gamma_{\|}.$$

4.5.2. Steady-State Solution in the Spatially Uniform Field Approximation: The Threshold for Laser Spectrum Splitting

When the inversion is spatially uniform over the entire volume of the active medium, Eqs. (4.100) can be written in a simpler form:

$$\frac{\mathrm{d}m_\lambda}{\mathrm{d}\tau}=Gm_\lambda\left[\int_0^{\infty}\frac{n\,\mathrm{d}\tilde{\omega}}{1+(\tilde{\omega}_\lambda-\tilde{\omega}_0)^2}-1\right], \tag{4.102a}$$

$$\frac{\partial n}{\partial\tau}=Ah(\tilde{\omega}_0)-n\left[1+\sum\frac{m_\lambda}{1+(\tilde{\omega}_\lambda-\tilde{\omega}_0)^2}\right]-\tilde{\Gamma}\left[n-h(\tilde{\omega}_0)\int_0^{\infty}n\mathrm{d}\tilde{\omega}_0\right] \tag{4.102b}$$

In the absence of lasing, the spectral density of the inversion follows the distribution function

$$n = Ah(\tilde{\omega}_0)\,. \tag{4.103}$$

Substituting Eq. (4.103) into Eq. (4.102a) and demanding $\mathrm{d}m(\tilde{\omega}_{00})/\mathrm{d}\tau > 0$, we find the self-excitation condition

$$A\int_0^{\infty} \frac{h(\tilde{\omega}_0)\mathrm{d}\tilde{\omega}_0}{1+(\tilde{\omega}_0 - \tilde{\omega}_{00})^2} > 1\,, \tag{4.104}$$

where $\tilde{\omega}_{00}$ denotes the frequency of the gain line centre.

If the inhomogeneous broadening is much larger than the homogeneous broadening, then the value $h(\tilde{\omega}_{00})$ can be removed from the integral sign, and the approximate self-excitation condition looks like

$$\pi A h(\tilde{\omega}_{00}) > 1\,. \tag{4.105}$$

For an exact calculation of the integral we need to know the shape of the distribution function. Here and elsewhere below we will use one of the following two functions:

The Lorentzian function

$$h_L(\tilde{\omega}_0) = \frac{q/\pi}{q^2 + (\tilde{\omega}_0 - \tilde{\omega}_{00})^2}\,, \tag{4.106}$$

and the Gaussian function

$$h_G(\tilde{\omega}_0) = \frac{1}{q}\left(\frac{\ln 2}{\pi}\right)^{1/2} \exp\left[-\frac{(\tilde{\omega}_0 - \tilde{\omega}_{00})^2 \ln 2}{q^2}\right]. \tag{4.107}$$

By adequately approximating the real shape of the inhomogeneous spectral lines these functions can yield the results in a rather simple form.

The quantity q is the halfwidth of the distribution function normalized to $\gamma_\perp$, and it is called the inhomogeneous broadening parameter. The total width of an inhomogeneous broadened line is related to the elementary homogeneous linewidth of a separate atom by $q+1$.

For the Lorentzian line (4.106) the self-excitation condition (4.105) reduces to

$$A(q+1)^{-1} > 1\,, \tag{4.108}$$

and for the Gaussian line (4.107) it reduces to

$$Aq^{-1}(\pi \ln 2)^{1/2} \exp(\ln 2/q^2)[1 - \Phi(\sqrt{\ln 2}/q)] > 1\,, \tag{4.109}$$

where $\Phi(x)$ is the probability integral. If the inhomogeneous broadening is large, then, instead of the last two inequalities, we have

$$\frac{A}{q} > \begin{cases} 1 & \text{for the Lorentzian line,} \\ (\pi \ln 2)^{-1/2} & \text{for the Gaussian line.} \end{cases} \qquad (4.143)$$

The equations describing the time-independent lasing can be obtained assuming $d/d\tau = 0$ in Eq. (4.102):

$$\int_0^\infty \frac{\bar{n}\, d\tilde{\omega}}{1+(\tilde{\omega}_\lambda - \tilde{\omega}_0)^2} - 1 = 0, \qquad (4.111a)$$

$$Ah(\tilde{\omega}_0) - \bar{n}\left[1 + \sum \frac{\overline{m}_k}{1+(\tilde{\omega}_k - \tilde{\omega}_0)^2}\right] - \tilde{\Gamma}\left[h\int_0^\infty \bar{n}\, d\tilde{\omega}_0 - \bar{n}\right] \qquad (4.111b)$$

The unknown $\int n\, d\omega_0$ entering Eq. (4.111b) is found by integration of both parts of this equation with respect to frequency:

$$\int_0^\infty \bar{n}\, d\tilde{\omega}_0 = A - \sum \overline{m}_k\,. \qquad (4.112)$$

The case of single-mode lasing is simplest for analysis. Making use of relation (4.112) we find from Eq. (4.111b) the frequency dependence of the inversion

$$\bar{n}(\tilde{\omega}_0) = \frac{[(\tilde{\Gamma}+1)A - \tilde{\Gamma}\overline{m}]}{\tilde{\Gamma}+1+\overline{m}[1+(\tilde{\omega}_0 - \tilde{\omega}_{00})^2]^{-1}} h(\tilde{\omega}_0)\,. \qquad (4.113)$$

Substituting the result into Eq. (4.111a) yields an equation defining the radiation intensity

$$[A(\tilde{\Gamma}+1) - \tilde{\Gamma}\overline{m}]\int_0^\infty \frac{h\, d\omega_0}{(\tilde{\Gamma}+1)[1+(\tilde{\omega}_0 - \tilde{\omega}_{00})^2] + \overline{m}} - 1 = 0\,. \qquad (4.114)$$

When the inhomogeneous broadening is large, so that we can assume $h = h(\tilde{\omega}_{00})$, Eq. (4.114) reduces to

$$s^2 + \frac{1}{\pi\tilde{\Gamma}h(\tilde{\omega}_{00})} s - \frac{\tilde{\Gamma}+A}{\tilde{\Gamma}} = 0\,. \qquad (4.115)$$

For brevity we introduce, in place of $\overline{m}$, a quantity

$$s = \sqrt{1+\overline{m}/(\tilde{\Gamma}+1)}\,. \qquad (4.116)$$

Constraints on the inhomogeneous broadening need not be used in case of the Lorentzian distribution function: Eq. (4.114) is transformed, irrespective of q, into

$$s^2+\frac{q}{\tilde{\Gamma}+1}s-\frac{\tilde{\Gamma}+A}{\tilde{\Gamma}+1}=0. \tag{4.117}$$

Only one of two solutions is physically relevant, and it yields

$$\overline{m}=A-1+\frac{q^2}{2(\tilde{\Gamma}+1)}-q\left[\frac{q^2}{4(\tilde{\Gamma}+1)^2}+\frac{\tilde{\Gamma}+A}{\tilde{\Gamma}+1}\right]^{1/2} \tag{4.118}$$

In the limiting cases of strong ($\tilde{\Gamma}>>A$) and weak ($\tilde{\Gamma}<<1$) cross-relaxation the expression for the steady-state intensity is simplified:

$$\overline{m}=A-q-1 \quad \text{for } \tilde{\Gamma}>>A, \tag{4.119a}$$

$$\overline{m}=(A/q)^2-1 \quad \text{for } \tilde{\Gamma}<<1. \tag{4.119b}$$

The Eq. (4.119b) can be obtained directly from Eq. (4.114) if the cross-relaxation is neglected from the very beginning. The relation (4.119a) means that in the presence of strong cross-relaxation the laser medium behaves as if it is were homogeneous broadened with the larger linewidth.

The single-frequency regime is stable when the saturated gain does not reach the loss level at any of modal frequencies except for the line centre frequency. Provided that

$$\tilde{G}(\tilde{\omega}_k)=\int_0^\infty\frac{\overline{n}\,d\tilde{\omega}_0}{1+(\tilde{\omega}_k-\tilde{\omega}_0)^2}-1\geq 0$$

multifrequency operation of the laser will be observed. Substituting for $\overline{n}$ from Eq. (4.113) and assuming that the distribution function is Lorentzian, we integrate and obtain an inequality

$$q\frac{A-\tilde{\Gamma}(s^2-1)}{q^2-s^2}\left\{\frac{(q+1)^2(q-1)}{q[\Delta_k^2+(q+1)^2]}-\frac{(s+1)^2(s-1)}{s[\Delta_k^2+(s+1)^2]}\right\}\geq 1. \tag{4.120}$$

Using Eq. (4.117), we can transform Eq. (4.120) to a simpler form

$$\Delta_k^2\left\{\Delta_k^2-(q-1)\left[s^2+(q+2)s-\frac{(q+1)^2}{q-1}\right]\right\}\leq 0. \tag{4.121}$$

The value $\Delta_k=\tilde{\omega}_k-\tilde{\omega}_{00}$ can satisfy this inequality only if

$$(q-1)\left[s^2+(q+2)s-\frac{(q+1)^2}{q-1}\right]>0.$$

The binomial in the square brackets has a positive root

$$s_1 = -\frac{q+2}{2} + \sqrt{\frac{(q+2)^2}{4} + \frac{(q+1)^2}{q-1}} \tag{4.122}$$

provided

$$q > 1. \tag{4.123}$$

By definition (4.116), the quantity s cannot be less than unity. If the distribution function is the Gaussian, then in place of Eq. (4.123) we have the inequality $q > \sqrt{\ln 2}$ [378].

The first condition for multifrequency lasing (4.123) requires a definite inhomogeinity of the spectral line, while the second,

$$s > s_1, \tag{4.124}$$

means that a finite excess of pump level above the laser threshold is necessary. The parameter s_1 that characterizes the critical pumping is large only for $q \approx 1$ and falls rapidly as q grows. This is seen from Eq. (4.122) and illustrated by the table of values that satisfy this equation:

q	1.1	2.0	10
s_1	6.5	1.6	1.04

Equating the s values, defined by Eqs. (4.122) and (4.117), we can find the pumping parameter A at the multimode threshold. For $q \gg 1$ [98]

$$A = q + 1 + \frac{4}{q} + \tilde{\Gamma}\frac{8}{q^2}. \tag{4.125}$$

Returning to inequality (4.121), we see that it indicates that if the conditions (4.123) and (4.124) are satisfied then the single-frequency solution is unstable with respect to mode excitation in a finite band near the line centre. The function $\tilde{G}$ has the form shown in Fig. 4.10*b*. Figure 4.10*b* differs from Fig. 4.10*a*, which corresponds to the case of stable single-mode operation, in the character of the central extremum. Thus, at the boundary of instability domain, not only $\partial\tilde{G}(\tilde{\omega} = \tilde{\omega}_{00})/\partial\tilde{\omega} = 0$ but also $\partial^2\tilde{G}(\tilde{\omega} = \tilde{\omega}_{00})/\partial\tilde{\omega}^2 = 0$. General information on the steady-state spectrum is drawn from the properties of the function $\tilde{G}(\tilde{\omega})$ [379]. This function is analytical and it can turn to zero in a finite interval at a finite number of points. Hence, it is inferred that the spectrum is discrete. The laser spectrum consists of only one central frequency while the pumping less than the critical value. Two frequencies will occur directly above the critical value, and in the case of symmetric gain line they will be symmetric about the line centre. Above the critical value the laser action at the central frequency is terminated, since at the critical pumping $\partial^2\tilde{G}(\tilde{\omega} = \tilde{\omega}_{00})/\partial\tilde{\omega}^2 = 0$ and above it the gain is a minimum rather than a maximum at the central

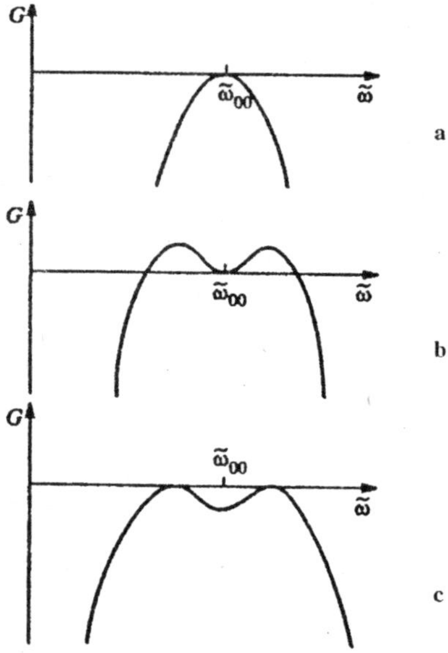

Fig. 4.10. Form of inhomogeneously broadened gain line of the laser medium in cases of (*a*) stable single-frequency, (*b*) unstable single-frequency and (*c*) stable double-frequency laser action.

frequency (Fig. 4.10*c*).

In the domain of two-frequency laser action the function $\tilde{G}(\tilde{\omega})$ remains symmetric with respect to the line centre although it changes in form as pumping further increase. Therefore, the second derivative is zero only at point $\tilde{\omega} = \tilde{\omega}_{00}$. This means that when the pumping reaches the next critical value, the spectrum will multiply by producing a new line rather than splitting the existing lines as at the first stage. More specifically, the central mode will again be excited and the distance between the sideband modes will be of the order of halfwidth of the elementary luminescence band. This problem is discussed in detail in the book [380]. The striped structure, originated from the spectral competition of modes in an inhomogeneously broadened laser medium, is observed in the output spectrum of a neodimium glass laser (see Section 1.2.3). The evolution of the spectrum, as pumping grows, which is followed in Fig. 1.16, is in qualitative agreement with the theory. A similar dependence of the radiation spectrum structure on the pumping parameter is exhibited by dye lasers (Fig. 1.9). Obviously, the spectrum splitting for small excess pumping above threshold, described in Refs. [59–61], indicates the same features of inhomogeneously broadened laser media are inherent in dye solutions.

4.5.3. Connection between the Spatial Mode Structure and the Spectral Structure of Laser Emission

The spatial nonuniformity of modes in the active medium leads to simultaneous participation of many modes with close frequencies in the laser action. For homogeneous broadening this was shown in Section 4.1. Inhomogeneous broadening does not remove the spatial competition of modes. Thus, it can be expected that each component of the large-scale spectral structure, mentioned in Section 4.5.2, will broaden to a band. If the width of a separate band is comparable to the interband spacing, then the discrete structure due to competition among the modes will not be apparent in the experiment.

For a very rough estimate of the spectral width of the laser emission due to spatial competition of the modes we ignore the spectral inhomogeinity of the medium and make use of Eq. (4.23). We should remember that the pumping parameter definitions in Section 4.1 and here are different and A should be replaced by $A\,(q+1)^{-1}$ in Eq. (4.23). No other changes are required if we assume $\Delta = \Delta\omega/\delta\omega_{\text{inh}}$, where $\delta\omega_{\text{inh}}$ is the halfwidth of the gain line at halfheight. Thus, in place of Eq. (4.23) we have the expression

$$\Delta_{\text{las}} = \left(3\Delta\frac{A-q-1}{8A}\right)^{1/3}. \tag{4.126}$$

Substituting the pumping value (4.125), which corresponds to spectrum splitting in the uniform field model, the lower limit on spectral width at the critical point is

$$2\Delta^*_{\text{las}} = q^{-1}[12\Delta(q+2\tilde{\Gamma})]^{1/3}. \tag{4.127}$$

This equation is valid for $q >> 1$ and $\tilde{\Gamma} << q^3$. It makes no sense to consider the case $\tilde{\Gamma} > q^3$, since the spectral splitting threshold will be much higher than the attainable pumping parameter values.

The strips in the laser spectrum will, of course, overlap, if $\Delta^*_{\text{las}} > 1/(2q)$ and, therefore, if the intermode interval satisfies the condition

$$\Delta > \Delta^* = \frac{1}{12(q+2\tilde{\Gamma})}. \tag{4.128}$$

For estimates we assume that the cavity length exceeds 10 cm and, consequently, the interval between the neighbouring longitudinal modes $\Delta\omega \le 10^{10}\text{s}^{-1}$. Assigning $\delta\omega_{\text{inh}} = 2{\cdot}10^{13}\text{s}^{-1}$ (the luminescence linewidth of neodymium glass) we obtain $\Delta < 5\cdot10^{-4}$. Substituting this in the left-hand side of Eq. (4.128) we find that the neodymium glass spectrum can be structureless for $q > 100$ ($\tilde{\Gamma} << q$). This degree of inhomogeneous broadening is achievable only for extreme cooling of the glass matrix [381].

When the intermode spacing does not satisfy the inequality (4.128) as, for example, in a neodymium glass laser at room temperature, the radiation spectrum must possess a structure. There are two options discussed below. Both options satisfy the time-independent laser equations. Therefore, the problem reduces to analyzing them for stability. This problem has not been yet solved and so it is impossible to specify more accurately the conditions needed for this or that spectrum.

1. Condition (4.128) is not met. The spectrum consists of individual bands spaced on the order of the halfwidth of the homogeneous part of the line broadening.

2. Condition (4.128) is met. This is possible when not all adjacent modes are generated. The spectrum consists of narrow lines and their spacing is greater than the intermode spacing and less than the homogeneous part of the line broadening.

Both possibilities satisfy the time-independent laser equations. Therefore, the problem reduces to their stability analysis. This problem is not solved and that is why we cannot indicate more definitely the condition needed for realization of that or other spectrum. Intuitively, it can be expected that the first type 'banded spectra' should be observed with greater probability in lasers with a spherical cavity, and the second type 'line spectra' in lasers with a plane-parallel cavity. Any small-scale structure in the spectrum is most likely caused by frequency-selected losses in the cavity[1]. The latter type of structure of the time-independent spectra is most often observed in practice.

The large number of modes and the rather small intermode frequency spacing facilitate calculation of the total width of the line type spectrum [382]. Owing to these features, the same approximate method of the space integral calculation, described in Section 4.1.2, is applicable, and the summation over all modes is replaced by integration with respect to frequency.

The case without cross-relaxation is the simplest. The radiation spectrum coincides in shape with the gain line. The spectrum boundaries are defined by the condition of balance between the unsaturated gain and loss.

In the presence of cross-relaxation, calculation of the steady-state spectral width is formally complicated by the term $\int n \mathrm{d}\tilde{\omega}$ in Eq. (4.100b). Putting $\mathrm{d}/\mathrm{d}\tau = 0$, integrating Eq. (4.100b) over the cavity volume and making use of Eq. (4.100a) yields

$$\iint \overline{n} \mathrm{d}\tilde{\omega}_0 \mathrm{d}v = A - \sum \overline{m}_k \,. \tag{4.129}$$

It is pertinent to recall that multimode lasing is accompanied by a spatially uniform distribution of the inversion throughout the laser medium. Thus, we can remove $\overline{n}$ from the integral sign and reduce Eq. (4.129) to

[1]Intracavity laser spectroscopy is based on this phenomenon [383–386].

$$\int \bar{n} \mathrm{d}\tilde{\omega}_0 = A - \sum \bar{m}_k \,.$$

By elimination of superfluous unknown we resolve Eq. (4.100b) with respect to $\bar{n}$:

$$\bar{n} = h(\tilde{\omega}_0) \frac{A(\tilde{\Gamma}+1) - \tilde{\Gamma}\sum \bar{m}_k}{\tilde{\Gamma}+1+\sum \tilde{L}_k \psi_k^2 \bar{m}_k},$$

and substitute the resultant expression into Eq. (4.100a). The inequality

$$\sum \tilde{L}_k \psi_k^2 \bar{m}_k << \tilde{\Gamma}+1$$

is equivalent to a small exceeds over threshold only in the case of weak cross-relaxation. For strong cross-relaxation we can use $(\tilde{\Gamma}+1)^{-1}$ as a small parameter and make the expansion

$$\frac{1}{\tilde{\Gamma}+1+\sum \tilde{L}_k \psi_k^2 \bar{m}_k} \approx \frac{1}{\tilde{\Gamma}+1} - \frac{\sum \tilde{L}_k \psi_k^2 \bar{m}_k}{(\tilde{\Gamma}+1)^2} \,.$$

After the simple integration we arrive at the equality

$$Z \int_o^{\infty} \left(1 - \frac{\sum \tilde{L}_q \bar{m}_q + \frac{1}{2} \tilde{L}_q \bar{m}_q}{\tilde{\Gamma}+1} \right) \tilde{L}_k h(\tilde{\omega}_0) \mathrm{d}\tilde{\omega}_0 = 1, \tag{4.130}$$

in which

$$Z = A - \tilde{\Gamma}(\tilde{\Gamma}+1)^{-1} \sum \bar{m}_k \,.$$

The further transformations are based on the assumption of a large value for the inhomogeneous broadening parameter, $q \gg 1$, and a broad radiation spectrum, $\Delta_{\text{las}} >> 1$. These transformations lead to the equality

$$Z \left[1 - \frac{\pi \bar{m}_k}{(\tilde{\Gamma}+1)\Delta} \right] = \frac{1}{\pi h(\tilde{\omega}_k)} \,. \tag{4.131}$$

Since $\bar{m}_k = 0$ at the spectrum edge, from Eq. (4.131) we get the equation $Z^{-1} = \pi h(\tilde{\omega}_{00} \pm \Delta_{\text{las}})$ to find the spectral width of the laser emission. The quantity Z can be found by summing both parts of Eq. (4.131) over all excited modes. In the case of Lorentzian line shape the problem reduces to the equation ($Y = \Delta_{\text{las}} / q$)

$$Y^4 + \frac{4\tilde{\Gamma}}{3\pi} Y^3 - \left(\frac{A}{q} - 2 \right) Y^2 - \left(\frac{A}{q} - 1 \right) = 0 \,. \tag{4.132}$$

Its solutions for some values of $\tilde{\Gamma}$ are given in Fig. 4.11. It is seen that the spectral width of the laser emission diminishes as cross-relaxation grows.

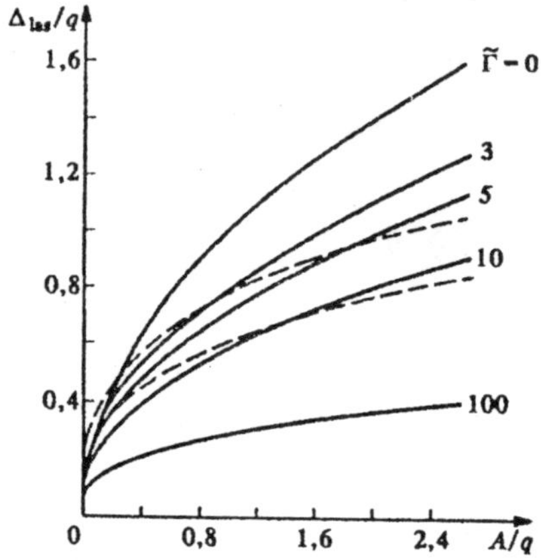

Fig. 4.11. Spectrum width of an inhomogeneously broadened laser versus pumping level at different cross-relaxation rates $\tilde{\Gamma}$.

The function $\Delta_{las}(A)$ is simple only in the case $\tilde{\Gamma} >> 1$ at $\Delta_{las} << q$:

$$\Delta_{las} = q\left[\frac{3\pi}{4\tilde{\Gamma}}\left(\frac{A}{q}-1\right)\right]^{1/3} \tag{4.133}$$

The curves calculated using Eq. (4.133) are shown by dash lines in Fig. 4.11. In the limit $\tilde{\Gamma} \to \infty$, Eq. (4.133) is not valid, nor is the calculation as a whole, since the condition $\Delta_{las} >> 1$ is violated. For very large $\tilde{\Gamma}$ the line can be considered homogeneously broadened, so that Eq. (4.126) can be used for estimation of the spectral width of the laser emission.

4.5.4. Transients in the Presence of Cross-Relaxation

By providing migration of excitations within the inhomogeneously broadened line contour, the cross-relaxation mechanism tends to cancel any deviation in the spectral distribution of active centres from the equilibrium one. A decrease of the number of active centres in some local part of the spectrum due to induced emission by a narrow frequency band produces a directed inflow of excitations exactly to this part of the spectrum. In this sense, the cross-relaxation acts as supplementary source of pumping. Correspondingly, the cross-relaxation influences any transients, leading to faster arrival at a steady state. This mechanism is the most effective in the case of narrow linewidth signal generation when an appreciate part of the active medium is not directly involved in laser action, playing the role of an energy reservoir. Below we consider a single-frequency time-dependent regime.

We now linearize Eqs. (4.102) near the fixed point $\bar{n},\bar{m}$, defined by equalities (4.113) and (4.118), and introduce the variables

$$\delta m(\tau) = m - \bar{m}, \quad \delta n(\tau,\tilde{\omega}_0) = n - \bar{n}.$$

Assuming solutions in the form $\delta m = \delta m_0 e^{\lambda\tau}$, $\delta n = \delta n_0(\tilde{\omega}_0)e^{\lambda\tau}$, we arrive at

$$\lambda \delta m_0 = G\overline{m}\int_0^{\infty} \frac{\delta n_0 \mathrm{d}\tilde{\omega}_0}{1+(\tilde{\omega}_0 - \tilde{\omega}_{00})^2} . \tag{4.134a}$$

$$\delta n_0\left[\lambda + \tilde{\Gamma} + 1 + \frac{\overline{m}}{1+(\tilde{\omega}_0 - \tilde{\omega}_{00})^2}\right] = \frac{\overline{n}}{1+(\tilde{\omega}_0 - \tilde{\omega}_{00})^2} + \tilde{\Gamma}h\int_0^{\infty} \delta n_0 \mathrm{d}\tilde{\omega}_0 . \tag{4.134b}$$

Integrating Eq. (4.134b) with respect to $\tilde{\omega}_0$ and making use of Eq. (4.134a) we find

$$\int_0^{\infty} \delta n_0 \mathrm{d}\tilde{\omega}_0 = -\delta m_0 (G+\lambda)/G(\lambda+1) .$$

Thus, from Eq. (4.134b) we can define the relation between δn_0 and δm_0. Substituting this relation into Eq. (4.134a) leads to the desired characteristic equation

$$\lambda = -G\overline{m}\left\{\int_0^{\infty} \frac{\overline{n}[1+(\tilde{\omega}_0 - \tilde{\omega}_{00})^2]^{-1}\mathrm{d}\tilde{\omega}_0}{\lambda+\tilde{\Gamma}+1+\overline{m}[1+(\tilde{\omega}_0 - \tilde{\omega}_{00})^2]^{-1}} + \frac{(G+\lambda)\tilde{\Gamma}}{G(\lambda+1)}\int_0^{\infty} \frac{h[1+(\tilde{\omega}_0 - \tilde{\omega}_{00})^2]^{-1}\mathrm{d}\tilde{\omega}_0}{\lambda+\tilde{\Gamma}+1+\overline{m}[1+(\tilde{\omega}_0 - \tilde{\omega}_{00})^2]^{-1}}\right\}. \tag{4.135}$$

Generally, it is clear that the damping of transient oscillations will be accelerated with an increase in $\tilde{\Gamma}$ only for relatively weak cross-relaxation. For very large $\tilde{\Gamma}$ the laser medium behaves as homogeneously broadened one thus obeying the laws stated in Section 3.2. It is reasonable to consider the limiting cases separately.

1. *Relatively weak cross-relaxation*

$$\Omega_1 >> |\theta_1|, \tilde{\Gamma}, \overline{m} . \tag{4.136}$$

Using these inequalities we expand the integrand in a series of the small quantity $1/\Omega_1$, and then find the approximate values of the frequency and damping rate of small oscillations.

For the Lorentzian distribution function (4.106) and $q >> 1, \tilde{\Gamma} >> 1$ the approximate relations

$$\Omega_1^2 \approx G\overline{m}(A-\overline{m})/(2q), \tag{4.137}$$

$$\theta_1 = -\tilde{\Gamma}/2, \tag{4.138}$$

are valid. Here $\overline{m}$ is defined by Eq. (4.118). So, the damping of transient oscillations increases with increasing cross-relaxation when the latter is not too large [387].

Using Eqs. (4.137) and (4.138) we can write the inequalities (4.136) in a more explicit form. By virtue of $G \gg 1$, a small excess over the laser threshold is sufficient to satisfy the inequality $\Omega_1 >> \overline{m}$. The other two inequalities of (4.136) coincide and can be reduced, using Eqs. (4.137) and (4.118), to

$$\tilde{\Gamma}^2 << \frac{1}{2}G(A-q). \tag{4.139}$$

2. *Strong cross-relaxation*

$$\tilde{\Gamma} >> \Omega_1, |\theta_1|, \overline{m}. \tag{4.140}$$

In this case the small parameter for the series expansion of the integrand in Eq. (4.135), is $1/\tilde{\Gamma}$. The characteristic equation (4.135) is reduced, provided $q >> 1$, to the simple form

$$2q\tilde{\Gamma}\lambda^2 + [G\overline{m}q + 2\tilde{\Gamma}(q+\overline{m})]\lambda + 2G\tilde{\Gamma}\overline{m} = 0.$$

Solving this quadratic equation we find

$$\Omega_1^2 \approx G(Aq^{-1} - 1), \tag{4.141}$$

$$\theta_1 = -\frac{Gq\overline{m} + 2\tilde{\Gamma}(q+\overline{m})}{4\tilde{\Gamma}q}. \tag{4.142}$$

Unlike the case of weak cross-relaxation, the damping rate $|\theta_1|$ diminishes as $\tilde{\Gamma}$ grows, so that as $\tilde{\Gamma} \to \infty, |\theta_1| = A/2q$. This means that the medium behaves as if it were homogeneously broadened in this limit.

Substituting Eqs. (4.142) and (4.118) into Eq. (4.140) we see that the three inequalities are satisfied when

$$\tilde{\Gamma}^2 >> \frac{1}{4}G(A-q). \tag{4.143}$$

From the general dependence $\theta_1(\tilde{\Gamma})$ it is inferred that in the intermediate domain, defined by $G(A-q)/4 < \tilde{\Gamma}^2 < G(A-q)/2$, the damping of the relaxation oscillations is a maximum, and $|\theta_1| \approx \tilde{\Gamma}$.

4.6. Dynamical Instability of Steady State of a Multimode Travelling-Wave Laser (Risken–Nummedal–Graham–Haken Theory)

The conditions for laser instability with respect to perturbation of the oscillating mode itself found in Section 3.4 are quite clear from the physical point of view. First, the nonlinear distortion of the gain line must ensure the mode splitting effect. Second, the sideband components, thus having the right to exist, must not be suppressed by the cavity; therefore, a wide passband of the mode ($\tilde{\kappa} > 1$) is required. This last condition is no longer necessary if we speak about the instability of single-mode lasing with respect to the excitation of other modes. The mode, which is tuned to the centre (the so-called resonant mode) ensures the laser self-excitation at the carrier frequency while the nonresonant modes are used for the side-band components. Formulated like this, the dynamical instability problem was first considered by Risken and Nummedal [388, 389], Graham and Haken [390] and discussed more recently by many authors [215, 289, 391-397]. The consideration is not limited now by class C lasers.

The main features of the phenomenon should be apparent if we turn to the one-dimensional model of a travelling-wave laser with uniformly distributed (over the cavity perimeter) loss and gain

$$\frac{\partial F}{\partial \tau}+\frac{\partial F}{\partial \zeta}=\tilde{\kappa}(P-F), \tag{4.144a}$$

$$\frac{\partial P}{\partial \tau}=nF-P, \tag{4.144b}$$

$$\frac{\partial n}{\partial \tau}=\tilde{\gamma}(A-n-PF). \tag{4.144c}$$

These equations should be supplemented with cycling condition

$$F(\tau,\zeta)=F(\tau,\zeta+l). \tag{4.145}$$

Besides the notation previously adopted we have introduced $\zeta = z\gamma_\perp / c$, a dimensionless coordinate along the cavity axis, and $l = L\gamma_\perp / c$, a dimensionless cavity length. By using equations in the form of (4.144) we mean that one mode is resonant.

Without the spatial variation of the variables, Eqs. (4.144) are identical to Eq. (3.85); thus, the steady-state solution of (4.144) coincides with Eqs. (3.86) and (3.87). There is no need to return to the laser self-excitation condition which again has the form $A > 1$. However, the linear stability analysis is rather specific in this model.

Assuming that small deviations of all variables from the steady state $\overline{F}_b^2 = \overline{P}_b^2 = A-1$, $\overline{n}_b = 1$ evolve as $\exp(\lambda\tau - i\mu\zeta)$ and linearizing Eqs. (4.144) be these deviations we arrive at a characteristic equation

$$\lambda^3+(\tilde{\kappa}+\tilde{\gamma}+1-i\mu)\lambda^2+[\tilde{\gamma}(A+\tilde{\kappa})-i\mu(1+\tilde{\gamma})]\lambda+2\tilde{\gamma}\tilde{\kappa}(A-1)-i\mu\tilde{\gamma}A=0 \tag{4.146}$$

where μ is a real quantity and $\lambda = i\Omega+\theta$, which corresponds to the Hopf type of the assumed bifurcation at the second laser threshold. We then use the familiar method: only the term linear with respect to θ are related in Eq. (4.146). Equation (4.146) is split into two real ones, which, after elimination of μ, reduce to

$$\theta = -\tilde{\kappa}\frac{\Omega^4 - \tilde{\gamma}[3(A-1)-\tilde{\gamma}]\Omega^2 + 2\tilde{\gamma}^2 A(A-1)}{\Omega^4 + [\tilde{\kappa}(2\tilde{\gamma}+1)+(\tilde{\gamma}+1)^2 - 2\tilde{\gamma}A]\Omega^2 + 2\tilde{\gamma}\tilde{\kappa}(\tilde{\gamma}+1)(A-1)+\tilde{\gamma}^2 A(A+\tilde{\kappa})}. \tag{4.147}$$

The boundary between the domains of stable and unstable solutions of Eqs. (4.144) can easily be found assuming $\theta = 0$. The roots of the resultant quadratic equation for the case are given by

$$2\Omega^2/\tilde{\gamma} = 3(A-1) \pm \sqrt{9(A-1)^2 - 8A(A-1)}. \tag{4.148}$$

They are real if $A \geq A_{cr} = 9$. Thus, the critical value of the pumping parameter for a multimode travelling-wave laser coincides with that obtained for a single-mode laser under the most favourable conditions $\tilde{\gamma} << 1, \tilde{\kappa} = 3$.

The dependence expressed by Eq. (4.148) is given in Fig. 4.12. Its asymptotes for $A \gg 1$ are represented by the lines

$$\Omega^2_{max}/\tilde{\gamma} = 2A, \quad \Omega^2_{min}/\tilde{\gamma} = A. \tag{4.149}$$

The potential instability domain is shown hatched in Fig. 4.12. It should be borne in mind, however, that the instability could develop only if one. frequency of the intermode beats enters this domain (the cycling condition). This additional (to $A > A_{cr}$) requirement is a compensation for no limit being imposed on the cavity Q-factor. Roughly speaking, it is necessary that the normalized intermode frequency spacing $\Delta = 2\pi c / L\gamma_\perp$ does not exceed the frequency $\Omega_{max} = (2A\tilde{\gamma})^{1/2}$. Thus, a limit is placed on the cavity perimeter:

$$L > L_{cr} = 2\pi c(\gamma_\parallel \gamma_\perp A)^{-1/2}. \tag{4.150}$$

Numerical estimation for a Nd:YAG laser ($\gamma_\parallel = 10^4$ s^{-1}, $\gamma_\perp = 10^{12}$ s^{-1}) for A = 10 yields $\omega_{max} = \gamma_\perp \Omega_{max} = 4 \cdot 10^8$ s^{-1} and $L_{cr} \approx 6$ m. In principle, the fibre optical delay permits design of laser cavities with a perimeter of a few hundreds of meters [86], but these are rare in practice. So, this type of instability is not an urgent problem for solid-state laser. However, it is easy to reach these values in is retained. If we take into account the existing spatial nonuniformity, caused, for example, by the loss localization on the mirrors, then the second threshold is still higher [396, 397].

Specific features of a multimode laser are essential to pulsations above the second threshold. Since the spectrum of allowed pulsation frequencies is rigidly subject to the discrete spectrum of the cavity modes, the transient process from amplitude modulation to beats, shown in Fig. 3.6, is highly problematic. Nevertheless, the numerical investigation of Eqs. (4.144) shows that higher-order bifurcations can occur. The pulse train

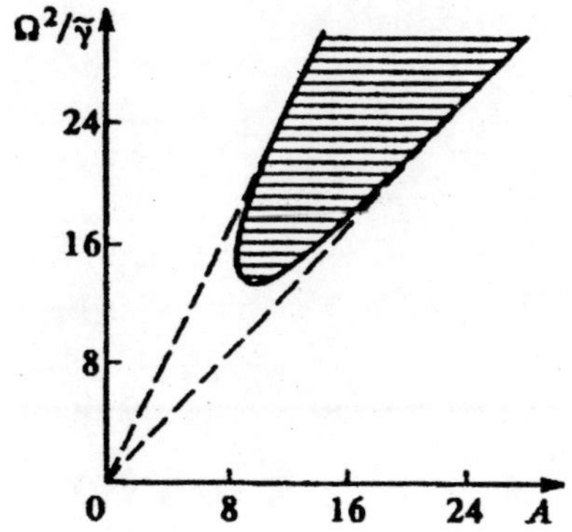

Fig. 4.12. Phase diagram of a travelling wave multimode laser in the control parameter plane (A, Ω) plotted according to Eq. (4.148). The hatching denotes the instability domain.

envelope in the laser instability domain can be both regular and chaotic [391, 395].

There is no strong experimental confirmation of the Risken–Nummedal–Graham–Haken theory until now. Information about observation of undamped pulsations in an erbium fibre laser is reported in Ref. [398]. The period of observed pulsations is equal to the round trip time of the cavity. According to all parameters this regime corresponds to the theory predictions. Only the instability threshold is evidently lower than theoretically predicted. This fact forced us to think about the modification of the model [399] or about the presence of any casual nonlinearity in the experiment, for example, a weak saturable absorption. But one must bear in mind that the gain line can be slightly inhomogeneously broadened.

Chapter 5

Multimode Lasers with Quasi-Frequency-Degenerate Modes

From a physical point of view, the principal difference between large and small intermode frequency spacings is determined by their relation to the frequency of dynamical processes in lasers.

In Chapter 4 we have discussed models, in which

$$\Delta\omega >> \delta\omega_c$$

The intermode beat frequencies suggested frequencies that are much greater than relaxation oscillation frequencies in the class *B* lasers and are situated essentially higher than all the frequencies characteristic for the spectral dynamics of dye lasers. Therefore, we did not take into account the mode beats in the considered rate equation models. Combination tone mode-mode couplings, which occur via oscillating inversion gratings, also do not require the presence of amplitudes of these gratings among the variables.

Such an approach is incorrect if the modes are close in frequencies. In this case the amplitudes of oscillating inversion gratings are included in the variable collection, making the system's dimension higher and changing its physical properties.

In contrast to the simple rate equation models, an important role is played not only by the amplitudes but also by phases of the physical quantities that characterize the state of the field and the medium. That is way we use the term 'model with the phase-sensitive interaction'.

The phase-sensitive interaction takes place between a pair of levels and it is reasonable to study this phenomenon considering the two-mode models. It is important to indicate the type of the resonator. In Fabry-Perot cavities the modes are standing waves and the phase-sensitive interaction occurs here in addition to a more simple interaction, which is realized through saturation and cross-saturation of the active medium. Models of these lasers smoothly, without bifurcations, transform to rate equations, when a control parameter, for example, the intermode frequency spacing, is changing. In ring lasers, where modes are travelling waves, the phase-sensitive interaction is the main type of interaction, and a corresponding model is never reduced to rate equations.

5.1 Two-Mode Class *B* Laser with a Fabry-Perot Resonator

First papers devoted to the two-mode laser theory date back to the middle 1960-s [400–405]. However, we shall use an approach developed in more recent works [406, 407].

5.1.1 Equations of Class *B* Two-Mode Laser

Let us start with Eqs. (S.12), slightly generalizing them:

$$\frac{\mathrm{d}f_1}{\mathrm{d}t}=\frac{G}{2}\left\{-f_1+(\alpha'+i\alpha'')[f_1(n_0+n_1)+f_2(n_{12}^-+n_{12}^+)]\right\},$$

$$\frac{\mathrm{d}f_2}{\mathrm{d}t}=\frac{G}{2}\left\{-(1+\beta+i\Delta)f_2+(\alpha'+i\alpha'')[f_2(n_0+n_2)+f_1(n_{12}^-+n_{12}^+)]\right\},$$

$$\frac{\mathrm{d}n_0}{\mathrm{d}t}=A-n_0-\alpha'[n_0(|f_1|^2+|f_2|^2)+n_1|f_1|^2+n_2|f_2|^2+(f_1^*f_2+f_1f_2^*)(n_{12}^-+n_{12}^+)],$$

$$\frac{\mathrm{d}n_1}{\mathrm{d}t}=-n_1-\alpha'[n_1(|f_1|^2+|f_2|^2)+\frac{1}{2}n_0|f_1|^2+\frac{1}{2}(f_1^*f_2+f_1f_2^*)(n_{12}^-+n_{12}^+)],\qquad(5.1)$$

$$\frac{\mathrm{d}n_2}{\mathrm{d}t}=-n_2-\alpha'[n_2(|f_1|^2+|f_2|^2)+\frac{1}{2}n_0|f_2|^2+\frac{1}{2}(f_1^*f_2+f_1f_2^*)(n_{12}^-+n_{12}^+)],$$

$$\frac{dn_{12}^{+}}{dt} = -n_{12}^{+} - \alpha'[n_{12}^{+}(|f_1|^2 + |f_2|^2) + \frac{1}{2}n_{12}^{-}(|f_1|^2 + |f_2|^2) + \frac{1}{2}(f_1^* f_2 + f_1 f_2^*)(n_0 + n_1 + n_2)],$$

$$\frac{dn_{12}^{-}}{dt} = -n_{12}^{-} - \alpha'[n_{12}^{-}(|f_1|^2 + |f_2|^2) + \frac{1}{2}n_{12}^{+}(|f_1|^2 + |f_2|^2) + \frac{1}{2}(f_1^* f_2 + f_1 f_2^*)(n_0 + n_1 + n_2)].$$

The first two equations require an additional comment. On the one hand, the asymmetry in these equations is due to the inequality of the mode losses presented by β, and, on the other hand, due to the fact that the reference frequency is chosen equal to the eigenfrequency of the first mode, so that Δ_{c2} coincides with the intermode beat frequency D.

Rigorously speaking, equations (S.12) are obtained under the assumption of two-level approximation for the laser medium. Hence $\alpha' = \tilde{L} = 1/(1+\Delta_0^2)$; $\alpha'' = -\Delta_0 \tilde{L}$. However, the two-level approximation in the simplest form is not always defensible. Even doped dielectrics often do not satisfy the requirement $\alpha'' = 0$ at the centre of the gain line because the polarizabilities of atoms at different energy levels do not coincide [408-410]. However, some qualitative features of the dynamics of semiconductor lasers can be described using such approach with only little transformation and supplementation of Eqs. (5.1). First of all, it is necessary to take into account the diffusion of carriers, the velocity of which can be considerable in semiconductors. In the limiting case the diffusion leads to complete smoothing of the small-scale gratings with a step of the order of the wavelength, n_1, n_2, n_{12}^{+}, but almost does not affect the large-scale gratings. Such a situation was considered in [406]. In paper [407] the case of arbitrary diffusion rate was considered, which is interesting from the methodical point of view, since the diffusion rate is the only control parameter that permits to affect the grating amplitudes.

Taking the spatial diffusion into account reduces to adding a term $-D\nabla^2 n$ in the material equations. There are new terms $n_1 d_{dif}$, $n_2 d_{dif}$, $n_{12}^{+} d_{dif}$. It is possible to use the unified diffusion coefficient $d_{dif} = D\gamma^{-1}(\pi/\lambda)$ because the wavelengths of interacting modes are nearly the same.

The necessity of one more correction of the equations, as applied to semiconductor lasers, follows from the fact that the form of the gain line is far from being Lorentzian. This does not influence the value of α', which is close to unity, since the laser frequencies are in the immediate proximity from the gain line maximum. Meanwhile, unlike the two-level case, the value of α'' can strongly differ from the intermode beat frequency, reaching a few units [411, 412]. Since α'' defines the degree of the influence of amplitude fluctuations on frequency fluctuations, this parameter is generally referred to as line width enhancement factor (Henry's α-factor) of a semiconductor laser. But there is one more term expressing the fact of dependence of the oscillation frequency in a nonlinear system on its amplitude, and this term is 'nonisochronity'.

In view of all these comments, Eqs. (5.1) can be rewritten in a more general form ($\alpha \equiv \alpha''$):

$$\frac{df_1}{dt} = \frac{G}{2}\{- f_1 + (1+i\alpha)[f_1(n_0+n_1) + f_2(n_{12}^- + n_{12}^+)]\},$$

$$\frac{df_2}{dt} = \frac{G}{2}\{-(1+\beta+i\Delta_c)f_2 + (1+i\alpha)[f_2(n_0+n_2) + f_1(n_{12}^- + n_{12}^+)]\},$$

$$\frac{dn_1}{dt} = -n_1(1+d_{dif}+|f_1|^2+|f_2|^2) - \frac{1}{2}n_0|f_1|^2 - \frac{1}{2}(f_1^* f_2 + f_1 f_2^*)(n_{12}^- + n_{12}^+),$$

$$(5.2)$$

$$\frac{dn_2}{dt} = -n_2(1+d_{dif}+(|f_1|^2+|f_2|^2) - \frac{1}{2}n_0|f_2|^2 - \frac{1}{2}(f_1^* f_2 + f_1 f_2^*)(n_{12}^- + n_{12}^+)],$$

$$\frac{dn_{12}^+}{dt} = -n_{12}^+(1+d_{dif}+|f_1|^2+|f_2|^2) - \frac{1}{2}n_{12}^-(|f_1|^2+|f_2|^2) - \frac{1}{2}(f_1^* f_2 + f_1 f_2^*)(n_0+n_1+n_2),$$

$$\frac{dn_{12}^-}{dt} = -n_{12}^-(1+n_{12}^-|f_1|^2+|f_2|^2) - \frac{1}{2}n_{12}^+(|f_1|^2+|f_2|^2) - \frac{1}{2}(f_1^* f_2 + f_1 f_2^*)(n_0+n_1+n_2).$$

5.1.2 Steady-State Solutions and Relaxation Oscillations

The considered two-mode model with the phase-sensitive interaction has two single-mode steady-state solutions

$$\bar{f}_1 = \bar{m}_1^{1/2}\exp\left(-\frac{1}{2}iG\alpha\tau\right), \quad \bar{f}_2 = 0,$$

$$\bar{n}_0 = A - \bar{m}_1, \quad \bar{n}_1 = 1 + \bar{m}_1 - A, \quad \bar{n}_2 = 0, \quad \bar{n}_{12} = 0, \qquad (5.3)$$

$$\bar{m}_1 = \frac{1}{2}\{A - 2(2+d_{dif}) + \sqrt{A^2 + 4[(1+Ad_{dif}) + (1+d_{dif})^2]}\}$$

and

$$\bar{f}_1 = 0, \quad \bar{f}_2 = \bar{m}_2^{1/2}\exp\left\{\frac{1}{2}iG[\Delta_c + (1+\beta)\alpha]\tau\right\},$$

$$\bar{n}_0 = A-(1+\beta)\bar{m}_2, \quad \bar{n}_1 = 0, \quad \bar{n}_2 = (1+\beta)(1+\bar{m}_2)-A, \quad \bar{n}_{12} = 0,$$

$$\bar{m}_2 = \frac{1}{2}\left\{\tilde{A}-2(2+d_{\mathrm{dif}})+\sqrt{A^2+4[(1+\tilde{A}d_{\mathrm{dif}})+(1+d_{\mathrm{dif}})^2]}\right\} \tag{5.4}$$

where $\tilde{A} = A/(1+\beta)$. In the case $d_{\mathrm{dif}} = 0$ the expression for the steady-state intensity in Eqs. (5.3) coincide with Eq. (3.71).

The presence of oscillating factors in the expressions for the mode fields in Eqs. (5.3) and (5.4) only means that the laser frequencies do not coincide with the reference frequency, which, we remind, was chosen equal to the cavity eigenfrequencies. It is also apparent that the transition from the first to the second mode is accompanied by a change in laser frequency to a value close to the intermode frequency spacing.

Linearization of Eqs. (5.2) near any of the single-mode steady-state solutions leads to a characteristic equation of the form

$$\lambda^2 P_3(\lambda)P_4(\lambda) = 0, \tag{5.5}$$

where $P_j(\lambda)$ is a polynomial of power j. The cubic polynomial $P_3(\lambda)$ has one real root λ_1 and a pair of complex-conjugate roots $\lambda_{2,3} = \theta_1 \pm i\Omega_1$. It should be not surprising that the latter roots correspond to relaxation oscillations, which are available in the single-mode model too, since the same cubic characteristic polynomial can be obtained by completely ignoring the presence of the second mode.

The quartic polynomial $P_4(\lambda)$ has a pair of complex-conjugate roots $\lambda_{4,5} = \theta_A \pm i\Omega_A$, whereas the remaining two roots, λ_6 and λ_7, are real. Thus, the second type of relaxation oscillations exists, which exhibits noticeably different properties compared to the first type. This is clearly demonstrated in Figs. 5.1 and 5.2, which show the dependencies of the characteristic roots on pumping for different values of the diffusion coefficient. The dependences $\Omega_1 \sim \sqrt{A-1}$ and $\theta_1 \sim A$ prove to be nearly the same as those met above, as was to be expected. On the contrary, the behaviour of Ω_A and θ_A strongly depends on near which of the steady-state solutions the action takes place and on the value of the diffusion coefficient d_{dif}.

An important feature of the roots $\lambda_{4,5}$ is that $\mathrm{Re}\lambda_{4,5}$ reverses its sign for some bifurcation value (HB) of the control parameter. This corresponds to a Hopf bifurcation accompanied by the loss of stability of the single-mode solution and transition to the two-mode solution. The real root λ_6 can reverse its sign too (a corresponding point is marked as BP) and this also affects the stability of the single-mode solution. Multistability and instability of the steady states characterize the model with phase-sensitive interaction and these features are absent in the simplest rate equation model.

Figure 5.3 shows the phase diagram of a two-mode laser in the plane of

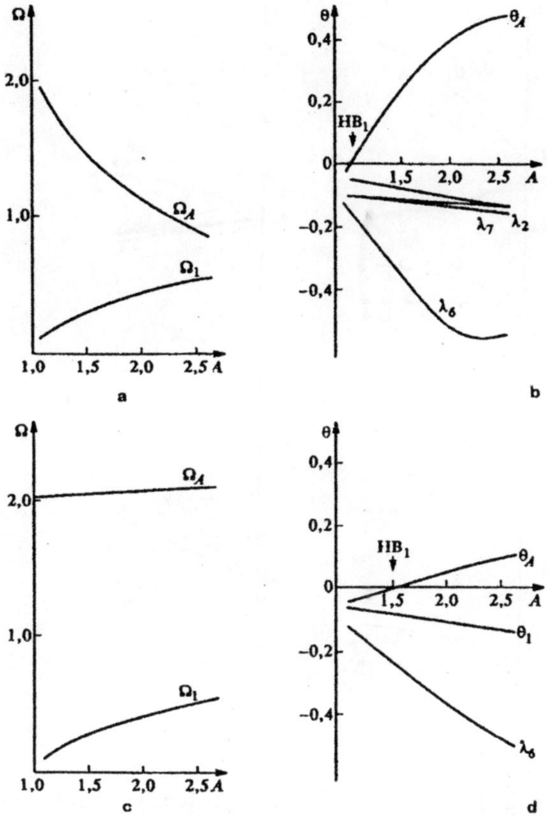

Fig. 5.1. Relaxation oscillation frequencies (*a,c*), their decrements and the real roots (*b,d*) of the characteristic equation (5.5) corresponding to the steady-state solution (5.3) as a function of the pumping level: $\alpha = 4.0$; $\Delta_c = 2.0$; $G = 20$; $\beta = 0.05$; $d = 0$ (*a,b*), 50 (*c,d*).

control parameters (A, d_{dif}). Two zones of stable steady-state solutions (SM_1 and SM_2) and the domains of two-mode solutions (TM and TM_S) are seen. The existence of stable solution for mode with higher losses (SM_2) is due to phase-sensitive interaction between the modes, mainly through a large-scale grating n_{12}^-. This solution is absent in ordinary rate equation models. This solution appears for $d_{dif} > d_1$ when the diffusion results in an appreciable smoothing of small-scale gratings. For smaller d_{dif} the competition between modes is largely weakened, so that two-mode operation dominates here. The boundary value $d_{dif} = d_1$ is not a bifurcation one, and it separates a domain of a relatively slow diffusion, in which the influence of small-scale gratings is determining, from a domain of high diffusion in which the mode-mode coupling through a large-scale structure is most significant. It should be noted that the value d_1 is varied within small limits when the control parameters Δ_c and α are changed.

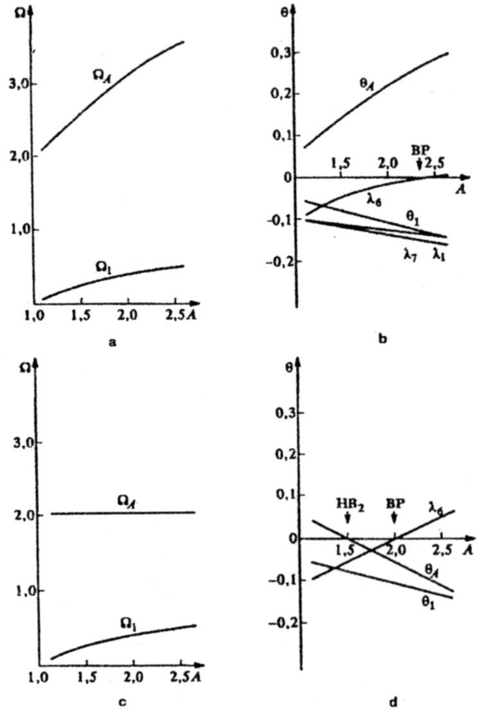

Fig. 5.2. The same as in Fig. 5.1 but for the steady-state solution (5.4) [407].

Zone TM_S in Fig. 5.3*a* corresponds to the steady-state two-mode solution with prevalence of the mode with highest losses, whereas this one, designated TM, is not steady-state. The depth of the self-modulation of mode amplitudes in this regime can be judged from Fig. 5.4. These diagrams are plotted on the basis of the numerical solutions of Eqs. (5.2), obtained for different values of the control parameters. In the particular case displayed in Fig. 5.4 the pumping parameter $A = 3.0$. In Fig. 5.3 this corresponds to a domain located noticeable higher than the boundary BP, and $m_2 > m_1$, i.e., the mode with higher losses dominates here. Attention is drawn to the very sharp change of the modulation depth when the diffusion coefficient passes through the value d_1.

The self-modulation frequency is smaller than intermode frequency spacing but approaches it with an increase in Δ_c. This fact allows to neglect the phase-sensitive mode coupling and use the rate-equation approximation for $\Delta_c > \max(1, \alpha)$. The time-dependent processes occurring in this case are represented by two relaxation oscillations: in-phase and antiphase oscillations of mode intensities, which have been considered in Chapter 4.

Let us summarize the properties of the model of two-mode laser with

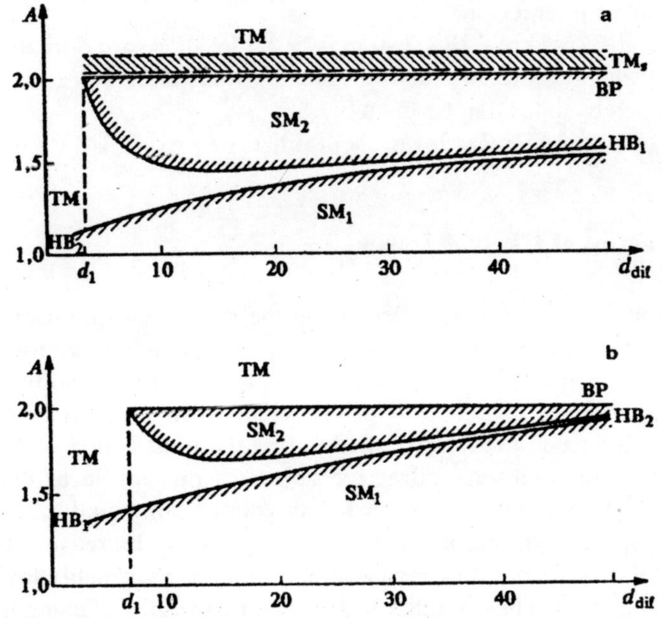

Fig. 5.3. Phase diagram for the laser model described by Eqs. (5.2) in the plane of control parameters (A, d_{dif}): $\alpha = 4.0$; $\Delta_c = 2.0$; $G = 20$; $\beta = 0.05$ (*a*), 0.09 (*b*).

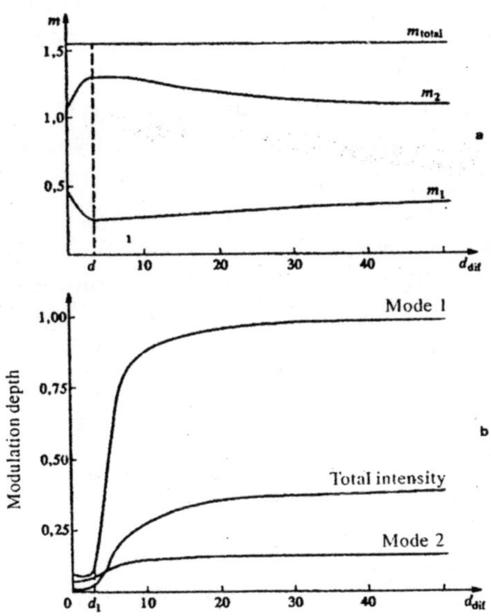

Fig. 5.4. Dependence of the mode intensities and the total intensity (*a*) and the self-modulation depth (*b*) on the diffusion coefficient: A=3.0; α = 4.0; Δ_c = 2.0; G = 20; β = 0.05 [407].

phase-sensitive interaction.

1. Generally speaking, the system is bistable: there are domains of the parameter values where two steady-state solutions are stable.

2. The steady states can be unstable.

3. The number of relaxation oscillations can exceed the number of modes.

5.2 Bidirectional Class *B* Laser

The specific feature of a ring laser from the dynamical point of view is conditioned by the fact that its modes are counter-running travelling waves. As a consequence there are no inversion gratings corresponding to individual modes in the presence of phase-sensitive interaction on the only one grating induced in the active medium by the joint action of modes.

The ring laser is a very attractive object of the nonlinear dynamics. Even the single-mode class *A* laser has the dimension equal to three, and that is enough for the complex dynamic behaviour. Increasing twice in class *B* lasers, the dimension, nevertheless, remains reasonably low for the modern computers. The complex regimes can be achieved using the ordinary combinations of parameters.

In the applied aspect at the first place stays the sensitivity of a ring laser to the rotation or, in more general sense, to the phase nonreciprocity of the cavity. As a rotation velocity sensor the preference have the atomic class *A* gas lasers. The gyroscopic potentialities of the class *B* ring laser it is hard to estimate at its true worth. What is nontrivial that information about phase nonreciprocity contains directly in the dynamic behaviour of the laser, and this is propaganda in favour of the inverse problems of laser dynamics.

The list of references at the subject given below though it is vast cannot pretend to be exhaustive. To fill this gap we can recommend the overview paper [413] authors of which have done much in the physics of the bidirectional ring lasers.

5.2.1 The Model of Single-Frequency Class *B* Ring Laser

We now use Eqs. (2.63) as the material equations. Assuming the field in the form (2.23) we approximate the material variables by using the series

$$\sigma = \sum \sigma_r e^{irkz}, \quad D = \sum D_r e^{irkz}.$$

Consideration of the elements D_r with $r \neq 0$ is dictated by the necessity of taking into account the counter-running wave interaction on a self-induced inversion grating. It is found, however, that the use of the harmonics $D_{\pm 2}$ is sufficient for this purpose, so the description of the inversion grating more thoroughly adds little to the dynamical characteristics of the laser system [414–416].

Termination of the series after the term D_r confines the polarization expansion to the terms $\sigma_{\pm 3}$. Assuming $\partial\sigma/\partial\tau = 0$ we come from Eq. (2.63) to

$$\sigma_{\pm 1} = -\frac{id(F_{\pm 1}D_0 + F_{\mp 1}D_{\pm 2})}{2\hbar[\gamma_\perp - i(\omega - \omega_0)]}, \quad \sigma_{\pm 3} = -\frac{idF_{\pm 1}D_{\pm 2}}{2\hbar[\gamma_\perp - i(\omega - \omega_0)]}. \tag{5.6}$$

Since the inversion is represented by slow variables, we find for

$$\frac{dD_0}{dt} + \gamma_\parallel(D_0 - D^{(0)}) = -\frac{i\beta_a}{2\hbar} d(F_1^*\sigma_1 - F_1\sigma_1^* + F_{-1}^*\sigma_{-1} - F_{-1}\sigma_{-1}^*), \tag{5.7}$$

$$\frac{dD_2}{dt} + \gamma_\parallel D_2 = -\frac{i\beta_a}{2\hbar} d(F_{-1}^*\sigma_1 - F_1\sigma_{-1}^* + F_1^*\sigma_3 - F_{-1}\sigma_{-3}^*).$$

The only thing left is to substitute the components $\sigma_\pm$ from Eq. (5.6) into Eqs. (5.7) and (2.38) bearing in mind that $P_{\pm 1} = N_S d\sigma_{\pm 1}$ we obtain a complete set of equations

$$\frac{dF_{\pm 1}}{dt} + [\kappa_\pm - i(\omega - \omega_c^\pm)]F_{\pm 1} = \frac{\pi\omega d^2 N_S}{\hbar[\gamma_\perp - i(\omega - \omega_0)]}(F_{\pm 1}D_0 + F_{\mp 1}D_{\pm 2}) + \frac{i}{2}\xi_\mp F_{\mp 1}$$

$$\frac{dD_0}{dt} + \gamma_\parallel(D_0 - D^{(0)}) = -\frac{\beta_a d^2\gamma_\perp}{2\hbar^2[\gamma_\perp^2 + (\omega - \omega_0)^2]}[(|F_1|^2 + |F_{-1}|^2)D_0 + F_1^*F_{-1}D_2 + F_1F_{-1}^*D_{-2}] \tag{5.8}$$

$$\frac{dD_2}{dt} + \gamma_\parallel D_2 = -\frac{\beta_a d^2\gamma_\perp}{2\hbar^2[\gamma_\perp^2 + (\omega - \omega_0)^2]}[(|F_1|^2 + |F_{-1}|^2)D_2 + F_1F_{-1}^*D_0].$$

Making use of the dimensionless notations introduced above and adding

$$\tau = t\gamma_\parallel, \quad \Delta_c^\pm = (\omega - \omega_c^\pm)/\kappa_0, \quad \tilde{F} = (1 - i\Delta_0)^{-1} = \tilde{L}(1 + i\Delta_0), \quad \rho_\pm = \xi_\pm/2\kappa_0,$$

we can write these equations in the form

$$\frac{df_\pm}{d\tau} - i\frac{G}{2}\Delta_c^\pm f_\pm = \frac{G}{2}[(\tilde{F}n_0 - 1 \mp \beta)f_\pm + (\tilde{F}n_{\pm 2} + i\rho_\mp)f_\mp],$$

$$\frac{dn_0}{d\tau} = A - [1 + \tilde{L}(|f_+|^2 + |f_-|^2)]n_0 - \tilde{L}(f_+^* f_- n_2 + f_+ f_-^* n_{-2}),$$

$$\frac{dn_2}{d\tau} = -[1 + \tilde{L}(|f_+|^2 + |f_-|^2)]n_2 - \tilde{L} f_+ f_-^* n_0. \tag{5.9}$$

5.2.2 Steady States in the Absence of Backscattering and Their Stability

At first we should ascertain the role of the Bragg gratings in the population inversion represented in Eqs. (5.9) by the variable . Omitting the linear coupling between waves and the nonreciprocity of the cavity we reduce Eqs. (5.9) to

$$\frac{df_\pm}{d\tau} = \frac{G}{2}[(\tilde{F}n_0 - 1) + \tilde{F}n_{\pm 2} f_\mp]. \tag{5.10}$$

This particular case is thoroughly investigated in [417, 418]. Besides the trivial steady state, the ring laser model possesses three other fixed points in the absence of linear coupling of waves. In terms of Eqs. (5.9) they are expressed by

$$|\bar{f}_+|^2 = A - \frac{1}{\tilde{L}}, \quad \bar{f}_- = 0, \quad \bar{n}_0 = \frac{1}{\tilde{L}}, \quad \bar{n}_2 = 0,$$

$$|\bar{f}_-|^2 = A - \frac{1}{\tilde{L}}, \quad \bar{f}_+ = 0, \quad \bar{n}_0 = \frac{1}{\tilde{L}}, \quad \bar{n}_2 = 0. \tag{5.11}$$

$$|\bar{f}_+|^2 = |\bar{f}_-|^2 \neq 0, \quad \bar{n}_2 \neq 0.$$

In Refs. [419–421] it is shown that the solution correspond to the standing wave mode is unstable. This is a characteristic feature of lasers with active media, which are not Doppler broadened.

Linearizing Eqs. (5.9) near each of the travelling wave steady states it is easy to obtain a characteristic equation, which can be decomposed into two parts. The first,

$$\lambda^2 + A\tilde{L}\lambda = G(A\tilde{L} - 1) = 0, \tag{5.12}$$

corresponds to introducing a perturbation into the excited wave (say, f_+) and into the uniform component of inversion. Owing to the condition $G \gg 1$, the roots of Eq. (5.12) have the form

$$\lambda \approx \frac{A\tilde{L}}{2} \pm i\sqrt{G(A\tilde{L} - 1)} = \theta_1 \pm i\Omega_1. \tag{5.13}$$

The frequency of relaxation oscillations exactly coincides with that for a single-mode travelling wave laser given by Eq. (3.20).

The reaction of the system to perturbation of the other pair of variables, f_- and n_2, which are equal to zero in the steady state being investigated, is indicated by the other part of the characteristic equation,

$$\lambda^2 + A\tilde{L}\lambda + \frac{1}{2}G(1+i\Delta_0)(A\tilde{L}-1) = 0. \tag{5.14}$$

This defines the additional relaxation oscillations frequencies

$$\Omega_{A,B} = \mathrm{Im}\,\lambda = \sqrt{G(A\tilde{L}-1)/2} = \Omega_1/\sqrt{2}, \tag{5.15}$$

characteristic of a bidirectional laser, which are doubly degenerate if the phase nonreciprocity is absent. They are the phase-sensitive relaxation oscillations, which have a decisive effect on the dynamic behaviour of this laser.

Under the condition Re $\lambda > 0$, which is equivalent to

$$\Delta_0^2 > \Delta_{cr}^2 = \frac{2}{G}\frac{(A\tilde{L})^2}{A\tilde{L}-1} \tag{5.16}$$

the oscillations of frequency $\Omega_1/\sqrt{2}$ are undamped. If the pumping in excess of threshold, $A\tilde{L}$, is assumed to be the control parameter, then Eq. (5.16) yields the critical detuning in explicit form. If we use the pumping parameter A_1, then Eq. (5.16) should be considered as an equation to solve with respect to A, from which we find

$$A_{1,2} = \frac{G}{4}\Delta_{cr}^2(1+\Delta_{cr}^2)\left(1 \pm \sqrt{1-\frac{8}{G\Delta_{cr}^2}}\right). \tag{5.17}$$

From this equation we get the necessary condition for instability $\Delta_{cr}^2 > 8/G$ and see the asymptotic behaviour $\Delta_{cr}^2 > 8/G$, namely $A_1 = G\Delta_{cr}^2(1+\Delta_{cr}^2)/2$ and $A_2 = 1 + \Delta_{cr}^2$ (the laser self-excitation threshold). The unstable domain in Fig. 5.5 lies between two branches of the curve (5.17).

The instability of the unidirectional steady-state solution with laser detuning was predicted in work [421]. The resulting time-dependent processes were calculated in [417, 422, 423]. Examples are given in Fig. 5.6. For small Δ_0, which nevertheless exceeds the critical value, the beam directions alternate with a characteristic time interval between switchings that greatly exceeds the period of relaxation oscillations. This interval is subject to considerable irregular fluctuations.

It is important to note that oscillations at the relaxation frequencies are involved in the dynamics and their presence becomes increasingly noticeable as the detuning increases. First, the onset of the self-modulation re-

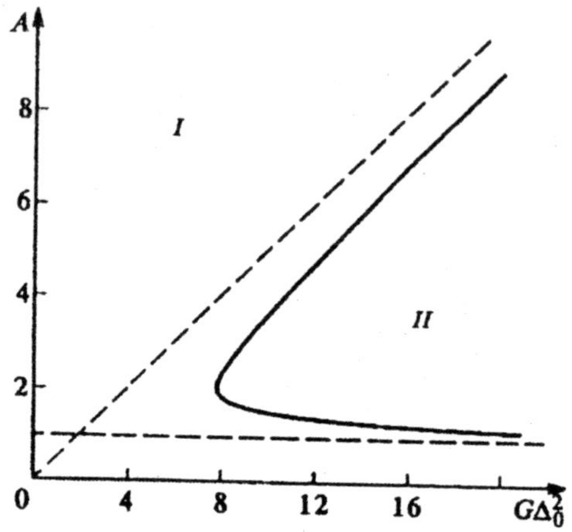

Fig. 5.5. Phase diagram of a bidirectional class B ring laser in the parameter plane $(G\Delta_0^2, A)$.

gime of the second kind, as it is called in Ref. [422], is due to the sign reversal of the damping rate of phase-sensitive relaxation oscillations (a Hopf bifurcation). Second, the coexistence of a several oscillatory processes with incommensurate frequencies is a prerequisite for deterministic chaos. The switching frequency grows with detuning, and the irregularity is enhanced as the switching frequency approaches Ω_1.

Experimental investigation of the time-dependent behaviour of a CO_2 ring laser with cavity detuning is reported in [155,156]. The results are reduced to the borderlines that are marked between the domains with different time-dependent processes in the parameter plane (discharge current, gas pressure), i.e., (pumping, atomic system relaxation rate). Besides the theoretically predicted self-modulation of the second kind, a regime of regular synchronous pulsations is revealed. Between these domains one can see a zone with pronounced irregular dynamics. According to [413] the necessary condition for existing the regime of regular synchronous pulsations is existence of asymmetric linear coupling of counter-running waves.

5.2.3 Influence of Fine Structure of the Gain Line on the Stability of the Steady States

The primary cause of the instability, which leads to self-modulation of the second kind, is the Bragg scattering of the waves from the wave-induced inversion grating. Dynamically, an important role is played by the phase shift of the scattered component. That is why the asymmetry of the gain line with respect to laser frequency is a necessary condition for the instability. In other words, the real part of the atomic system susceptibility at

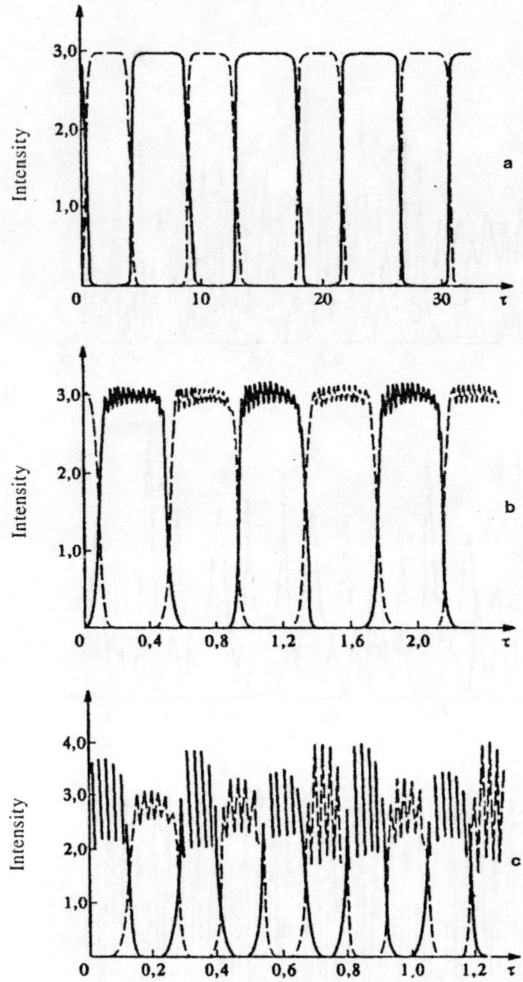

Fig. 5.6 a,b,c.

the laser frequency must be nonvanishing. In a homogeneously broadened laser this condition is ensured by controlling detuning of the cavity mode from the atomic line centre. This effect is easily achieved in low-pressure gas lasers where the gain line is narrow and the intermode frequency spacing can be large compared with the gain line. By contrast, solid-state lasers characteristically have broad gain lines, such that detuning within the limits required for instability is feasible only with use of special resonators containing controlled frequency selecting elements [424]. These a conditions under which the frequency dependence of the bidirectional ring Nd:YAG laser action was experimentally demonstrated [425].

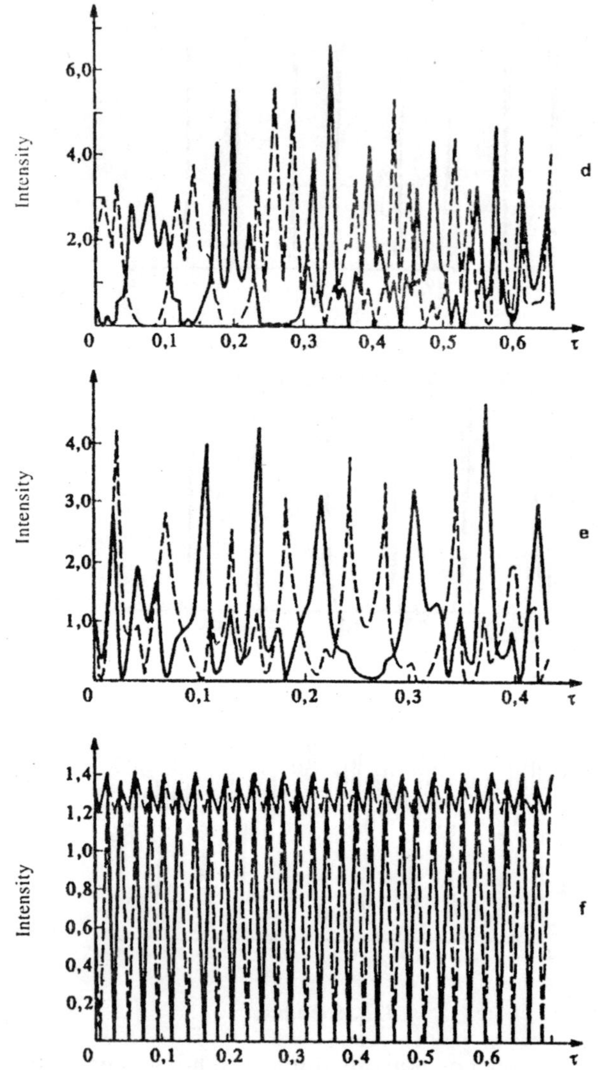

Fig. 5.6*a,b,c,d,e,f*. Examples of numerical solutions to Eqs. (5.9) for the parameter values: $G = 10^4$; $\beta_\pm = r_\pm = \Delta_c^\pm = 0$; $A = 4.0$; $\Delta = 0.05$ (*a*); 0.2 (*b*); 0.4 (*c*); 0.6 (*d*); 0.9; (*e*); 1.1 (*f*) [417].

However, Nd:YAG lasers exhibit unstable behaviour when operated with a nonselective resonator [113, 157, 158]. This requires a spectral asymmetry which is present here since the neodymium gain line is not homogeneously broadened but consists of two overlapping components. First of all, near $\lambda = 1.064$ μm there is located not one but two luminescence lines of the neodymium ions implemented in the YAG matrix, [43, 426]. The idea that the self-modulation regime of the second kind takes place due to the existence of fine structure of the gain line was first proposed in [427] and then developed in [422, 428]. At room temperature the distance between these components is of the order of the halfwidth of each line and the line intensities differ by about a factor of three. The asymmetry introduced by the weaker component plays, from the stability point of view, the same role as the cavity detuning. The laser frequency coincides with neither the centre frequency of the strong component, nor with the centre frequency of the additional spectral component. This impedes a simple separation of the functions of the two components in the sense that the strong component is responsible only for the gain while the weak one influences only the stability. Although there is a major contribution to the gain by the strong line, the wave interaction occurs through both inversion gratings, which makes the analysis somewhat more complicated.

Generalization of the class *B* ring laser equations to include a two-component medium is straightforward:

$$\frac{\mathrm{d}f_\pm}{\mathrm{d}\tau} - i\frac{G}{2}\Delta_c^\pm = \frac{G}{2}\left\{(\tilde{F}n_0 + \tilde{F}'n_0' - 1)f_\pm + [\tilde{F}n_{\pm 2} + \tilde{F}'n_{\pm 2}' + ir\exp(\mp i\vartheta_\mp)]f_\mp\right\},$$

$$\frac{\mathrm{d}n_0}{\mathrm{d}\tau} = A - n_0 - \tilde{L}[n_0(|f_+|^2 + |f_-|^2) + n_2 f_+^* f_- + n_{-2} f_+ f_-^*],$$

$$\frac{\mathrm{d}n_2}{\mathrm{d}\tau} = -n_2 - \tilde{L}[n_2(|f_+|^2 + |f_-|^2) + n_0 f_+ f_-^*], \tag{5.18}$$

$$\frac{\mathrm{d}n_0'}{\mathrm{d}\tau} = A' - n_0' - \tilde{L}'[n_0'(|f_+|^2 + |f_-|^2) + n_2' f_+^* f_- + n_{-2}' f_+ f_-^*],$$

$$\frac{\mathrm{d}n_2'}{\mathrm{d}\tau} = -n_2' - \tilde{L}'[n_2'(|f_+|^2 + |f_-|^2) + n_0' f_+ f_-^*].$$

The printed symbols indicate the quantities related to the weaker component of the gain line. Besides the variables, they include the parameters: $\Delta_0' = (\omega - \omega_0')/\gamma_\perp$ is the laser frequency detuning from the weak component centre, $\tilde{F}' = (1 - i\Delta_0')^{-1}$ is the function of the weak line form, $\Delta' = (\omega_0 - \omega_0')/\gamma_\perp$, $\tilde{L}' = \mathrm{Re}\,\tilde{F}'$, and $r = |\rho|$.

As for the homogeneously broadened model, Eqs. (5.18) have two time-independent solutions corresponding to the opposite travelling wave laser beam directions. In the limit $r = 0$ the following asymptotic form is valid

$$\bar{f}_+ = \bar{m}^{1/2}\exp(-i\dot{\varphi}_+\tau), \quad \bar{f}_- = 0, \quad \dot{\varphi}_+ = \frac{G}{2}\left(\Delta_c^+ + \frac{\bar{n}_0\Delta_0}{1+\Delta_0^2} + \frac{\bar{n}_0'\Delta_0'}{1+\Delta_0'^2}\right); \tag{5.19}$$

$$f_+ = 0, \quad f_- = \overline{m}^{1/2}\exp(-i\dot{\varphi}_-\tau), \quad \dot{\varphi}_- = \frac{G}{2}\left(\Delta_c^- + \frac{\overline{n}_0\Delta_0}{1+\Delta_0^2} + \frac{\overline{n}_0'\Delta_0'}{1+\Delta_0'^2}\right). \tag{5.20}$$

The expressions

$$\overline{m} = \frac{1}{2\tilde{L}\tilde{L}'}\left\{A + A')\tilde{L}\tilde{L}' - \tilde{L} - \tilde{L}' + \sqrt{[(A+A')\tilde{L}\tilde{L}' - \tilde{L} - \tilde{L}']^2 + 4\tilde{L}\tilde{L}'(A\tilde{L} + A\tilde{L}' - 1)}\right\} \tag{5.21}$$

$$\overline{n}_0 = A(1+\tilde{L}\overline{m})^{-1}, \quad \overline{n}_0' = A'(1+\tilde{L}'\overline{m})^{-1}, \quad \overline{n}_2 = \overline{n}_2' = 0,$$

do not depend on the direction of propagation. Linearization of Eq. (5.18) near one of the fixed points leads to a set of equations that can be decomposed into two closed subsets. Only that which corresponds to perturbation of the variables of zero steady states values contains a potential instability. Taking for specificity Eqs. (5.20) as the initial state we arrive at the linear subset of equations

$$\begin{aligned}
\frac{df_+}{d\tau} &= \frac{G}{2}(i\Delta_c^+ + \tilde{F}n_2 + \tilde{F}n_2')\bar{f}_-, \\
\frac{dn_2}{d\tau} &= -n_2(1+\tilde{L}m) - \tilde{L}n_0\bar{f}_-^* f_+, \\
\frac{dn_2'}{d\tau} &= -n_2'(1+\tilde{L}'m) - \tilde{L}'n_0'\bar{f}_-^* f_+.
\end{aligned} \tag{5.22}$$

This yields to a cubic characteristic equation $a_j\lambda^{3-j} = 0$ with the coefficients ($\Delta_c^\pm = 0$)

$$\begin{aligned}
a_0 &= 1, \quad a_1 = 2 + \overline{m}(\tilde{L} + \tilde{L}'), \quad a_2 = \frac{1}{2}G\overline{m}(\tilde{F}\tilde{L}\overline{n}_0 + \tilde{F}'\tilde{L}'\overline{n}_0'), \\
a_3 &= \frac{1}{2}G\overline{m}[\tilde{F}'\tilde{L}'\overline{n}_0'(1+\tilde{L}\overline{m}) + \tilde{F}\tilde{L}\overline{n}_0(1+\tilde{L}'\overline{m})]
\end{aligned} \tag{5.23}$$

Based on the assumption that the complex roots of the characteristic equation satisfy the condition $|\mathrm{Re}\lambda/\mathrm{Im}\lambda| \ll 1$, we calculate the approximate values of the relaxation oscillation frequencies

$$\Omega_{A,B} = \mathrm{Im}\lambda_{1,2} = \sqrt{\frac{1}{2}G\overline{m}(\tilde{L}^2\overline{n}_0 + \tilde{L}'^2 n_0')} \tag{5.24}$$

and find the instability condition of the travelling wave solution

$$\Omega_{A,B}^2\,\mathrm{Re}a_1 + \Omega_{A,B}\,\mathrm{Im}a_2 - \mathrm{Re}a_3 \geq 0. \tag{5.25}$$

At $\Delta'=0$ the expressions (5.24) and (5.25) coincide with Eqs. (5.15) and (5.16).

The boundaries of the instability domains in the phase plane Δ_0,Δ', calculated by Eqs. (5.24) and (5.25), are given in Fig. 5.7. The position of the resulting gain line maximum is also indicated. It almost coincides with the steady-state laser frequency since the longitudinal mode, the nearest to the maximum, is generated.

In the situation, shown in Fig. 5.7*a*, the instability is absent. Nevertheless, an instability can be obtained by a relatively small variation of the parameters. It should be borne in mind that the parameter G can be determined from experiments accurate to within 30 per cent [429] and the ratio A'/A is estimated as 1/3 with still a higher accuracy [43, 426]. The spectral linewidths are known with lower accuracy. The ratio $\gamma_\perp/\gamma'_\perp$ is evaluated as 1.06–1.14 using the experimental data from [426] and as 1.24 from [430].

Figure 5.7b shows the result of calculation of instability domains with the unequal $\gamma_\perp$ and $\gamma'_\perp$. The remainder of the parameters are chosen the same as in Fig. 5.7a. The time-independent solution is unstable in the range Δ'= 0.4-0.8. The measured value Δ'= 0.6 enters exactly this range [426]. Thus, the proposed interpretation of the self-modulation regime of the second kind of a Nd:YAG laser as resulting from the interaction of the counter-running waves by two self-induced inversion gratings seems quite realistic. This is confirmed by the form of the solution (Fig. 5.8), which is similar to that obtained in the presence of detuning in a homogeneously broadened medium (see Fig. 5.6).

There is another, more universal mechanism, which is equivalent to detuning of laser frequency from the gain line centre. This mechanism is based on the distinction in polarizabilities of the impurity ions being on different energy levels. We have mentioned about this phenomenon in Sec-

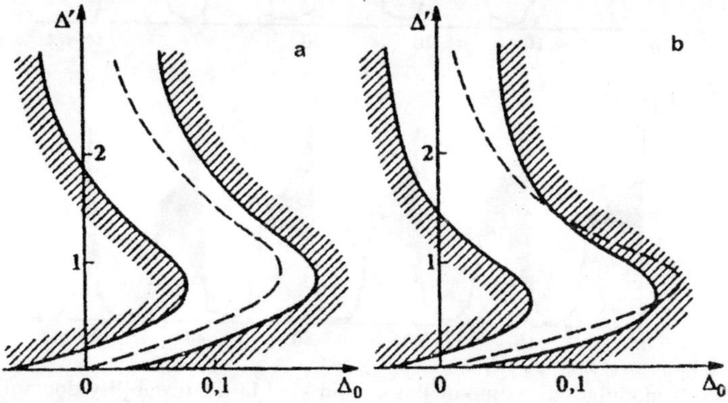

Fig.5.7. Phase diagram for the model of ring laser with a two-component gain line: $G=5000$; $A=1.2$; $A'=0.4$; $\gamma_\perp/\gamma'_\perp=1$ (*a*), 1.3 (*b*) [428].

tion 5.1. An important contribution to this phenomenon is made also by the transitions that are situated from the upper laser level at a distance much greater than the linewidth. These transitions influence the refractive index but do not change the gain.

In semiconductors this circumstance plays a very important role, but in crystals such as Nd:YAG we also cannot ignore it.

In the theory of the ring laser, the α-factor is taken into account in the same manner as in the theory of the Fabry–Perot laser, and the corresponding model has the form [431]

$$\frac{df_+}{d\tau} = \frac{1}{2}G[(1+i\alpha)(f_+n_0 + f_-n_2) - f_+], \tag{5.26a}$$

$$\frac{df_-}{d\tau} = \frac{1}{2}G[(1+i\alpha)(f_-n_0 + f_+n_2^*) - f_-], \tag{5.26b}$$

$$\frac{dn_0}{d\tau} = A - (1 + |f_+|^2 + |f_-|^2)n_0 - f_+^* f_- n_2 - f_+ f_-^* n_2^*, \tag{5.26c}$$

$$\frac{dn_2}{d\tau} = -(1 + |f_+|^2 + |f_-|^2)n_2 - f_+ f_-^* n_0 - d_{\text{dif}} n_2. \tag{5.26d}$$

Equation (5.26) contains a term responsible for the diffusion of the active

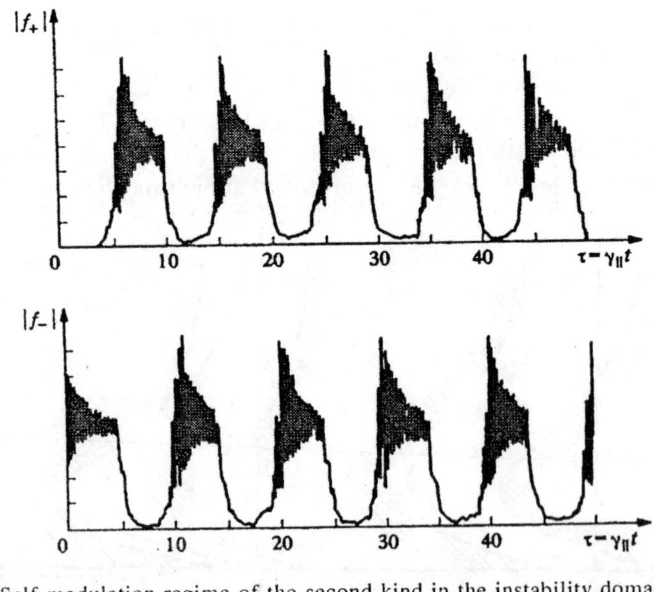

Fig.5.8. Self-modulation regime of the second kind in the instability domain of a laser with a two-component gain line, found by numerically solving Eqs. (5.18): $G = 5000$; $A = 1.2$; $\Delta' = 1.0$; $r = 10^{-3}$; $\vartheta = 0$.

centres, which makes it possible to influence (at least theoretically) the inversion grating amplitude.

In addition to two nontrivial steady-state travelling wave solutions

$$\bar{f}_+ = \overline{m}^{1/2}\exp(-i\dot{\varphi}\tau), \quad \bar{f}_- = 0$$
$$\bar{f}_+ = 0, \quad \bar{f}_- = \overline{m}^{1/2}\exp(-i\dot{\varphi}\tau), \tag{5.27}$$
$$\dot{\varphi} = \frac{1}{2}G\alpha, \quad \bar{n}_0 = 1, \quad \bar{n}_2 = 0, \quad \overline{m} = A - 1,$$

we must write one more solution corresponding to the regime of generation of two counter-running waves

$$\bar{f}_+ = \overline{m}^{1/2}, \quad \bar{f}_- = \overline{m}^{1/2}\exp(-i\dot{\varphi}\tau), \quad \bar{n}_2 = |\bar{n}_2|\exp(-i\psi),$$
$$\bar{n}_0 = A - 2\overline{m}, \quad |\bar{n}_2| = \bar{n}_0 - 1, \quad \varphi - \psi = \pi, \tag{5.28}$$
$$\overline{m} = \frac{1}{4}\left[A - 2d_{\text{dif}} - 4 + \sqrt{A^2 + 4Ad_{\text{dif}} + 4d_{\text{dif}}^2 + 8d_{\text{dif}} + 8}\right].$$

For zero values of the diffusion coefficient and the α-factor the steady state corresponding to the standing wave solution, as was mentioned above, is unstable [419, 420].

The stability analysis of each travelling wave solution gives results that generalize the results obtained in Section 5.2.1. Besides the main relaxation oscillation, there are two phase-sensitive oscillations with the following frequencies and damping rates:

$$\Omega_{A,B}^2 = \frac{1}{2}\left\{\frac{1}{2}\Omega_1^2 - \frac{1}{4}(A+d_{\text{dif}})^2 + \sqrt{\left[\frac{1}{2}\Omega_1^2 - \frac{1}{4}(A+d_{\text{dif}})^2\right]^2 + \frac{\alpha\Omega_1^2}{2}}\right\}, \tag{5.29}$$

$$\theta_{A,B} = -\frac{1}{2}(A + d_{\text{dif}}) \pm \frac{\alpha\Omega_1^2}{4\Omega_A}. \tag{5.30}$$

Both phase-sensitive relaxation oscillations are frequency-degenerate. They lose their stability when the inequality

$$\alpha > \alpha_{\text{cr}} = \frac{2(A + d_{\text{dif}})}{\sqrt{2G(A-1)}} \tag{5.31}$$

is fulfilled. For $d_{\text{dif}} = 0$ Eqs. (5.29) and (5.30) transform in Eqs. (5.15) and (5.16), respectively.

Dependences expressed by Eqs. (5.29) and (5.30) are shown in Fig. 5.9 for two values of the α-factor. For $\alpha = 0$, the phase-sensitive relaxation oscillations are completely frequency-degenerate in the whole domain of their existence. Exceeding of the diffusion coefficient of some critical value $d^{(0)}_{1cr}$ results in disappearance of these oscillations with transformation of the complex roots in the real ones. The picture drastically changes when $\alpha \neq 0$: the degeneration in the dumping rate is eliminated regardless of the value of the diffusion coefficient, while the frequency degeneration of the relaxation oscillations remains. In the critical point

$$d^{(1)}_{1cr} = \frac{1}{2}\alpha_{cr}\sqrt{G(A-1)} - A \tag{5.32}$$

the quantity θ_A reverses in sign. In the domain $d_{dif} < d^{(1)}_{1cr}$ the unidirectional steady state becomes unstable and gives way to either the undamped pulsations or the bidirectional generation.

The linear stability analysis of the steady state (5.28) leads to a characteristic equation of the seventh order, which can be investigated only by numerical methods. Examples of calculated dependencies of frequencies and damping rates are given in Fig. 5.10. Only two relaxation oscillations are inherent to the bidirectional regime of operation. The frequency of the main oscillation is close to W_1 changing a little depending on the values of d_{dif} and a. The phase-sensitive relaxation oscillation belongs to antiphase dynamics. Its frequency decreases when the diffusion coefficient grows

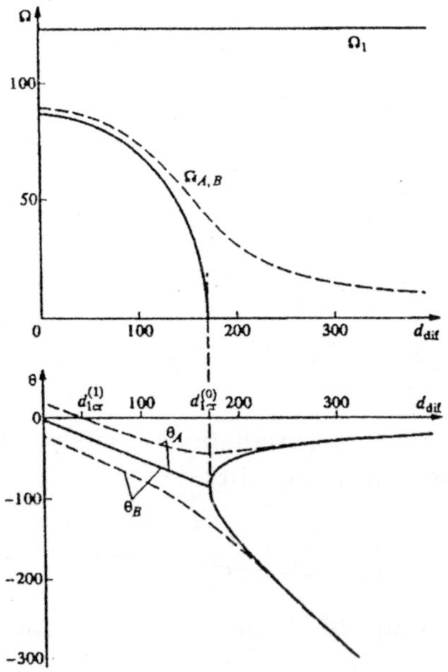

Fig.5.9. Dependence of the frequencies and decrements of relaxation oscillations near the steady-state solution (5.27) on the diffusion coefficient at G = 5000, A = 4. Solid lines correspond α = 0, while strip lines to α = 0.2 [431].

and becomes zero in a point $d = d_{2cr}^{(0)}$ where the complex characteristic roots transform in the real ones. Decrement θ becomes zero at $d_{2cr}^{(3)}$. Below this point the decrement is positive, which corresponds to undamped antiphase pulsations of counter-running wave intensities. In the interval $d_{2cr}^{(2)} < d_{dif} < d_{2cr}^{(3)}$ the regime of bidirectional operation remains stable but has only one relaxation oscillation.

Figure 5.11 represents the regionalization of the α, d_{dif} parameter plane into domains with different laser behaviour. Under the solid straight line (5.31) there is the region of stable generation of a travelling wave. The region of two-wave generation is situated in the acute angle between two dashed curves. The vertical orientation of the left border means that $d_{2cr}^{(1)}$ is practically independent on α. The transition to instability in the moment of the border intersection is realized through the Hopf bifurcation. Hatching on the phase diagram indicates zones of bistability: depending on the initial conditions, either single-wave or two-wave operation is realized to the right from $d_{2cr}^{(1)}$, whereas to the left from this point the stable unidirectional and unstable bidirectional operations co-exist.

From given in this section consideration it follows that the parameters of the inversion gratings burned in the active medium by the joint action of the interacting modes influence in some sense the dynamical behaviour of the laser. Theoretically, it is possible to influence the grating amplitude

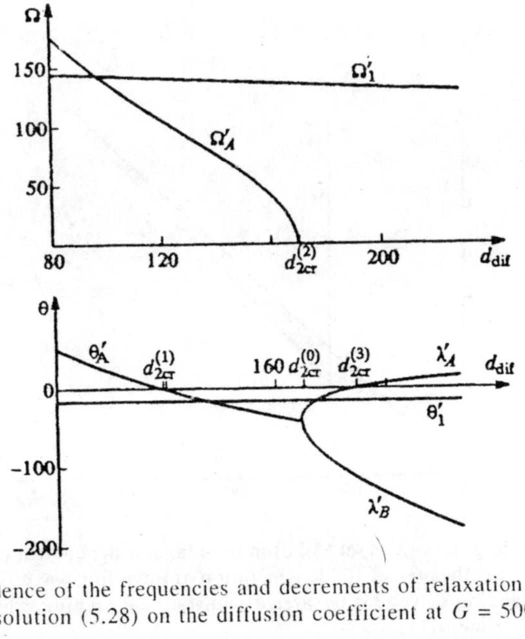

Fig.5.10. Dependence of the frequencies and decrements of relaxation oscillations near the steady-state solution (5.28) on the diffusion coefficient at G = 5000, A = 4, α = 2 [431].

by means of the diffusion coefficient, but the practical possibility of this method is very problematic. More simple is to control the α-factor using its dependence on the detuning.

5.2.4 Steady States in the Presence of Backscattering and Their Stability

In this section the form of the equations rewritten in terms of the real amplitude and phase is more useful. The transition to this form from Eqs. (5.9) is made by way of the relations

$$f_{\pm} = F_{\pm}\exp(i\varphi_{\pm}), \quad n_2 = n_{2r}\exp(i\varphi_2), \quad \Delta_{\rm NR} = \Delta_{\rm c}^{+} - \Delta_{\rm c}^{-},$$

$$\Phi_1 = \varphi_{+} - \varphi_{-} + \vartheta_{+}, \quad \Phi_2 = \varphi_{+} - \varphi_{-} - \varphi_2, \quad \vartheta = \vartheta_{+} - \vartheta_{-}.$$

The resulting equations are given by

$$\frac{dF_+}{d\tau} = \frac{G}{2}[(\tilde{L}n_0 - 1 - \beta_+)F_+ + \tilde{L}n_{2r}F_-(\cos\Phi_2 + \Delta_0\sin\Phi_2) + r_-F_-\sin(\Phi_1 - \vartheta)],$$

$$\frac{dF_-}{d\tau} = \frac{G}{2}[(\tilde{L}n_0 - 1 - \beta_-)F_- + \tilde{L}n_{2r}F_+(\cos\Phi_2 - \Delta_0\sin\Phi_2) - r_+F_+\sin\Phi_1],$$

(5.33a)

$$\frac{dn_0}{d\tau} = A - n_0 - \tilde{L}[(F_+^2 + F_-^2)n_0 + 2n_{2r}F_+F_-\cos\Phi_2],$$

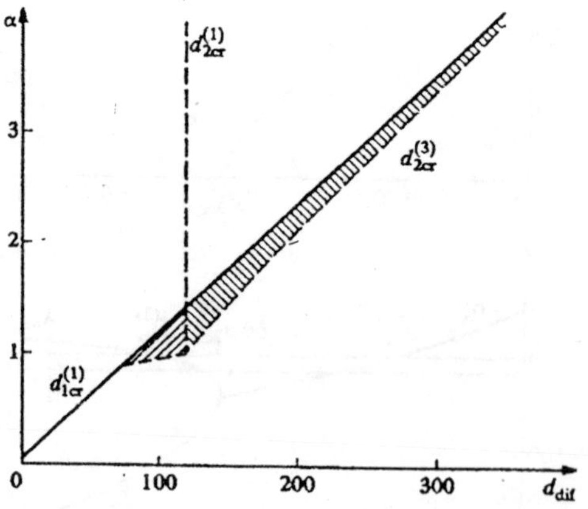

Fig.5.11. Phase diagram of the set (5.26) in the plane of the control parameters $d_{\rm dif}$, α, at $G = 5000$, $\alpha = 4$: *1* is the area of stable generation of travelling wave: *2* is the area of stable generation of standing wave; *3* is the area of nonstationary regimes; hatching corresponds to the bistability zone [431].

$$\frac{dn_{2r}}{d\tau} = A - n_{2r} - \tilde{L}[(F_+^2 + F_-^2)n_{2r} + n_0 F_+ F_- \cos\Phi_2],$$

$$\frac{d\Phi_1}{d\tau} = -\frac{G}{2}\left\{\Delta_{NR} - \tilde{L}n_{2r}\left[\left(\frac{F_-}{F_+} - \frac{F_+}{F_-}\right)\Delta_0 \cos\Phi_2 + \left(\frac{F_-}{F_+} + \frac{F_+}{F_-}\right)\sin\Phi_2\right]\right.$$

$$\left. + r\left[\frac{F_-}{F_+}\cos(\Phi_1 - \vartheta) - \frac{F_+}{F_-}\cos\Phi_1\right]\right\}, \tag{5.33b}$$

$$\frac{d\Phi_2}{d\tau} = \frac{d\Phi_1}{d\tau} + \frac{n_0}{n_{2r}} F_+ F_- \sin\Phi_2 .$$

We now simplify our consideration by confining ourselves to the particular case of neither detuning nor nonreciprocity [158–160, 213, 214, 432, 433]. Owing to linear coupling, the steady states cannot exist in the form of pure travelling waves. However, the deviations from these ideal solutions for $r \ll 1$ are small enough and the time-independent solutions can be represented in the form of series expansion in terms of the small parameter r. The presence of weak coupling is unnecessary for the conclusion that the regimes with equal representation of both waves are unstable. The steady-state solutions with considerably unequal mode amplitudes (for definiteness, $F_+ >> F_-$) for $\Delta_{NR} = \Delta_0 = 0, \beta_+ = \beta_- = 0, r_+ = r_- = r$ are expressed by

$$\overline{F}_+^2 = F_0^2 - r^2 A[\eta + (1 - F_0^2)/F_0^2],$$

$$\overline{F} = r_- A/F_0, \quad \overline{n}_0 = 1 + \eta r^2, \quad \overline{n}_{2r} = r, \tag{5.34}$$

$$\cos\overline{\Phi}_2 = -1, \quad \sin(\overline{\Phi}_1 - \vartheta) = -1 .$$

Here $F_0^2 = A - 1, \eta = A(1 + \cos\vartheta)/F_0^2$.

Linearization of Eqs. (5.33) near the steady state (5.34) leads to a characteristic equation

$$U_2(\lambda)[V_2(\lambda)W_2(\lambda) + r^2 A^{-1} G F_0^2 Z(\lambda)] = 0, \tag{5.35}$$

where $U_2(\lambda)$, $V_2(\lambda)$, $W_2(\lambda)$ and $Z_2(\lambda)$ represent quadratic polynomials:

$$U(\lambda) = \lambda^2 + \left(A - \frac{G}{2}\eta r^2\right)\lambda + \frac{G}{2}F_0^2 + \frac{G^2 r^2}{4F_0^2}(3 + \cos\vartheta),$$

$$V(\lambda)=\lambda^2+\left(A-\frac{G}{2}\eta r^2\right)\lambda+\frac{G}{2}F_0^2-\frac{G}{2}A\eta r^2,$$

$$W(\lambda)=\lambda^2+\left(A-\frac{G}{2}\eta r^2\right)\lambda+GF_0^2-\frac{G}{2}A\eta r^2, \tag{5.36}$$

$$Z_2(\lambda)=(5A-4)\lambda^2+(A^2-6A+6)\lambda-4AF_0^2(A^2-3A+1).$$

The characteristic equation (5.35) is decomposed into two parts. The first equation has roots

$$\lambda_{1,2}=-\frac{A}{2}\left[1-\frac{A}{2(A+1)}\frac{r^2}{r_{cr}^2}\right]\pm i\frac{\Omega_1}{\sqrt{2}}. \tag{5.37}$$

The fourth-order equation

$$V_2(\lambda)W_2(\lambda)+r^2A^{-1}GF_0^2Z_2(\lambda)=0$$

adds two more pairs of complex roots:

$$\lambda_{3,4}=-\frac{A}{2}\left(1-\frac{r^2}{r_{cr}^2}\right)\pm i\frac{\Omega_1}{\sqrt{2}}, \tag{5.38}$$

$$\lambda_{5,6}=-\frac{A}{2}\left[1+(A+1)\frac{r^2}{r_{cr}^2}\right]\pm i\Omega_1. \tag{5.39}$$

The parameter

$$r_{cr}^2=\frac{A}{G(A+1)(1+\cos\vartheta)}F_0^2 \tag{5.40}$$

is a minimum for $\vartheta=0$ and reaches ∞ if $\vartheta=\pm\pi$.

From Eqs. (5.37)–(5.39) it is seen that oscillations with a frequency Ω_1 are damped in any case, while those with frequency $\Omega_1/\sqrt{2}$ are undamped for $r>r_{cr}$. Instability developed at $r>r_{cr}$ leads to antiphase modulation of the counter-running wave intensities (Fig. 5.12), which we call the self-modulation regime of the first kind. Near the instability threshold the oscillation frequency $\Omega\approx\Omega_1/\sqrt{2}$ and it grows proportionally to r for $r\gg r_{cr}$. This regime of pulsations is always regular [434]. It has been observed experimentally in Nd:YAG lasers (see Section 1.2.3).

In a class B ring laser under the condition $r<r_{cr}$ both symmetric steady states corresponding to unidirectional modes are stable. The transient process is taking place in oscillatory manner. So, the relaxation oscillations belong to a definite steady state. However, for $r>r_{cr}$ the self-modulation

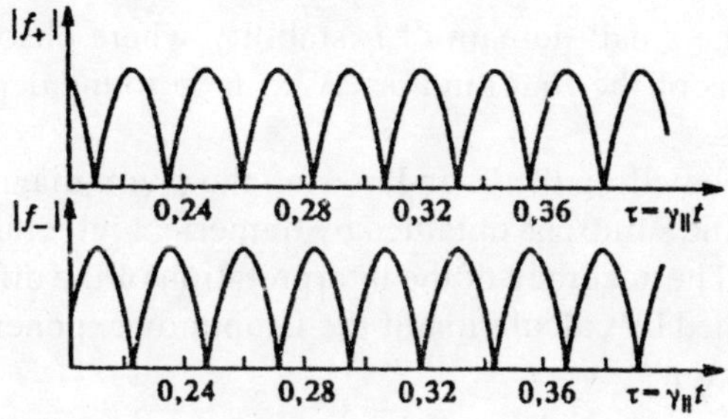

Fig.5.12. Example of numerical solution to Eqs. (5.33) for the self-modulation regime of the first kind: $G = 5000$; $A = 1.2$; $r = 0.05$; $\vartheta = 0$; $\Delta_{NR} = \Delta_0 = 0$

regime of the first kind sets in, which is also stable. In the phase space it is represented by a limit cycle. The transient process near this singular trajectory is considered in [435] where it is shown that it also has oscillatory behaviour. In this case one also has right to speak about relaxation oscillations.

5.2.5 Competition of two instability mechanisms

In the previous sections we have considered two ideal situations. In one of them detuning is absent and the laser behaviour is governed exclusively by the coupling of the waves via spontaneous scattering due to microinhomogeneities. In the other, there is no spontaneous scattering but the nonlinear wave scattering via the wave-induced inversion grating (induced scattering) becomes an important dynamical factor in the presence of detuning. In general, both factors are present, so that a competition (or reinforcement) between two instability mechanisms occurs [434, 436]. This general case is shown by the phase diagram in Fig. 5.13 in the plane of the control parameters (r, Δ_0). Note, first of all, the solid line, above which there is the domain of time-dependent solutions. To the right from the dash-dotted line there is the domain of self-modulation of the first kind with regular (periodic) solutions. The dash line giving the boundary between the domains of the second kind of instability and of chaotic behaviour is approximate since there is no sharp transition from one to the other.

The overlapping domains of laser bistability (hatched regions) add to the complexity. Stable steady solutions and self-modulation processes of the first kind coexist in the domain between s, r_0 and r_{cr}, chaotic and regular pulsations coexist in a narrow vertical band, chaos and steady-state solutions coexist in the region on the left of the sr_0 line. Self-modulation of the second kind smoothly connects to stable unidirectional solution since the switching frequency of the beam direction decreases continuously and infinitely as Δ_0 approaches Δ_{cr} and grows tending to Ω_0 as $r \to r_0$. Of further interest is the small domain of tristability where chaotic pulsations, regular pulsations of the first kind or CW can be found depending on the initial conditions.

The basis for labelling the boundaries on the phase diagram (Fig. 5.13) was the form of the solutions obtained by numerical integration of the ring laser equations. The accuracy of the interpretation of the different regimes has been confirmed by calculation of the Lyapunov exponents and dimension of the attractor.

5.2.6 Role of Phase Nonreciprocity in the Dynamics of a Class *B* Ring Laser

The prospect of making account of all three factors contributing to ring laser dynamics – detuning, backscattering and phase nonreciprocity –makes an analytical investigation of the model unrealistic. We therefore neglect detuning in the following treatment. If phase nonreciprocity is included in the analysis, then the steady-state solutions increase in complexity and are transformed from Eq. (5.34) into

$$\bar{F}_+^2 = F_0^2 - r^2 A\left(\eta + F_0^2 \frac{1-F_0^2}{F_0^4 + A^2\Delta_{NR}^2}\right), \quad \bar{F}_- = r\frac{AF_0}{(F_0^4 + A^2\Delta_{NR}^2)^{1/2}},$$

$$\bar{n}_0 = 1 + r^2\eta, \quad \bar{n}_{2r} = r\frac{F_0^2}{(F_0^4 + A^2\Delta_{NR}^2)^{1/2}}, \quad \cos\bar{\Phi}_2 = -1, \tag{5.41}$$

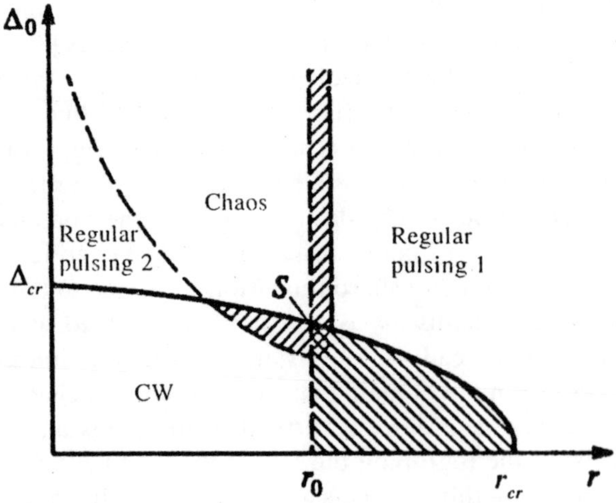

Fig. 5.13. Phase diagram of a ring laser plotted in the control parameter plane (r, Δ_0): $G = 5000$; $A = 1.2$; $\vartheta = 0$; $\Delta_{NR} = 0$.

$$\sin(\overline{\Phi}_1-\vartheta)=-\frac{F_0^2}{(F_0^4+A^2\Delta_{NR}^2)^{1/2}}\left[1-r^2\frac{A^3\Delta_{NR}}{(F_0^4+A^2\Delta_{NR}^2)^{1/2}}(F_0^2\sin\vartheta-A\Delta_{NR}\cos\vartheta)\right],$$

$$\eta=-\frac{A}{F_0^4+A^2\Delta_{NR}^2}\left[F_0^2(1+\cos\vartheta)-A\Delta_{NR}\sin\vartheta\right].$$

The generalized characteristic equation (5.35) takes the form

$$U_2(\lambda)\left[V_2(\lambda)W_2(\lambda)+r^2\frac{GF_0^2}{A}Z_2(\lambda)\right]+\frac{G^2\Delta_{NR}^2}{4}\left\{(A+\lambda)^2W_2(\lambda)\right.$$

$$\left.+\frac{r^2}{2}\frac{GA^2}{F_0^4+A^2\Delta_{NR}^2}\left[U_3(\lambda)+\left(\cos\vartheta-\frac{F_0^2}{A\Delta_{NR}}\sin\vartheta\right)U_4(\lambda)\right]\right\}=0.$$

(5.42)

Here $U_3(\lambda)$ and $U_4(\lambda)$ are third- and fourth-order polynomials, respectively. In the limit $r\to 0$ Eq. (5.42) again decomposed into two equations. The roots of the quadratic equation $W_2(\lambda)=0$ are known:

$$\lambda_{5,6}=-\frac{A}{2}\pm i\Omega_1.$$

The remaining fourth-order equation

$$\left(\lambda^2+A\lambda+\frac{G}{2}F_0^2\right)^2+\left[\frac{G}{2}\Delta_{NR}(A+\lambda)\right]^2=0 \qquad (5.43)$$

yields two pairs of conjugate complex roots

$$\lambda_{1,2}=\theta_A\pm i\Omega_A,\quad \lambda_{3,4}=\theta_B\pm i\Omega_B. \qquad (5.44)$$

Neglecting the weak damping, the relaxation oscillation frequencies are given by

$$\Omega_{A,B}=\sqrt{\frac{G}{2}F_0^2+\left(\frac{G}{4}\Delta_{NR}\right)^2}\pm\frac{G}{4}\Delta_{NR}. \qquad (5.45)$$

Thus, the presence of phase nonreciprocity removes the two-fold degeneracy of the relaxation oscillation spectrum. Phase nonreciprocity also influences the damping rates of relaxation oscillations

$$\theta_{A,B}=-\frac{A}{2}\left\{1\mp\frac{G}{4}\Delta_{NR}\left[\frac{G}{2}F_0^2+\left(\frac{G}{4}\Delta_{NR}\right)^2\right]^{-1/2}\right\}. \qquad (5.46)$$

The resultant dependences are illustrated in Fig. 5.14.

Eq. (5.45) shows the new possibilities offered by class B lasers for gyroscopic applications. It appears that information on the phase nonreciprocity (rotation, in particular) of the laser cavity is contained not only in the laser field optical spectrum but also in the low-frequency relaxation oscillation spectrum. Hence, the rigid requirements for equal representation of both waves are removed. However, the problem arises: How can this information be retrieved from the laser output?

Unfortunately, there is no simple answer to this question. It seems most natural to make use of the modulation transfer function, i.e., the laser response to modulation of one parameter. Such a characteristic must have resonant features at the relaxation frequencies (for more details see Chapter 6). The resonant peaks also should be observed in the intensity fluctuation spectrum. The resonance effect for the frequency Ω_1 was theoretically predicted in [352] and experimentally substantiated in the solid-state laser investigations [429, 437]. Besides the peak at the intensity fluctuation spectrum contains peaks at other frequencies of specific relaxation oscillations of the weak wave. However, the observation of such peaks requires some additional conditions to be satisfied [173, 174, 438].

The difficulty is that the transfer function has resonant features for the

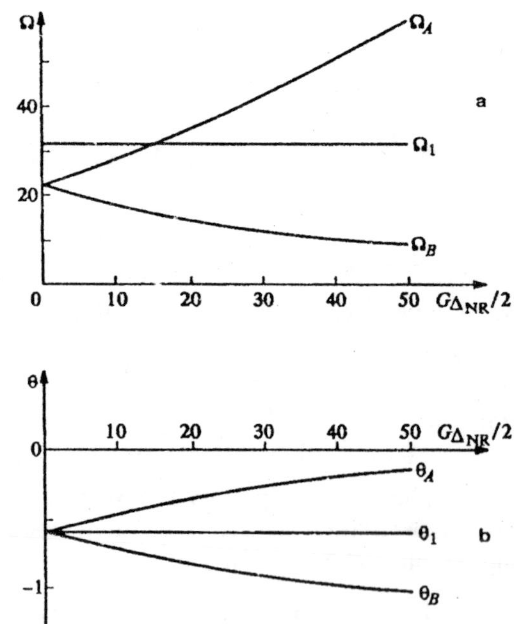

Fig.5.14. Dependence of (*a*) frequencies and (*b*) damping rates of relaxation oscillation on phase nonreciprocity: $G = 5000$; $A = 12$; $r = \beta = 0$.

frequencies given in Eq. (5.45) but they are present with strength proportional to the small coefficient r^2. This makes practical observations of such peaks more difficult. This problem is completely unresolvable in the laser far from the instability threshold. However, regenerative amplification of oscillations at the system eigenfrequencies is evident near the instability threshold, leading, as is well known, to an infinite growth of resonance peaks with the simultaneous infinite narrowing. It is the regenerative noise amplification, which makes it possible to observe the resonance features in their intensity fluctuation spectrum of a solid-state ring laser. Owing to the line narrowing, however, it is not easy to see the resonance peaks on the transfer function, since there are sharply increased stability requirements imposed both on the laser itself and on the modulation signal.

The roots of the characteristic equation (5.42) cannot be found analytically when $\Delta_{\mathrm{NR}} \neq 0$ and $r \neq 0$. In this case, however, the laser dynamics exhibits interesting features [174, 434, 436]. In the vicinity of the point $G\Delta_{\mathrm{NR}} = \Omega_1$, where two branches of relaxation oscillations, Ω_1 and Ω_A, intersect, their damping rates are subject to abrupt change. From the diagrams in Fig. 5.15, obtained by numerical methods, it is seen that there is a sign reversal in the damping rate and, therefore, a loss of stability of the time-independent solution for $r < r_{cr}$. Numerical integration of Eqs. (5.9) indicates that the self-modulation of the first kind arises. However, introducing a weak (within 10^{-4}–10^{-3}) amplitude nonreciprocity $\beta_{\mathrm{NR}} = \beta_+ - \beta_-$ converts the regular process to a chaotic one. The general situation is explained by the phase diagram in the plane of the control parameters $(G\Delta_{\mathrm{NR}}, \beta_{\mathrm{NR}})$ in Fig. 5.16 [439].

The behaviour of a Nd:YAG laser under these conditions is illustrated by the intensity fluctuation spectra obtained experimentally. Stable unidirectional lasing modulated by a weak noise is depicted in Fig. 5.17*a*. The presence of peaks other than just at Ω_1 indicates the laser is close to the instability threshold, since a single peak is observed for over most of the stable single-mode domain. A transition to the unstable domain is accompanied, as seen in Fig. 5.17*b*, by broadening of the resonance peak owing to convergence of Ω_1 and Ω_A. Such a broadening is characteristic for chaotic pulsations.

It is important to note that the strong and weak waves carry different information about the laser processes. In the intensity fluctuation spectrum of a strong wave, the peak dominates at Ω_1 and the peaks at Ω_A and Ω_B are suppressed because of smallness of the parameter r^2 and are, therefore, exhibited only near the instability threshold. By contrast, the peak at is absent in the weak wave spectrum while the conditions under which the other two peaks are manifested in the weak wave are the same as those described above.

The ring laser is sensitive not only to the absolute value but also to the

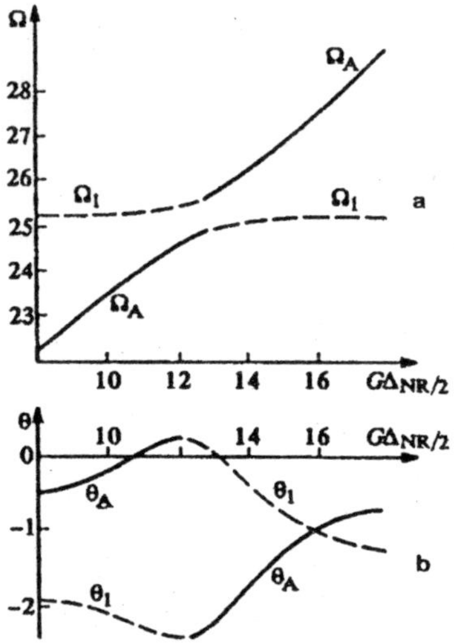

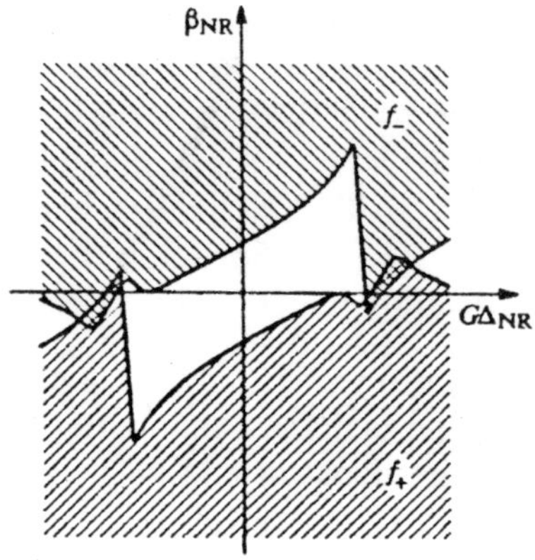

Fig.5.15. Frequencies (*a*) and damping rates (*b*) of relaxation oscillations as a function of phase nonreciprocity in the vicinity of the intersection point of branches Ω_I and Ω_A: $G = 5000$; $A = 1.2$; $r = 0.02$.
Fig.5.16 (right). Phase diagram of a ring laser in Δ_{NR}, β_{NR} plane: $G = 5000$; $A = 1.2$; $A' = 0.4$; $r = 0.005$; $\Delta_0 = 0$; $\Delta' = 0.6$; $\gamma_\perp/\gamma_\perp' = 1.3$; $\vartheta = 0$; $\Delta_0 = 0.173$.

sign of phase nonreciprocity. This property is apparent only if the gain line is asymmetric with respect to the laser frequency, as in the case of two-component gain line. If the relaxation oscillation frequency is followed by moving along the stability boundary in the phase diagram, then the result is the curve $\Omega(\Delta_{NR})$ presented in Fig. 5.18 [439]. The main feature of this dependence is the smooth transition from one branch of relaxation oscillations (Ω_A) to another (Ω_B) at point $\Delta_{NR} = 0$.

5.2.7 Frequency Dynamics of a Bidirectional Ring Laser

According to Eq. (5.11), the ring laser for $r = 0$ has two equivalent steady states in the form of travelling waves. The phases of the time-independent solutions are given by the expressions

$$\dot{\varphi}_\pm = \frac{G}{2}(\Delta_0 + \Delta_c^\pm),$$

which determine their frequencies

$$\omega_\pm = \omega_0 + \frac{\gamma_\parallel}{\tilde{\kappa}}\dot{\varphi}_\pm.$$

In the absence of phase nonreciprocity these formulas offer the same frequency shift from ω_c toward ω_0 for the two single mode solutions because of the

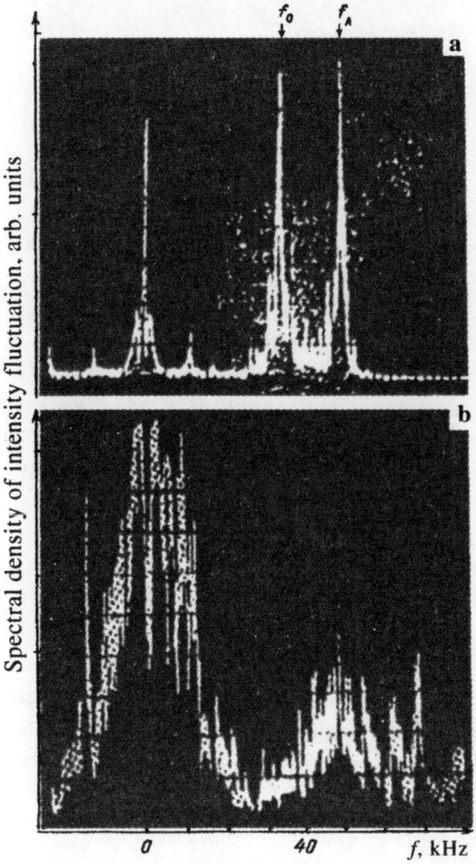

Fig.5.17. Experimental spectra of the intensity fluctuations of a ring Nd:YAG laser below (*a*) and above (*b*) instability threshold of stationary generation [174, 436].

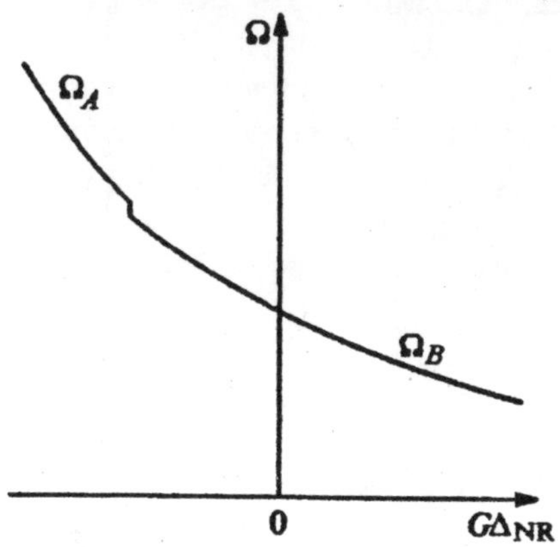

Fig.5.18. Plot of the dependence of the relaxation oscillation frequency on phase nonreciprocity to illustrate the smoothness of the transition from branch Ω_A to branch Ω_B at point $\Delta_{NR} = 0$. The dip on the plot corresponds to the point of intersection of branches Ω_A and Ω_1 [439].

linear frequency pulling.

The behaviour of frequencies under self-modulation of the second kind, which arises for $\Delta_0 > \Delta_{cr}$, can be investigated by numerical integration of Eqs. (5.9) [440]. Figure 5.19 shows the behaviour of the wave intensities and frequencies in the absence of nonreciprocity and scattering ($\Delta_{NR} = r = 0$). It is seen that as the direction is changed, the laser frequency retains its position corresponding to the steady-state single-mode solution. At the time when the wave becomes strong it forces the competing counter-running wave to a position further from the line centre. The frequency difference is $\Omega_A(\Omega_B)$ until the weak wave reaches a minimum in its intensity. At this moment its frequency changed by $\Omega_A + \Omega_B$, so that the weak wave (while it is growing) is closer to the line centre than the strong wave.

Ω_A and Ω_B are the frequencies of phase sensitive relaxation oscillations of the ring laser. These frequencies coincide in the absence of phase nonreciprocity. Note also that both waves have completely equal rights and each wave dominates exactly half the self-modulation period.

According to [440], nonzero phase nonreciprocity produces a dominant wave, the duration of which exceeds half the self-modulation period (Fig. 5.20). Simultaneously, the symmetry is broken in the frequency dynamics. In Fig. 5.20*b* it is seen that the laser frequency is closer to the line centre during the time interval when the dominant wave is strong (dashed line). In this case $\Omega_A \neq \Omega_B$ and $\Omega_A - \Omega_B = G\Delta_{NR}/2$.

Switching of the weak wave frequency by $\Omega_A + \Omega_B$ is nonmonotonic. It is preceded by growth of oscillations near the initial frequency. The frequency damping is ended by damped oscillations near the new value. The frequencies of the transient oscillations found by numerical computation coincide with $\Omega_A + \Omega_B$. It should be mentioned that a particular combina-

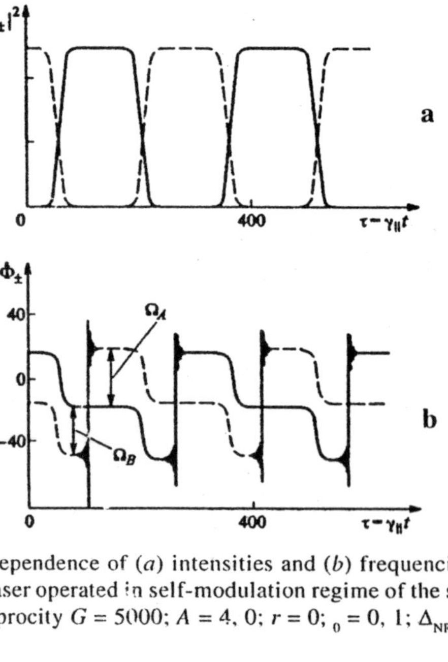

Fig.5.19. Time dependence of (*a*) intensities and (*b*) frequencies of the counter-running waves of a ring laser operated in self-modulation regime of the second kind in the absence of phase nonreciprocity $G = 5000$; $A = 4, 0$; $r = 0$; $_0 = 0, 1$; $\Delta_{NR} = 0$ [440].

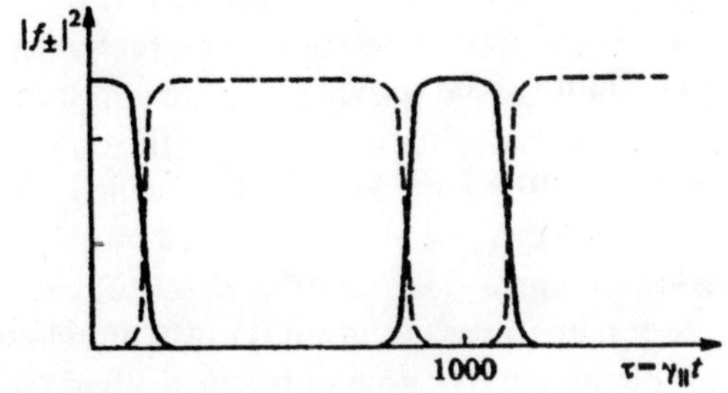

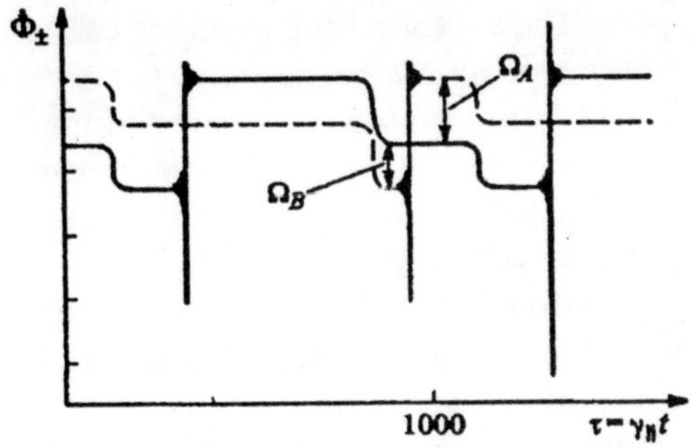

Fig. 5.20. The same as Fig. 5.19 but in presence of phase nonreciprocity ($G = 5000$; $A = 4, 0$; $r = 0$; $_0 = 0, 1$; $\Delta_{NR} = 30$) [445].

tion of the signs of Δ_{NR} and Δ_0 is not important to the dynamic process described except that it defines which wave will dominate.

Summing up our consideration of a bidirectional class *B* ring laser we would like to focus on its most typical properties:

- there are two more specific (phase-sensitive) types of relaxation oscillations in addition to the ordinary oscillations;
- there are two different instabilities of the unidirectional lasing and there are, correspondingly, two forms of self-modulation processes reached by the specific relaxation oscillations;
- there is a strong influence of both amplitude and (which is of greater interest) phase nonreciprocity on the laser dynamics and the radiation characteristics in both stable and unstable domains.

It should be emphasized that all of the conclusions of the theoretical analysis in this section have found experimental confirmation and almost all the experimental facts have received theoretical interpretation. The qualitative agreement between theory and experiment means that the laser models described here are adequate to describe the real laser systems. Thus, the class *B* ring laser has priority as one of the most promising laser systems for complex behaviour with quantitative agreement between theory and experiment. This is important both for verification of the fundamental theoretical postulates of nonlinear dynamics and for the development and verification of the new methods of retrying information about the laser parameters from the dynamic behaviour.

5.3 Vector Model of a Fibre Laser

Until the polarization remains a fixed characteristic of the radiation field the scalar model is adequate to the situation. However, when the polarization plays the role of independent degree of freedom, then the theory would be made more complex. The vector model assumed existence of two states of the field with orthogonal, generally – elliptic polarizations. So, all lasing modes are separated in two groups according to their polarization. In its turn, each such an ensemble manifested some dynamic properties and this allows using the term *polarization mode*. There are only two polarization modes. They compete like ordinary longitudinal or transverse modes through the cross-saturation of the active medium. Indeed, hole burning means deformation of the angular profile and this is possible if the distribution of the dipole moments of active centres is different from δ-function. A uniform angular distribution is peculiar to the glasses and optical fibres. Something like this takes place in semiconductor lasers with vertical cavities. The impurity ions in crystals can take the places in different lattice points, which lead to different orientations of the transition dipole moments.

So, the main features of polarization dynamics can be obtained using the vector model, which is in many aspects similar to the ordinary two-mode model. Presence of a weak birefringence prevents the frequency degeneration. Nevertheless, the theoretical model would take into attention the phase-sensitive interaction of polarization modes.

There is a vast literature devoted to polarization dynamics including [441–449], which are devoted to class B lasers. It is impossible to present a complete list of existing publications. We will refer to the Special Issue of the Quantum and Semiclassical Optics journal (v.10, #1, 1998).

Creating the vector model adequately describing the main features of the fibre laser polarization dynamics, we will proceed from the existence of two orthogonal states of elliptical polarization of the field but neglecting the variation of these states along the fibre axis. The vector field inside the cavity can be represented as a superposition of two orthogonal polarized components:

$$\mathbf{E} = (f_1\mathbf{U}_1 + f_2\mathbf{U}_2)\exp(i\omega t), \tag{5.47}$$

where

$$\mathbf{U}_{1,2} = \sqrt{2}\mathbf{n}^0_{1,2}\sin(\pi k_{1,2}\zeta) \tag{5.48}$$

are the cavity eigenfunctions, and

$$\mathbf{n}_m^0 = \frac{\mathbf{x}^0 + \gamma_m \mathbf{y}^0}{\sqrt{1 + |\gamma_m|^2}} \tag{5.49}$$

are the unit vectors, which are determined by the Cartesian components $\mathbf{x}^0$, $\mathbf{y}^0$ and the complex parameters $\gamma_{1,2}$ satisfying the relation $\gamma_1\gamma_2^* = -1$ and determining the ellipticity of the eigenfunctions

$$\varepsilon_{1,2} = \left| \tan \arcsin \frac{2\,\mathrm{Im}\,\gamma_{1,2}}{1 + |\mathrm{Im}\,\gamma_{1,2}|^2} \right|. \tag{5.50}$$

The dipole moments of the active centre transitions are suggested linearly polarized and uniformly distributed over all azimuthal angles ψ in the plane normal to the fibre axis. Interaction of the elliptically polarized field with the ensemble of randomly oriented dipoles leads to hole burning in the azimuthal inversion distribution, which can be presented in the form

$$n = n^0 + 2n^c \cos(2\psi) + 2n^s \sin(2\psi) + \dots . \tag{5.51}$$

Each angular component of the inversion n^i ($i = 0, \mathrm{c}, \mathrm{s}$) in its turn consists of the uniform along the cavity axis component and spatial harmonics:

$$n^i = n_0^i + 2n_{1,2}^i \cos[\pi(k_1 - k_2)\zeta]. \tag{5.52}$$

Only large-scale components of spatial expansion of the field are kept in Eq. (5.52) because the number of longitudinal modes in the fibre lasers is very large, which lead to smoothing of the small-scale structures in the distribution and absence of antiphase relaxation oscillations, which have been mentioned above. The difference in wave numbers k_1 and k_2 for the polarization modes can be caused by a weak birefringence in a fibre due to local mechanical tensions.

The developed ideas lead to the following fibre laser model [446]:

$$\frac{df_1}{d\tau} - i\frac{G}{2}\Delta = \frac{G}{2}[f_1(n_0^0 + c_1 n_0^c + d_1 n_0^s - 1) + f_2(Pn_{12}^c + Ln_{12}^s)],$$

$$\frac{df_2}{d\tau} + i\frac{G}{2}\Delta = \frac{G}{2}[f_2(n_0^0 + c_2 n_0^c + d_2 n_0^s - 1) + f_1(Pn_{12}^c + L^* n_{12}^s)],$$

$$\frac{dn_0^0}{d\tau} = A^0 - n_0^0(|f_1|^2 + |f_2|^2) - n_0^c(c_1|f_1|^2 + c_2|f_2|^2) - n_0^s(d_1|f_1|^2 + d_2|f_2|^2)$$

$$- n_{12}^c P(f_1^* f_2 + f_1 f_2^*) - n_{12}^s(L^* f_1 f_2^* + L f_1^* f_2),$$

$$\frac{dn_0^c}{d\tau} = A^c - n_0^c(1 + |f_1|^2 + |f_2|^2) - \frac{1}{2}n_0^0(c_1|f_1|^2 + c_2|f_2|^2) - \frac{1}{2}n_{12}^0 P(f_1^* f_2 + f_1 f_2^*),$$

$$\frac{dn_0^s}{d\tau} = A^s - n_0^s(1+|f_1|^2+|f_2|^2) - \frac{1}{2}n_0^0(d_1|f_1|^2+d_2|f_2|^2) - \frac{1}{2}n_{12}^0(L^* f_1 f_2^* + L f_1^* f_2,) \tag{5.53}$$

$$\frac{dn_{12}^0}{d\tau} = -n_{12}^0(1+|f_1|^2+|f_2|^2) - n_{12}^c(c_1|f_1|^2+c_2|f_2|^2) - n_{12}^s(d_1|f_1|^2+d_2|f_2|^2),$$

$$-\frac{1}{2}n_0^c P(f_1^* f_2 + f_1 f_2^*) - \frac{1}{2}n_0^s(L^* f_1 f_2^* + L f_1^* f_2),$$

$$\frac{dn_{12}^c}{d\tau} = -n_{12}^c(1+|f_1|^2+|f_2|^2) - \frac{1}{2}n_{12}^0(c_1|f_1|^2+c_2|f_2|^2) - \frac{1}{2}n_0^0 P(f_1^* f_2 + f_1 f_2^*),$$

$$\frac{dn_{12}^s}{d\tau} = -n_{12}^s(1+|f_1|^2+|f_2|^2) - \frac{1}{2}n_{12}^0(d_1|f_1|^2+d_2|f_2|^2) - \frac{1}{2}n_0^0(L^* f_1 f_2^* + L f_1^* f_2).$$

For simplicity we used here the following notation:

$$c_m = \frac{1-|\gamma_m|^2}{1+|\gamma_m|^2}, \quad d_m = \frac{2\,\mathrm{Re}\,\gamma_m}{1+|\gamma_m|^2},$$

$$P = \frac{1}{\sqrt{(1+|\gamma_1|^2)(1+|\gamma_2|^2)}}, \quad L = \frac{\gamma_1^* + \gamma_2}{2}P, \tag{5.54}$$

$$A^0 = \frac{1}{\pi}\int_0^\pi A\,d\psi, \quad A^c = \frac{1}{\pi}\int_0^\pi A\cos(2\psi)\,d\psi, \quad A^s = \frac{1}{\pi}\int_0^\pi A\sin(2\psi)\,d\psi.$$

The last line contains quantities describing the pump polarization produced by a semiconductor laser beam, which is focused through the end inside the fibre.

The set of Eqs. (5.53) describes the main features of dynamical behaviour of the neodymium doped fibre laser: the dependence of parameters of the output radiation on the pump polarization, presence of the resonance peaks, corresponding to antiphase relaxation oscillations, in the power spectrum of the polarization modes and its absence in the total intensity. The measured ratio of the frequencies of relaxation oscillations can be reproduced in theory by appropriately choosing the degree of ellipticity of the polarization.

Chapter 6

Lasers with Time-Dependent Parameters

Parameter modulation is an effective method to control the laser behaviour. A relatively weak periodic modulation is capable of producing a much stronger response in the laser output. Owing to the system nonlinearity this response can be not only regular but also chaotic.

Due to the high sensitivity of lasers to external modulation, uncontrolled parameter variations during the laser operation can produce random spiking.

Thus, there is considerable interest in nonautonomous laser models both for control of the emission and for interpretation of spontaneous time-dependent phenomena.

6.1 Lasers with Periodic Parameter Modulation

We will focus our discussion on the response of class B lasers to weak periodic modulations of parameters [96, 450-457]. There are four important parameters accessible for modulation in the models we have considered: $\kappa, A, \sigma_{\text{tr}}$ and $\gamma_{\parallel}$. However, the last parameter is relatively insensitive to variation of the external conditions, therefore, it is generally more interesting to study the laser behaviour when one of the other three parameters is modulated.

6.1.1 Linear Response of a Single-Mode Laser to Low-Frequency Modulation

Our discussion is based on the rate equations (3.11):

$$\frac{dm}{d\tau} = Gm(\tilde{L}n - 1)\,. \tag{6.1a}$$

$$\frac{dn}{d\tau} = A - n(\tilde{L}m + 1)\,. \tag{6.1b}$$

1. Assume first that the cavity losses are modulated harmonically and, therefore, Eqs. (6.1) are replaced by

$$\frac{dm}{d\tau} = Gm\left[\tilde{L}n - 1 - \beta_{loss}\cos(\Omega\tau)\right], \tag{6.2a}$$

$$\frac{dn}{d\tau} = A - n(\tilde{L}m + 1)\,, \tag{6.2b}$$

where $\beta_{\rm loss}$ is the depth of loss modulation.

Since Eqs. (6.2) contain time explicitly, they have no time-independent solutions. Only in the limit $\beta_{\rm loss} \to 0$ do the solutions of these equations tend to the solutions of the unperturbed Eqs. (6.1):

$$\overline{m} = \frac{A-1}{\tilde{L}}, \quad \overline{n} = \frac{1}{\tilde{L}}\,. \tag{6.3}$$

Considering the loss modulation as a perturbation we will seek Eqs. (6.2) for solution in the form

$$m = \overline{m} + \tilde{m}\exp(i\Omega\tau) + \text{c.c.}, \quad n = \overline{n} + \tilde{n}\exp(i\Omega\tau) + \text{c.c.} \tag{6.4}$$

Inserting Eqs. (6.4) into Eqs. (6.2) and neglecting all the nonlinear combinations of $\beta_{\rm loss}, \tilde{n}, \tilde{m}$ we find the transfer functions

$$\frac{\tilde{m}}{\overline{m}} = \frac{G\beta_{\rm loss}}{2}\,\frac{i\Omega + A\tilde{L}}{\Omega^2 - \Omega_1^2 - iA\tilde{L}\Omega}, \tag{6.5a}$$

$$\frac{\tilde{n}}{\overline{n}} = \frac{G\beta_{\rm loss}}{2}\,\frac{\overline{m}\tilde{L}}{\Omega^2 - \Omega_1^2 - iA\tilde{L}\Omega}\,. \tag{6.5b}$$

The quantity

$$K_{\rm loss} \equiv \frac{2}{\beta_{\rm loss}}\left|\frac{\tilde{m}}{\overline{m}}\right| = G\left|\frac{i\Omega + AL_{\rm loss}}{\Omega^2 - \Omega_1^2 - iA\tilde{L}\Omega}\right|, \tag{6.6}$$

which represents the ratio of the modulation depth of the output beam to the modulation depth of the cavity losses will be called the loss modulation amplification factor.

The perturbation can be assumed to be weak and the linear approximation to be valid if $|\tilde{m}/\overline{m}| \ll 1, \tilde{n}/\overline{n} \ll 1$, which is equivalent to

$$G\beta_{\text{loss}}\left|\frac{i\Omega+A\tilde{L}}{\Omega^2-\Omega_1^2-iA\tilde{L}\Omega}\right|<<1. \tag{6.7}$$

Relations (6.5) and (6.6) have the form of resonance. The loss modulation amplification factor reaches a maximum at $\Omega=\Omega_1=\sqrt{G(A\tilde{L}-1)}$, i.e., when the modulation frequency coincides with the frequency of relaxation oscillations. Under the condition $\Omega>>A\tilde{L}$ the relations (6.6) and (6.7) at resonance transform to

$$K_{\text{loss}}(\Omega_1)=\frac{G}{A\tilde{L}} \tag{6.8}$$

and

$$\beta_{\text{loss}}<<\beta_{\text{loss}}^{\text{cr}}=\frac{A\tilde{L}}{G}=\frac{1}{K_{\text{loss}}}, \tag{6.9}$$

respectively. For solid-state lasers the typical values are $G\approx10^4$ and $A\tilde{L}$ is of the order of unity. This means that the order of magnitude of K_{loss} is 10^4, so that a loss modulation depth $\beta_{\text{loss}}\approx10^{-4}$ is sufficient for the onset of large-amplitude nonlinear pulsations.

2. If the pumping is modulated harmonically, then Eqs. (6.1) are replaced by

$$\frac{dm}{d\tau}=Gm(\tilde{L}n-1) \tag{6.10a}$$

$$\frac{dn}{d\tau}=A[1+\beta_{pump}\cos(\Omega\tau)]-n(\tilde{L}m+1). \tag{6.10b}$$

An established solution is again sought in the form of Eqs. (6.4). Putting this solution into Eqs. (6.10) and linearizing yields

$$\frac{\tilde{m}}{\overline{m}}=\frac{G\beta_{\text{pump}}}{2}\frac{A\tilde{L}}{\Omega^2-\Omega_1^2-iA\tilde{L}\Omega}, \tag{6.11a}$$

$$\frac{\tilde{n}}{\overline{n}}=-\frac{G\beta_{\text{pump}}}{2}\frac{i\Omega\tilde{L}}{\Omega^2-\Omega_1^2-iA\tilde{L}\Omega}. \tag{6.11b}$$

The modulation amplification factor is represented in this case by the quantity

$$K_{\text{pump}}=\frac{2}{\beta_{\text{pump}}}\left|\frac{\tilde{m}}{\overline{m}}\right|=G\left|\frac{A\tilde{L}}{\Omega^2-\Omega_1^2-iA\tilde{L}\Omega}\right|. \tag{6.12}$$

Under exact resonance

$$\frac{\tilde{m}}{\overline{m}} = -\frac{iG\beta_{\text{pump}}}{2\Omega_1}, \tag{6.13}$$

$$K_{\text{pump}}(\Omega_1) = G/\Omega_1. \tag{6.14}$$

The condition for small response is now expressed by

$$\beta_{\text{pump}} << \beta_{\text{pump}}^{\text{cr}} = \Omega_1/G = \sqrt{(A\tilde{L}-1)/G} \tag{6.15}$$

The values $G = 10^4$, $A\tilde{L} = 2$ correspond to $\beta_{\text{pump}}^{\text{cr}} = 10^{-2}$.

3. The transition cross-section σ_{tr} is included in Eqs. (6.1) and elsewhere below in the line shape function $\tilde{L}$. Modulation of cross-section is thus equivalent to laser frequency (detuning) modulation. Replacing $\tilde{L}$ by $\tilde{L}[1+\beta_{\text{c-s}}\cos(\Omega\tau)]$ in Eqs. (6.1) we arrive at

$$\frac{dm}{d\tau} = Gm\{n\tilde{L}[1+\beta_{\text{c-s}}\cos(\Omega\tau)]-1\}. \tag{6.16a}$$

$$\frac{dn}{d\tau} = A - n\{m\tilde{L}[1+\beta_{\text{c-s}}\cos(\Omega\tau)]+1\}. \tag{6.16b}$$

repeating our search for linear response to weak harmonic modulation of the parameter ($\tilde{L}$ in this case) we get

$$\frac{\tilde{m}}{\overline{m}} = -\frac{G\beta_{\text{c-s}}}{2}\frac{1+i\Omega}{\Omega^2-\Omega_1^2-iA\tilde{L}\Omega}, \tag{6.17a}$$

$$\frac{\tilde{n}}{\overline{n}} = -\frac{G\beta_{\text{c-s}}\overline{m}}{2}\frac{i\Omega+G}{\Omega^2-\Omega_1^2-iA\tilde{L}\Omega}. \tag{6.17b}$$

In the limit $\Omega >> A\tilde{L}$ the relation (6.17a) coincides, except for the sign, with Eq. (6.5a) and, therefore, $\beta_{\text{c-s}}^{\text{cr}} = \beta_{\text{loss}}^{\text{cr}}$.

Of all these methods for resonance excitation of class B laser pulsations, pump modulation is the least effective. According to Eq. (6.14), the resonant pump modulation amplification factor is proportional to $\sqrt{G}$, whereas G appears in the other two cases. This is because the inversion decay rate in such lasers is noticeably less than the relaxation oscillation frequency; thus, pump modulation is transformed into inversion modulation with considerable attenuation.

It is worth noting that in a time-independent process the inversion is stabilized at a level given by the cavity losses. Low-frequency pump fluctuations are unable to disturb the inversion from its steady state value. This is seen from Eq. (6.11b), according to which $\overline{n} \to 0$ as $\Omega \to 0$.

6.1.2 Linear Response of a Multimode Laser to Low-Frequency Loss Modulation

If synchronous and proportional modulation of all modes is provided, then the result will be the same as that obtained in Section 6.1.1 for a single-mode model. Let us suppose that the losses of the separate modes can be modulated independently. The response of all of the mode intensities to the modulation of the loss of one selected mode will be considered below. This situation is described by the set of equations

$$\frac{dm_k}{d\tau} = Gm_k\left[\tilde{L}_k \int n\psi_k^2 dv - 1 - \delta_{k,p}\beta_p \cos(\Omega\tau)\right], \tag{6.18a}$$

$$\frac{\partial n}{\partial \tau} = A - n\left(1 + \sum \tilde{L}_q \psi_q^2 m_q\right), \tag{6.18b}$$

which are a direct generalization of Eqs. (4.10). If solutions are assumed in the form

$$m_k = \overline{m}_k + \tilde{m}_k \exp(i\Omega\tau) + \tilde{m}_k^* \exp(-i\Omega\tau),$$

$$n_k = \bar{n} + \tilde{n}\exp(i\Omega\tau) + \tilde{n}^* \exp(-i\Omega\tau),$$

then their substitution into Eqs. (6.18) leads to the linear response equations

$$i\Omega\tilde{m}_k - G\tilde{L}_k\overline{m}_k \int \tilde{n}\,\psi_k^2 dv - \frac{1}{2}G\overline{m}_k\beta_p\delta_{k,p} = 0, \tag{6.19a}$$

$$(i\Omega + A/\bar{n})\tilde{n} + \bar{n}\sum \tilde{L}_q\psi_q^2\tilde{m}_q = 0. \tag{6.19b}$$

We then follow the scheme given in Section 4.2.1. Expressing from Eq. (6.19b)

$$\tilde{n} = -\bar{n}\frac{\sum \tilde{L}_q\psi_q^2\tilde{m}_q}{i\Omega + A/\bar{n}}$$

we use it in Eq. (6.19a) and integrate.

Assuming $\bar{n} \approx 1$ and $\tilde{L} \approx 1$ we obtain, for large number of longitudinal modes ($N \gg 1$), the set of equations

$$\left[i\Omega(i\Omega + A) + \frac{3}{2}G\overline{m}_k\right]\tilde{m}_k + G\overline{m}_k\sum_{q\neq k}\tilde{m}_q = -\frac{1}{2}G\overline{m}_k(i\Omega + A)\beta_k\delta_{k,p}. \tag{6.20}$$

The solution is expressed by Kramer's rule: $\tilde{m}_k = D_k / D$. The determinants D and D_k are given, according to [330], by

$$D = \left(1 + 2\sum_{q=1}^{N} \frac{\Omega_q^2}{H_q}\right) \prod_q H_q , \tag{6.21}$$

$$D_k = -\left[\left(1 + 2\sum_{q \neq k} \frac{\Omega_q^2}{H_q}\right)\xi_k - 2\Omega_k^2 \sum_{q \neq k} \frac{\xi_q}{H_q}\right] \prod_{q \neq k} H_q , \tag{6.22}$$

where

$$H_q = \Omega_q^2 - \Omega^2 + iA\Omega, \quad \Omega_q^2 = \frac{1}{2} G\overline{m}_q, \quad \xi_q = \frac{1}{2} G\overline{m}_q (i\Omega + A)\beta_q \delta_{q,p} . \tag{6.23}$$

Since the losses of only one selected mode undergo modulation, it should be reasonable to give the responses of this and remaining modes separately:

$$\frac{\tilde{m}_p}{\overline{m}_p} = -\frac{G\beta_p S_p}{2} \frac{i\Omega + A}{\Omega_p^2 - \Omega^2 + iA\Omega} , \tag{6.24}$$

$$\frac{\tilde{m}_k}{\overline{m}_k} = \frac{1}{2} \frac{G\beta_p \overline{m}_p}{1 + 2\sum(\Omega_q^2 / H_q)} \frac{i\Omega + A}{(\Omega_p^2 - \Omega^2 + iA\Omega)(\Omega_k^2 - \Omega^2 + iA\Omega)} , \tag{6.25}$$

where

$$S_p = \frac{1 + 2\sum(\Omega_q^2 / H_q) - 2\Omega_p^2 / H_p}{1 + 2\sum(\Omega_q^2 / H_q)}$$

is a quantity, which differs slightly from unity when a large number of modes are excited. The expressions noticeably differ from each other.

Let us specify the relations (6.24) and (6.25) for two of most interesting types of laser: a solid-state laser and a dye laser. Characteristic of the first type is the large value of the parameter G and the relatively small number of excited modes, so that $\Omega_k^2 = G\overline{m}_k / 2 \cong Gm/(2N) >> 1$. At resonance ($\Omega = \Omega_p$) for the modulated mode we have

$$\left|\frac{\tilde{m}_p}{\overline{m}_p}\right| \approx \frac{G\beta_p}{2A} . \tag{6.26}$$

The criterion of applicability of the linear approximation is expressed by inequality

$$\beta_p << \beta_p^{\mathrm{cr}} = \frac{A}{G} , \tag{6.27}$$

which, formally, is the same as Eq. (6.9). Making the same assumptions for an unmodulated mode we obtain

$$\left|\frac{\tilde{m}_k}{\overline{m}_k}\right| \approx \frac{G\beta_p}{2AN}, \tag{6.28}$$

i.e., an N times smaller response.

Things are quite different in the case of a dye laser, since $G \leq 1$, while $N \gg 1$ and, consequently, $\Omega_k^2 << 1$. Assuming that the modulation frequency is in the range $\Omega_k << \Omega << 1$ we find $H_q \cong iA\Omega$ and $\sum(\Omega_q^2) \cong Gm/(2iA\Omega)$, which leads to

$$\left|\frac{\tilde{m}_p}{\overline{m}_p}\right| \approx \frac{G\beta_p}{2\Omega}, \quad \left|\frac{\tilde{m}_k}{\overline{m}_k}\right| \approx \frac{G\beta_p}{2N\Omega}. \tag{6.29}$$

Here again the modulation amplification factor of the selected mode, $K_{\text{loss}} \cong G/\Omega$, can reach large values, but this time it is due to the small value of Ω rather than to a large value of G. The condition for validity of the linear approximation is

$$\beta_p << \beta_p^{\text{cr}} \approx \frac{\Omega}{G}. \tag{6.30}$$

Experimental investigations of an Nd:YAG laser with a small number of excited longitudinal modes shows that the transfer function for the total intensity contains remarkable resonance features at the frequencies of antiphase relaxation oscillations [458]. Meanwhile, as was noted in the Section 4.2.2, no such features have been found in the power spectra of total intensity.

An interesting regularity is traced in distribution of the phases of the intensity oscillations of individual modes [458, 459]. Let us be consistent with the principle of mode numeration in accordance with decreasing steady-state intensities and relaxation oscillation numbering in accordance with decreasing eigenfrequencies adopted in Chapter 4. If the modulation frequency $\Omega = \Omega_p$, then modes are divided into two groups: the amplitudes of modes with numbers $k = 1,\ldots, p-1$ oscillate in phase as well as the modes with numbers $k = p,\ldots,N$. However, relative to each other these two groups oscillate in opposite phases. Such are the results of numerical calculations of modules and phases of the transfer functions using the model (6.10). The experiment confirming the fact of clusterization shows, however, some different distribution of modes into groups [458]. The reason could be a difference in length of the cavity and the active element that does not take into account in the model but can influence the laser dynamics as have been shown in Chapter 4.

There are also discrepancies between the results of theoretical calculation and experimental data in positions of the peaks with maximal intensity on the transfer functions of individual modes. Calculations give a monotonic shift of the dominate peak in the high-frequency side when one move from the weak modes to the more intensive one [167, 458], while the experiment does not confirm such a regularity [458].

It is possible to give one more example of the sensitivity of transfer functions to symmetry breaking in a laser. When the modes are situated symmetrically with respect to the gain line centre and all the mode losses are equal, then the transfer functions for symmetrical pairs of modes are doubly degenerate (Fig. 6.1*a*). Introducing an extra losses in one of side modes (in the mode with $k = 2$ in this case) removes the degeneration, which results in splitting of transfer functions of modes with $k = 2$ and $k = 3$ (Fig. 6.1*b*). This circumstance can be used for indication of selective absorption in the intracavity medium.

6.1.3 Nonlinear Response of a Single-Mode Laser to Periodic Loss Modulation

The laser response can be nonlinear even if the loss modulation is rather weak. This follows from the preceding discussion. To clarify how the nonlinear response deviates from the linear response with an increase in the modulation amplitude, it is reasonable to transform Eqs. (6.2) as follows.

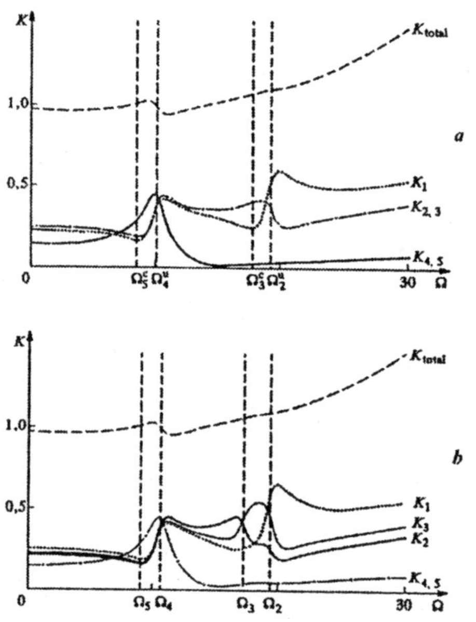

Fig. 6.1. Transfer functions at symmetrical positions of laser modes relative to the gain line centre and equal losses (*a*), at the presence of additional losses in one of modes (*b*) [350]

First, eliminating the variable n, we pass from Eqs. (6.2) to one second-order equation

$$\frac{d^2m}{d\tau^2}-\frac{1}{m}\left(\frac{dm}{d\tau}\right)^2+(m+1)\frac{dm}{d\tau}+Gm(m-\overline{m})=Gm\beta_{loss}[\Omega\sin(\Omega\tau)-(m+1)\cos(\Omega\tau)]$$

Then, by a change of variable $m=\overline{m}e^x$, where $\overline{m}=A-1$, we arrive at the desired form [460]

$$\frac{d^2x}{d\tau^2}+(\overline{m}e^x+1)\frac{dx}{d\tau}+\Omega_1^2[e^x(1+\beta_{loss}\cos\Omega\tau)-1]=G\beta_{loss}[\Omega\sin(\Omega\tau)-\cos(\Omega\tau)]. \tag{6.31}$$

In the absence of loss modulation Eq. (6.31) reduces to the nonlinear oscillator equation

$$\frac{d^2x}{d\tau^2}+(\overline{m}e^x+1)\frac{dx}{d\tau}+\Omega_1^2(e^x-1)=0, \tag{6.32}$$

where $\Omega_1^2=G(A-1)$ is the frequency of the relaxation oscillations. This equation describes anharmonic oscillations of a particle in the field with Toda's potential $V(x)=e^x-x$ in the presence of nonlinear damping [461, 462].

The loss modulation changes the situation in two ways. First, the term

$$G\beta_{loss}[\Omega\sin(\Omega\tau)-\cos(\Omega\tau)],$$

which represents the driving term, appears on the right-hand side. Second, the last term on the left-hand side now contains an additional component $\beta_{loss}\Omega_1^2e^x\cos(\Omega\tau)$, which is responsible for the parametric excitation of the nonlinear oscillator.

A general idea of the properties of a forced nonlinear oscillator can be gained by defining the form of the so-called 'skeleton curve'. This curve is the geometrical site of the resonance contour extremes and yields the frequency-amplitude dependence of the eigenmode of the corresponding conservative system [463]. Thus, we substitute the solution $x=a+b_1\cos(\Omega\tau)$ and obtain, with the damping term neglected,

$$-\frac{\Omega^2}{\Omega_1^2}b_1\cos(\Omega\tau)+\exp[a+b_1\cos(\Omega\tau)]=1. \tag{6.33}$$

Representing $\exp[b_1\cos(\Omega\tau)]$ as a Taylor expansion series, we rewrite Eq. (6.33) in the form

$$e^{-a}\left[1+\frac{\Omega^2}{\Omega_1^2}b_1\cos(\Omega\tau)\right]=1+\sum_{k=1}^{\infty}\frac{b_1^k\cos^k(\Omega\tau)}{k!}.$$

Equating the time-independent terms on the right-hand and left-hand sides of this equation, we find

$$e^{-a} = 1 + \sum_{k=1}^{\infty} \frac{b_1^{2k}}{2^k (k!)^2} . \tag{6.34}$$

Equating the coefficients of $b_1 \cos(\Omega\tau)$ we get

$$\frac{\Omega^2}{\Omega_1^2} e^{-a} = 1 + \sum_{k=2}^{\infty} \left(\frac{b_1}{2} \right)^{2k-2} (k-1)!k! . \tag{6.35}$$

Using (6.34) in (6.35) yields the equality

$$\frac{\Omega^2}{\Omega_1^2} = \frac{1 + \sum_{k=2}^{\infty} b_1^{2k-2} [2^{2k-2} (k-1)!k!]^{-1}}{1 + \sum_{k=2}^{\infty} b_1^{2k} [2^k (k!)]^{-1}} , \tag{6.36}$$

which defines implicitly the skeleton curve $b_1(\Omega)$.

In the limit $b_1 \to \infty$ we have $\Omega \to 0$. This means that for this type of nonlinearity the skeleton curve and, correspondingly, the resonance contours of the oscillator are inclined towards the low frequency side as shown in Fig. 6.2. The form of the resonance curve indicates that there are two stable solutions (bistability) and that the system manifests the hysteresis behaviour as either the frequency or the modulation index is varied. Two branches of stable solutions exist in the domain $\Omega < \Omega_1$. These branches radically differ from each other by the high contrast of the solutions, i.e., in the ratio of the maximum and minimum intensity during a modulation period or in the difference $\delta x = x_{max} - x_{min}$.

Since Eq. (6.31) is strongly nonlinear, the system may have, besides the main resonance, resonances on overtones and undertones of the driving force [212]. Each of them is represented by a resonance curve like that shown in Fig. 6.2. In order to obtain the general amplitude-frequency characteristic (transfer function) of the investigated laser, one should assume the solutions of Eq. (6.31) in the form

$$\sum b_i \cos[(m/k)_i \Omega\tau + \varphi_i ,$$

where m and k are mutually prime numbers, and make use of the harmonic balance condition. The implementation of this program requires cumbersome calculations even to find the skeleton curves. The only thing easy to ascertain is that all of these curves are inclined to the low frequency side.

The anticipated form of amplitude-frequency characteristic is illustrated in Fig. 6.3. The possible number of branches of stable solutions exceeds two; thus the laser may exhibit multistability properties where several branches overlap. The result, in particular, is a sophisticated pattern that includes not one, as in Fig. 6.2, but several hysteresis loops. It should be

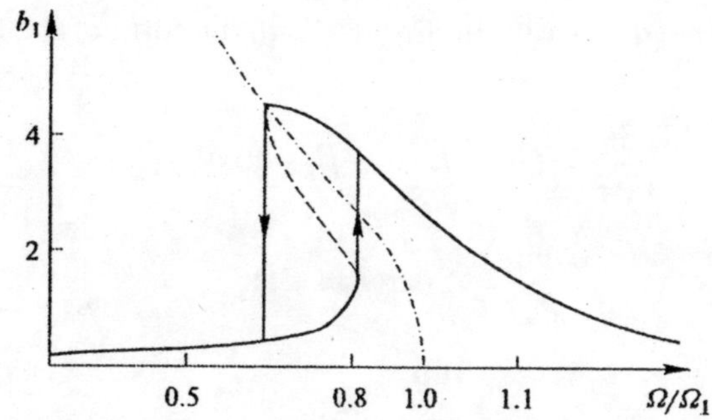

Fig. 6.2. Nonlinear resonance response of the laser to harmonic loss modulation. The dash lines show the unstable branch and the dash-dotted line indicate the skeleton curve. The arrows mark the boundaries of the hysteresis region [460]

emphasized that the complex profile of the potential is not the reason for multistability in this case, since the Toda potential possesses the unique extremum.

6.1.4 Bifurcations and Chaos

Consideration of the general properties of a nonlinear oscillator with the Toda potential is beyond the scope of this book. We briefly discuss only the data obtained directly in investigations of lasers. First of all, we note that although the class B laser is described by a second-order set of equations, introducing a parameter modulation provided not simply time-dependent behaviour but even chaotic behaviour. The point is that the nonautonomous system possesses an additional degree of freedom. Actu-

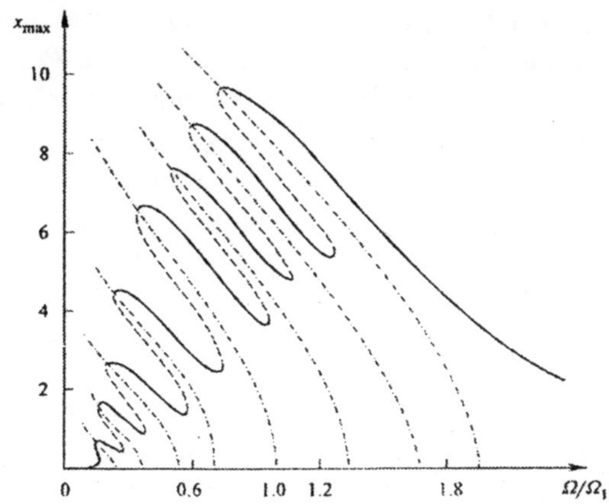

Fig. 6.3. Sample view of the amplitude-frequency characteristic of the system described by Eq. (6.31) [460]

ally, assigning $\Omega\tau = \Phi$ we rewrite Eqs. (6.2) in the form of the third-order ser of equations

$$\frac{dm}{d\tau} = Gm(n\tilde{L} - 1 - \beta_{\text{loss}} \cos\Phi) ,$$

$$\frac{dm}{d\tau} = A - n(\tilde{L}m + 1) , \tag{6.37}$$

$$\frac{d\Phi}{d\tau} = \Omega .$$

A three-dimensional phase space is large enough for a strange attractor (chaos) to occur.

Particular data on lasers with modulated loss, pumping or detuning, obtained by numerical calculations are in good agreement with those achieved in experiments [23, 460, 464-471]. This should be clearly apparent using a Nd:YAG laser as an example and comparing the calculated (Fig. 6.4) and experimental (Fig. 6.5) results.

According to calculations, the linear response to weak sinusoidal loss modulation (Fig. 6.4*a*) transforms to more complicated modulation of the output with the harmonics of Ω, as the modulation depth increases above $\beta_{\text{loss}} = 0.003$. In the range $0.003 < \beta_{\text{loss}} < 0.018$ the modulation of the solution remains rather shallow (Fig. 6.4*b,c*). At $\beta_{\text{loss}} = 0.021$ there is jump-like passage to short pulse generation with a pulse repetition rate Ω (Fig. 6.4*d*). A further increase in β_{loss} yields a sequence of period doubling bifurcations (Fig. 6.4*e* shows the regime with a period 2*T*), which leads to chaos at $\beta_{\text{loss}} = 0.027$ (Fig. 6.4*f*). At $\beta_{\text{loss}} = 0.035$ the laser response is again periodic but with period 4*T*. The abrupt passage, from chaotic to regular pulsations, without intermediate state, with an increase in β_{loss} is of particular interest. The minimum intensities between pulses are very different in these domains.

Results of an experimental investigation of the dynamics of a Nd:YAG ring laser are presented in Fig. 6.5. The set of parameters is about the same as that adopted in the calculations. Unidirectional lasing was ensured by use of a glass-based Faraday isolator with a large Verde constant. The laser was 20% above threshold, corresponding to a relaxation frequency $\nu_1 = 27\ \text{kHz}$. The modulation frequency was set in the range 8–15 kHz, where the choice of a particular value was not crucial to the realization. A LiNb crystal (the *z*-section) to which an alternating electric field applied across the laser beam was used as a loss modulator. On comparing Figs. 6.4 and 6.5 we notice that the scenarios coincide almost in all details, being indicative of the appropriateness of the mathematical model adopted.

These regularities fit well into the ideas developed in Section 6.1.3 and

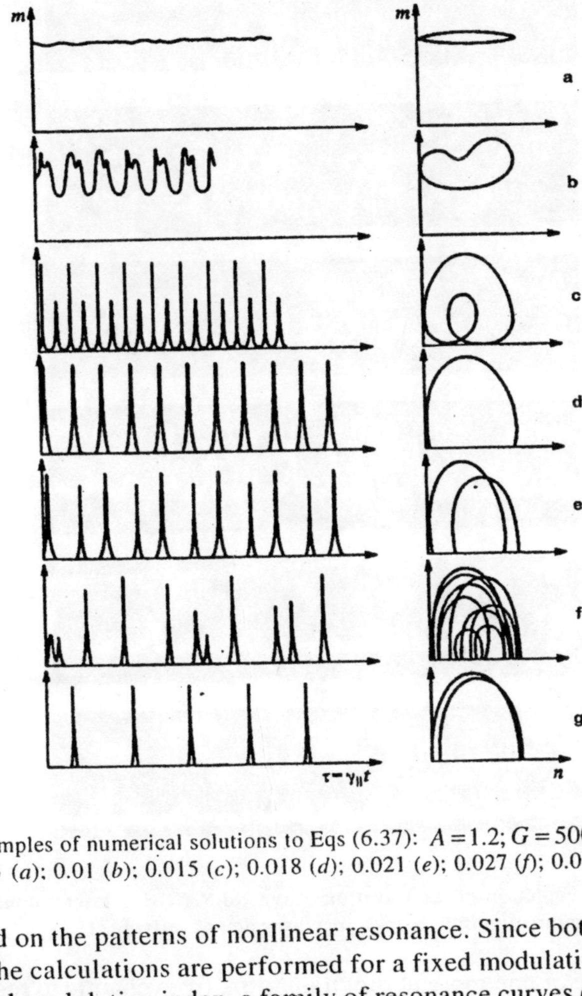

Fig. 6.4. Examples of numerical solutions to Eqs (6.37): $A = 1.2$; $G = 5000$; $\Omega = 0.4\Omega_1$; $\beta_{loss} = 0.003$ (*a*); 0.01 (*b*); 0.015 (*c*); 0.018 (*d*); 0.021 (*e*); 0.027 (*f*); 0.035 (*g*) [471].

those based on the patterns of nonlinear resonance. Since both the experiment and the calculations are performed for a fixed modulation frequency and a varied modulation index, a family of resonance curves dependent on β_{loss} as a parameter (Fig. 6.6) is useful for interpretation of the results. In this figure a change in β_{loss} corresponds to passage from one resonance curve to another. Suppose the modulation frequency Ω is set to the fixed value less than Ω_1 indicated by the vertical dash line. At the smallest β_{loss} the oscillator nonlinearity is not apparent and, therefore, the response to harmonic loss modulation is also harmonic. Nonlinear distortions of the resonance smoothly grow while the operating point remains on the lower branch of the resonance curves. Once $\beta_{loss} = \beta_5$ is achieved, the system jumps over into the state on the upper branch, which is exhibited as an abrupt transition to pulsed laser action. If β_{loss} is decreased from a value larger than β_5 the jump to the lower branch occurs at $\beta_{loss} = \beta_3 < \beta_5$ (hysteresis).

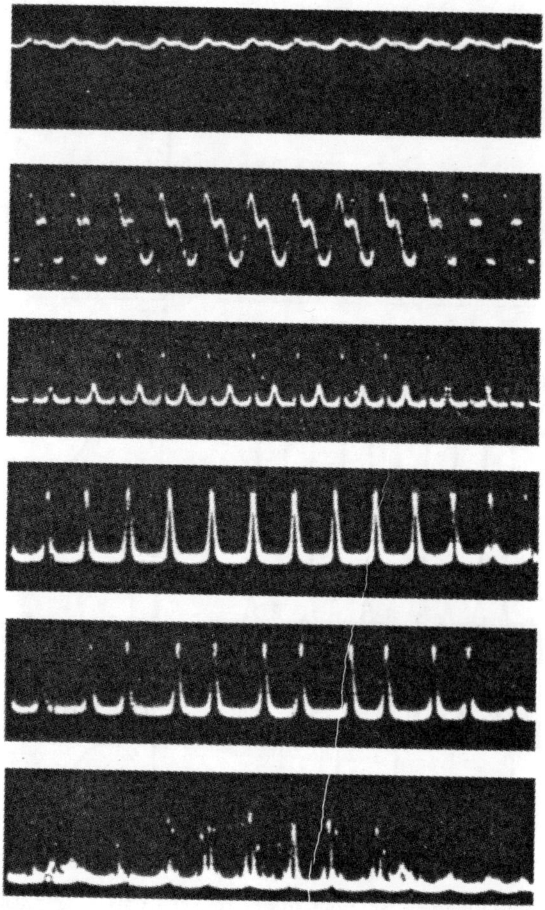

a
b
c
d
e
f

Fig. 6.5. Time evolution of the travelling wave Nd:YAG ring laser output intensity for different loss modulation indices (β_{loss} increases downwards) [471].

The presence of one more abrupt transition, from chaotic to regular pulsations with a period $4T$ at $\beta_{loss} = 0.0355$, is an indication that, besides the main resonance shown in Fig. 6.6, there is another higher resonance at higher amplitudes. Reproducing the whole complex structure of nonlinear resonances by numerical integration of Eqs. (6.2) or (6.31) is rather time-consuming. However, enough fragments have been found by this method to confirm the conception as a whole [460].

There are others examples of numerical solutions in combination with the experimental ones. Thorough investigations of CO_2 laser behaviour under periodic loss and frequency modulation are described in Refs. [468-470], which followed the seminal paper [465]. The succession of different regimes with an increase in the parameter modulation index differs from Fig. 6.4 in its details but, undoubtedly, there are common features. The

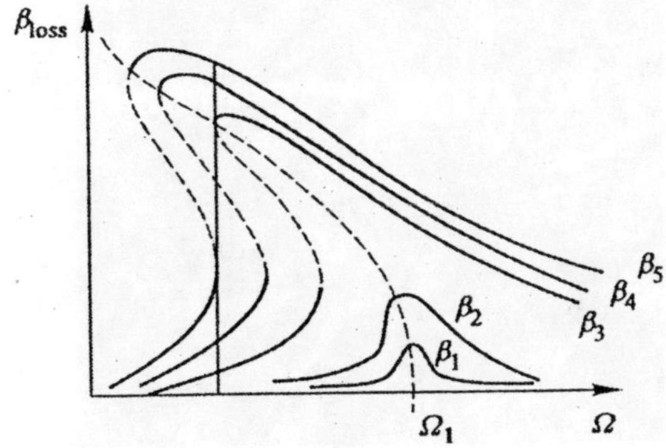

Fig. 6.6. Family of resonance curves of a nonlinear oscillator dependent on β_{loss} as a parameter [471].

CO_2 laser exhibits both bistability and hysteresis. Bistability can be generalized to cases when one of both coexisting regimes is chaotic and the corresponding attractors in the phase space are strange attractors. It is found that chaos is reached via a cascade of period doubling bifurcations and escaped via an inverse cascade.

Windows of regular behaviour with a fundamental period $3T$ instead of T may occur in the domain of chaotic behaviour. The situation is illustrated in Fig. 6.7. The diagram shows two branches of a resonance structure. On the lower branch chaos is due to the chain of bifurcations *T–2T–4T–C*, while on the upper branch the succession *3T–6T–C* takes place. Thus, the window of regular response with period $3T$ is not inside some domain of chaos but between the domains belonging to the different branches. Figure 6.7 confirms that changing of regimes may occur either in a 'soft' manner within one branch of a resonance structure or in a 'hard' manner as the result of switching from one branch to another.

Let us pay attention to one more regime, which is also shown in Fig. 6.7. It is called the attractor crisis. Without going into the details we note that the crisis is due to the coalescence between two attractors or between an attractor and an unstable fixed point or periodic orbit. The phenomenon is exhibited as a sudden expansion, contraction or disappearance of the attractors as the control parameter is varied. The behaviour of the attractor at a crisis is displayed in Fig. 6.8. The loss modulation index is plotted on the horizontal axis and the set of output intensities sampled at a fixed time in every period of modulation is shown on the vertical axis. The point of splitting of the curve corresponds to a period doubling bifurcation and the diffuse curve corresponds to chaotic dynamics. The weak diffusion means that although the regularity is lost, the difference in spike amplitudes and time intervals between spikes is relatively small. The dimension of the

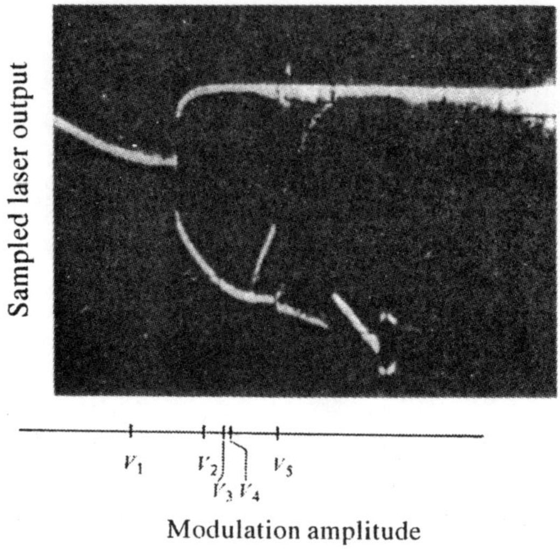

Fig. 6.7. Bifurcation diagram of a CO_2-laser with periodic combined modulation of cavity frequency and Q-factor [468]. Intensities sampled at time intervals equal to modulation period (vertical axis) are displayed versus the modulation amplitude (horizontal axis). Modulating voltage V_1 corresponds to a bifurcation T—$2T$, voltage V_2 corresponds to a bifurcation $2T$—$4T$, and V_3 corresponds to a transition to chaos, and at V_5 chaos is replaced by period $3T$ pulsations, which enter, via $6T$, the new domain of chaos.

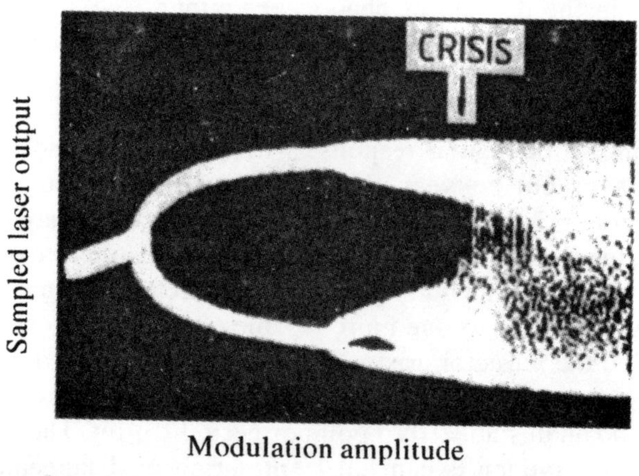

Fig. 6.8. The attractor crisis phenomenon in the response of CO_2-laser to periodic parameter modulation [468]. Here the crisis is a boundary crisis in which the attractor suddenly expands

strange attractor at the crisis and the resulting dispersion of the pulsation characteristics increase until the branches overlap.

We do not claim to have explained the physical reasons for attractor crises in this particular case, but we can point to two excitation mechanisms of the Toda oscillator presented in Eq. (6.31), which may contribute to crises. These are external forcing and parameter modulation. The second is effective for modulation frequencies exceeding the relaxation oscillation frequency. Phenomena such as attractor crises may occur when these two mechanisms compete.

The complex dynamics of a laser subject to periodic parameter modulation indicates the relationship between eigenoscillations and forced oscillations. Variation of the control parameter, while influencing the interaction efficiency, by no means affects the properties of the relaxation oscillations themselves. Therefore, factors such as detuning and inhomogeneous broadening never exert such a strong influence on the stability of modulated lasers as they do for autonomous lasers [472].

An important circumstance for the response of a laser to internal periodic modulation is the mode spectrum, since the latter defines the spectrum of relaxation oscillations [473, 474]. Generally, the number of degrees of freedom increases proportional to the number of modes only if the mode locking is absent. For this reason, we should differentiate between the phase diagrams under periodically modulated losses with locked and unlocked modes (Fig. 6.9). In the latter case the experimental phase diagram displays several minima at the boundary between the domains of chaotic and regular behaviour. The positions of these minima are identified with the fundamental relaxation oscillation frequency (70 kHz), its subharmonic (35 kHz) and the highest frequency of antiphase relaxation oscillations (22 kHz). The lower relaxation oscillation frequencies are to close to each other to be distinguished in the diagram. One more minimum is located near the second harmonic of the fundamental relaxation oscillation frequency, which is natural, given the parametric mechanism of excitation of pulsations. The diagram in Fig. 6.9a shows that the domain of chaos has a smooth boundary, which does not contain the additional relaxation oscillation frequencies and which does not appear near these frequencies.

Both phase diagrams, presented in Fig. 6.9, were obtained in experiments with Nd:YAG lasers, in which an electrooptical modulator was used for low-frequency loss modulation. Active mode locking was provided in the first case as well. The number of modes excited in the experiment is estimated to be 10–15.

The attractor dimensions, found by processing of both the experimental results [23,469,470] and the numerical simulations [474,475], are rather low. Only in one case, which identified with a five-mode model, the fractal dimension is slightly larger than 3 when the laser response to modulation is chaotic. For all the chaotic regimes of a single-mode laser the integer part of the attractor dimension does not exceed 2.

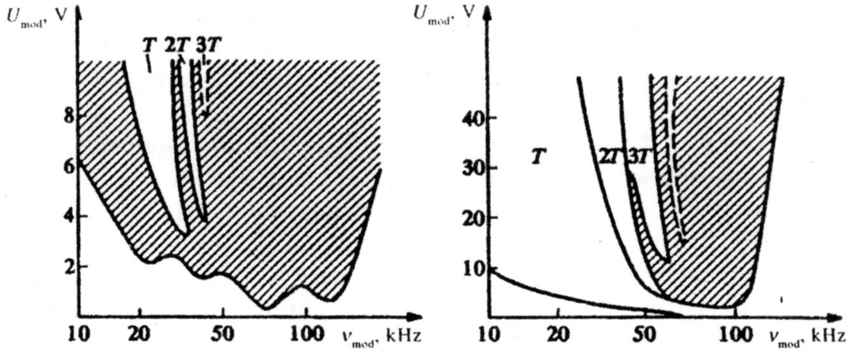

Fig. 6.9. Experimental phase diagrams of a periodically modulated multimode Nd:YAG-laser in the control parameter space (modulation frequency ν_{mod}, modulator driving voltage amplitude U_{mod}) (a) with and (b) without mode locking [474].

6.2 Monotonic Adiabatic Variation of Parameters

The premises developed in this section are directly related to the theory of spectrally swept lasers they are a key to explaining the spontaneous pulsations of solid-state lasers and they are useful in the analysis of the condition for giant pulse formation. Our discussions are based on Refs. [24, 476, 477].

6.2.1 Sweeping of the Losses

Without specifying the loss variation method and the form of the loss modulation we can write the set of equations

$$\frac{dm}{d\tau} = Gm[n-1-\beta(\tau)], \tag{6.38a}$$

$$\frac{dn}{d\tau} = A - n(m+1). \tag{6.38b}$$

Our interest is in the case of losses, which diminish with time, $d\beta/d\tau = \dot{\beta} < 0$, since it offers richer laser dynamics.

To find the characteristics of a quasistationary regime, at which the amplitude adiabatically follows the varied losses, we assume $\dot{m} = 0$ as a first-order approximation. Hence, the quantity

$$\tilde{n} = 1 + \beta \tag{6.39}$$

is varied during the lasing at a rate

$$\frac{d\tilde{n}}{d\tau} = \frac{d\beta}{d\tau} = \dot{\beta}. \tag{6.40}$$

Putting Eq. (6.40) into Eq. (6.38b) leads to the following expression for laser intensity:

$$\tilde{m} = \frac{A - 1 - \beta - \dot{\beta}}{1 + \beta}. \tag{6.41}$$

The next level of approximation provides a correction to the inversion, $n - 1 - \beta = \dot{\tilde{m}} / G\tilde{m}$, which is small compared to $\tilde{n}$ if

$$\frac{\dot{\tilde{m}}}{\tilde{m}} = -\frac{\dot{\beta}(A - \beta)}{(1+\beta)(A - 1 - \beta - \dot{\beta})} << G(1+\beta). \tag{6.42}$$

It is useful to introduce the notion of slow and fast sweep rates with the boundary between them specified by

$$\dot{\beta}_{\text{s-f}} = A - 1 - \beta. \tag{6.43}$$

In the domain of slow sweeping, $|\dot{\beta} << A - 1|$, the laser dynamics is fully determined by the pumping rate. In the domain of fast sweeping, $|\dot{\beta} >> A - 1|$, the rate, at which the losses are decreased, is of primary importance. Correspondingly

$$\tilde{m} = \begin{cases} (A - 1 - \beta)/(1 + \beta\} & \text{for } A >> |\dot{\beta}|, \\ -\dot{\beta}(1+\beta) & \text{for } A << |\dot{\beta}|. \end{cases} \tag{6.44}$$

We now linearize Eqs. (6.38) in the vicinity of the quasistationary values $\tilde{m}, \tilde{n}$ and obtain a generalization characteristic equation (3.16) for the present case of adiabatically varied losses. The roots of this equation for $G \gg 1$ have the form

$$\lambda = -\frac{A + \dot{\beta}}{2(1+\beta)} \pm i\sqrt{G(A - 1 - \beta - \dot{\beta})}. \tag{6.45}$$

At slow sweep rates, the expressions for the frequency and damping of relaxation oscillations look like those for the autonomous model:

$$\Omega_1 = \sqrt{G(A - 1 - \beta)}, \quad \theta_1 = -\frac{A}{2(1+\beta)}, \tag{6.46a}$$

while for fast sweep rates,

$$\Omega_1 = \sqrt{G|\dot{\beta}|}, \quad \theta_1 = -\frac{\dot{\beta}}{2(1+\beta)}. \tag{6.46b}$$

For the linearly decreasing losses (as $\beta = \beta_0 - u_{sw}\tau$, where $\dot{\beta} = -u_{sw} = \text{const}$), the results are the simplest. The dependence $\tilde{m}(\tau)$, expressed by Eq. (6.41), is shown by a solid line in Fig. 6.10 (curve 2). It is important to note that the curve $\tilde{m}(\tau)$ begins at $u_{sw} > 0$ at the point with coordinates

$$\tau_{thr} = (\beta_0 + 1 - A)/u_{sw}, \quad \tilde{m}_{thr} = u_{sw}/A.$$

Therefore, the initial (fluctuation) intensity, when the laser threshold is crossed, is markedly different from the quasi-equilibrium value $\tilde{m}_{thr} = u_{sw}/A$ at the threshold, which means that transient pulsations will necessarily occur. The oscillatory process calculated by Eqs. (6.38) with the initial conditions

$$\beta(0) > A - 1, \quad m(0) << u_{sw}/A$$

is also presented in Fig. 6.10 (curve 3).

We would like to draw attention to a feature of the lasing process shown by curve 3. The delay of laser action with respect to the moment at which the self-excitation boundary is crossed is exactly the same as τ_{thr}, the time from the start of sweeping to the moment at which the laser threshold is crossed. This remarkable feature was ascertained in Ref. [476] for a class A laser with an aperiodic transient; however, this property is retained for other classes of lasers as well.

Information on the first spike after the laser threshold is passed can be obtained in a fashion similar to that used in Section 3.2.2. Disregarding the change of the inversion in the linear regime we have, by virtue of $A - 1 - \beta_0 = 0$,

$$\frac{dm}{d\tau} = Gmu_{sw}\tau. \tag{6.47}$$

Knowing the solution of the above equation

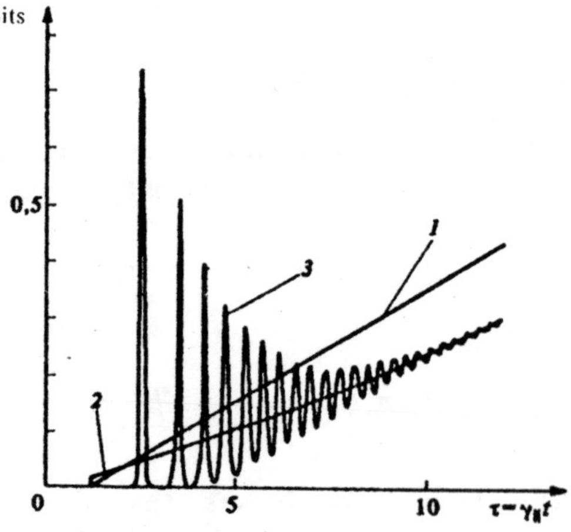

Fig. 6.10. Time dependence of (1) steady-state intensity, (2) quasi-steady-state intensity and (3) intensity that occurred in numerical simulation of Eqs. (6.38) during the uniform diminishing of cavity losses.

$$\ln\frac{m}{m_{\min}}=\frac{G}{2}u_{\mathrm{sw}}\tau^2$$

we can estimate

$$u_{\mathrm{sw}}\tau_{\mathrm{lin}}=\sqrt{\frac{2}{G}u_{\mathrm{sw}}\ln\frac{\tilde{m}}{m_{\min}}},\tag{6.48}$$

where τ_{lin} is the time the laser intensity reaches a quasistationary value $\tilde{m}$. Since $u_{\mathrm{sw}}\tau_{\mathrm{lin}}$ plays the role of $\eta_{\max}$, we have, by analogy with (3.33),

$$m_{\max}=u_{\mathrm{sw}}\ln\frac{\tilde{m}}{m_{\min}}.\tag{6.49}$$

According to (6.41), the quantity $\tilde{m}$ depends on sweeping rate but this does not lead to considerable deviations from the proportionality between $m_{\max}$ and u_{sw} in (6.49).

Consider now the following adiabatic criterion, which, according to [478], is expressed by the inequality

$$\tau_0\left|\frac{\mathrm{d}\beta}{\mathrm{d}\tau}\right|<<1+\beta.\tag{6.50}$$

We use τ_0 to denote the period of the slowest eigenoscillations, specifically the relaxation oscillations. Thus, the inequality (6.50) can be rewritten as

$$u_{\mathrm{sw}}<<\Omega_1(1+\beta).\tag{6.51}$$

Using Ω_1 as defined in (6.46b), we find the final form of the adiabatic following criterion

$$u_{\mathrm{sw}}<<G(1+\beta)^2.\tag{6.52}$$

Recalling inequality (6.42) and assuming $u_{\mathrm{sw}}=-\dot{\beta}>>A$, we see that this inequality is identical to (6.52).

Under the free running operation conditions, the mode sweeping through the occasional selection band of the resonator is the most probable. Then the losses are varied as

$$\beta(\tau)=\beta_0[1-g(\tau)],\tag{6.53}$$

where $g=1/(1+\Delta_{\mathrm{sel}}^2)$ and the time dependence is due to the fact that

$$\Delta_{\mathrm{sel}}=\frac{\nu-\nu_{\mathrm{sel}}}{\delta\nu_{\mathrm{sel}}}=\Delta_{\mathrm{init}}-u_{\mathrm{sw}}\tau.\tag{6.54}$$

Equations (6.40)–(6.46) are still valid bearing in mind that

$$\dot{\beta} = -\beta_0 \dot{g} = \beta_0 g' u_{sw},$$

where $g' = dg/d\Delta_{sel}$. This means that

$$u_{sw}^{s\text{-}f} = \frac{A - 1 - \beta_0 + \beta_0 g}{\beta_0 g'}, \tag{6.55}$$

and in the fast sweeping domain

$$\tilde{m} = -\frac{\beta_0 g' u_{sw}}{1 + \beta_0 (1 - g)}, \tag{6.56a}$$

$$\Omega_1 = \sqrt{-G\beta_0 g' u_{sw}}. \tag{6.56b}$$

For all these reasons, when the mode involved in the lasing process is swept through the selector band, an oscillatory transient process is excited. The selection depth restricts the limiting amplitude of the population difference oscillations: $\eta_{max} < \beta_0$. This amplitude will be reached if the mode is retuned into selection band halfwidth during the linear stage of lasing. In order to estimate the amplitude of the first spike, we make use of the expansion

$$g(\tau) = g(0) + \left(\frac{dg}{d\tau}\right)_{\tau=0} \tau = g(0) + g' u_{sw} \tau. \tag{6.57}$$

Using it in Eq. (6.38a) and bearing in mind that $A - 1 - \beta_0 [1 - g(0)] = 0$ we find

$$\frac{dm}{d\tau} = Gm\beta_0 g' u_{sw} \tau, \tag{6.58}$$

so that by analogy with Eqs. (6.47)–(6.49) we can write

$$u_{sw} \tau_{lin} = \sqrt{-\frac{2u_{sw}}{G\beta_0 g'} \ln \frac{\tilde{m}}{m_{min}}}, \tag{6.59}$$

$$m_{max} = -\beta_0 g' u_{sw} \ln \frac{\tilde{m}}{m_{min}}. \tag{6.60}$$

For the first spike of the transient to develop, the duration of the linear stage should not exceed the residence time of the mode in the selector band: $u_{sw} < 1/\tau_{lin}$ or, according to Eq. (6.59),

$$u_{sw} < \frac{G\beta_0 |g'|}{2 \ln(\tilde{m}/m_{max})}.$$

On the other hand, the pulsation intensity will be greater than in a laser with fixed tuning when the pumping is switched on only if the sweeping rate is high enough:

$$u_{sw} > \frac{A-1}{\beta_0 \,|\, g' \,|} O(1),$$

where

$$O(1) = \frac{\ln[(A-1)/m_{\min}]}{\ln(\tilde{m}/m_{\min})}$$

is the quantity of order of unity. This inequality follows from the comparison of Eq. (6.60) and Eq. (4.20). The necessary condition for occurrence of intense pulsations is given by

$$\beta_0 > \sqrt{\frac{2(A-1)\ln(\tilde{m}/m_{\min})}{Gg'^2} O(1)}\,. \tag{6.61}$$

Assigning $A = 10, G = 10^5$ and assuming $g' = 0.5, \ln(\tilde{m}/m_{\min}) = 25$, we find that the necessary condition (6.62) takes the form $\beta_0 = 0.1$.

The limit on the sweeping rate in the dimensional notations can be rewritten as

$$U_{sw} > \gamma_{\|}\delta_{sel} \frac{A-1}{\beta_0 \,|\, g \,|^2} O(1)\,. \tag{6.62}$$

Substituting these values, as well as $\gamma_{\|} = 10^3 \text{s}^{-1}$, $\delta v_{sel} = 0.1\,\text{cm}^{-1}$ yields $U_{sw} > 2 \cdot 10^{-2}\ \text{cm}^{-1}/\mu\text{s}$.

6.2.2 Sweeping of Detuning

For this discussion Eqs. (6.1) are now the reference equations, but the problem is in that $\Delta_0 = \Delta_0(\tau)$. The sequence of operations for analysis is very similar to those used in Section 6.2.1, so that no additional comments are needed.

Assuming in the first-order approximation $\mathrm{d}m/\mathrm{d}\tau = 0$ we find the quasistationary value of inversion

$$\bar{n} = \frac{1}{\tilde{L}}, \tag{6.63}$$

which is varied during lasing at the rate

$$\frac{\mathrm{d}\tilde{n}}{\mathrm{d}\tau} = -\frac{\tilde{L}'}{\tilde{L}^2} u_{sw}, \tag{6.64}$$

where $u_{sw} = \mathrm{d}\Delta_0/\mathrm{d}\tau$ is the dimensionless rate of the detuning sweeping.

Substituting Eq. (6.64) into Eq. (6.1a) leads to the expression for intensity

$$\tilde{m} = A - \tilde{L}^{-1} + \tilde{L}'\tilde{L}^{-2}u_{sw}. \tag{6.65}$$

Higher order corrections are, as previously, small if the sweeping rate is within the framework stipulated by the adiabatic following criterion.

Expression (6.65) shows that there are domains of fast and slow sweeping in the parameter plane (u_{sw}, Δ_0), which are separated by the line

$$u_{\rm sw}^{\rm s-f} = \frac{\tilde{L}}{\tilde{L}'}(A\tilde{L} - 1). \tag{6.66}$$

In the domain of slow sweeping the laser dynamics is determined by the pumping, while in the domain of fast sweeping it depends on the mode frequency movement towards the gain line centre (only this case is considered). The domains of fast and slow sweeping correspond to different laser intensities

$$\tilde{m} = \begin{cases} (A\tilde{L} - 1)/\tilde{L} & \text{for } u_{\rm sw} << u_{\rm sw}^{\rm s\text{-}f}, \\ (\tilde{L}'/\tilde{L}^2)u_{\rm sw} & \text{for } u_{\rm sw} >> u_{\rm sw}^{\rm s\text{-}f}. \end{cases} \tag{6.67}$$

Linearizing Eqs. (6.1) in the vicinity of the values $\tilde{m}, \tilde{n}$ we arrive at a quadratic characteristic equation with the roots

$$\lambda \approx -\frac{1}{2}(1 + \tilde{m}\tilde{L}) \pm i\sqrt{G\tilde{m}\tilde{L}}. \tag{6.68}$$

In the domain of fast sweeping Eq. (6.68) transforms to

$$\Omega_1^2 = G\frac{\tilde{L}'}{\tilde{L}}u_{sw}, \quad \theta_1 = -\frac{1}{2}\left(1 + \frac{\tilde{L}'}{\tilde{L}}u_{sw}\right). \tag{6.69}$$

The adiabatic following criterion is written as

$$\tau_0\left|\frac{d\tilde{L}}{d\tau}\right| << \tilde{L}. \tag{6.70}$$

Remembering that $d\tilde{L}/d\tau = \tilde{L}'u_{sw}$ we rewrite Eq. (6.70) as

$$u_{\rm sw} << u_{\rm sw}^{\rm cr} = \frac{\tilde{L}}{\tilde{L}'}\Omega_1. \tag{6.71}$$

In the domain of fast sweeping

$$u_{\rm sw}^{\rm cr} = \frac{\tilde{L}}{\tilde{L}'}G. \tag{6.72}$$

Comparing Eq. (6.72) with Eq. (6.66) it is easy seen that $u_{sw}^{cr} >> u_{sw}^{s\text{-}f}$ provided $G \gg 1$. Thus, the adiabatic theory applies to solid-state lasers in the whole domain of slow sweeping and in part of the domain of fast sweeping. Given in natural dimension the relation (6.72) becomes

$$U_{sw}^{cr} = \left(\frac{d\nu}{dt}\right)_{cr} = \frac{\tilde{L}}{\tilde{L}'}\frac{\delta\nu_0}{T_c}. \tag{6.73}$$

The excitation of a mode by dynamical tuning of the mode frequency will necessarily be accompanied by the growth of pulsations. This is because $A\tilde{L}=1$ at the laser threshold, and the threshold point lies at the boundary of the fast sweeping domain according to Eq. (6.66). As in the case of swept loss the quasistationary intensity at the time the laser threshold is passed has a finite value defined by Eq. (6.67), while $m(\Delta_{thr}) << \tilde{m}$.

Assuming that the period of the excited pulsations is much shorter than the time the mode is above the laser threshold we can make use of the expansion

$$\tilde{L}(\tau) = \tilde{L}(0) + \tilde{L}' u_{sw}\tau. \tag{6.74}$$

During the linear stage of the spike Eq. (6.1a) can be written, in view of Eq. (6.74), as

$$\frac{dm}{d\tau} = Gmu_{sw}\frac{\tilde{L}}{\tilde{L}'}\tau.$$

By analogy with Section 6.2.1, we find the maximum excess of the unsaturated gain above the threshold value that is achievable in the sweeping

$$\eta_{max} = A\tilde{L} - 1 = \sqrt{\frac{2\tilde{L}'}{G\tilde{L}}u_{sw}\ln\frac{\tilde{L}'u_{sw}}{\tilde{L}^2 m_{min}}}. \tag{6.75}$$

The spike amplitude can be determined by using conservative approximation equations, which only differ from Eq. (3.31) in that m is replaced by $m\tilde{L}$. Similar to Eqs. (3.33) and (6.62), for the spike amplitude we have

$$m_{max} = \frac{G\eta_{max}^2}{2\tilde{L}} = \frac{\tilde{L}'u_{sw}}{\tilde{L}^2}\ln\frac{\tilde{L}'u_{sw}}{\tilde{L}^2 m_{min}}. \tag{6.76}$$

Comparison of this value with the amplitude of the first spike, which is formed in a laser with fixed tuning when the pumping is switched on [see (3.33)], shows that the former is greater for

$$u_{sw} > \frac{\tilde{L}^2(A-1)}{\tilde{L}'}O(1), \tag{6.77}$$

where O(1) is the quantity of order of unity. In unnormalized units, the inequality (6.77) has the form

$$U_{sw} > \gamma_{\parallel}\delta\nu_0 \frac{A-1}{|\Delta_0|} O(1) . \tag{6.78}$$

Assigning $\gamma_{\parallel} = 10^{-3}\ \mu s^{-1}$, $\delta\nu_0 = 6\,cm^{-1}$, $A = 10$ and $|\Delta_0| = 0.5$ we obtain $U_{sw} > 0.1\,cm^{-1}/\mu s$.

6.3 Mechanisms of Undamped Pulsations in Solid-State Free-Running Lasers

In previous sections of this chapter we have considered two processes capable of destabilizing laser operation: weak periodic modulation of the parameters and their monotonous variation, owing to which one or more modes reach a threshold and begin to oscillate. The models for both processes are in excellent agreement with experiments. Laser dynamics with periodic loss modulation was investigated, for example, in the early papers [479-482]. Nevertheless, there are no controlled variations of the laser parameters in a free-running laser so we must consider the effects of parasitic modulation.

The hypothesis that the laser parameters were time dependent was first used to explain the spiking of two-level paramagnetic masers [234, 289]. The idea that the pulsations in solid-state lasers are due to the technical fluctuations was put forward in [483]. However, the assumption that a dominant role is played by inversion fluctuations produced by the pumping has not been confirmed in practice. Another approach, which relates cavity loss modulation to random cavity mirror displacements and to temperature variations, proves to be more realistic.

6.3.1 Origin of the Pulsations of a Two-Level Paramagnetic Maser

The specific feature of such a maser is that the pumping and the emission do not coincide in time. This predetermines the short duration of the emission stage. Analysis of the experimental data given in [484–488] shows the following features of the process;

– the amplitude of the maser output signal oscillates in time;

– the emission duration noticeably exceeds the relaxation times T_2 and T_c;

– when the emission is over, the active medium remains in the inverted state and can be used for the amplification.

The attempts to explain the spiking phenomenon in terms of autonomous maser theory have been a failure. These approach attribute the maser pulsations to magnetization nutation of the sample [489–492], but it is unclear why the pulse duration could then greatly exceed the relaxation times of the magnetization and of the field and how the atomic system could remain in the inverted state when emission is over.

All these features will find a natural explanation if we bear in mind the nonautonomous nature of the maser inherent in the method of the adiabatic rapid passage of the resonance, which is used for the production of the inversion [493]. Initially, the sample is placed in a magnetic field corresponding to a large detuning between the paramagnetic resonance and the pumping source frequencies. Thereafter the magnetic field is varied monotonically, so that these frequencies converge until they reach coincidence and than they diverge to a distant exceeding the paramagnetic resonance linewidth. If this process takes a time less than the relaxation times but is adiabatic, then population inversion, or reversal of magnetization, is achieved after the resonance passage.

On completing of the adiabatic rapid passage of the resonance the pumping source is switched off. The magnetic field then must be changed in the opposite way, since the frequencies ω_0 and ω_c need to be matched. For this reason, the induced emission process occurs under the conditions of sweeping of the detuning. Experiment showed [494] that the sweeping rate of the magnetic field influences the form of the output signal.

When choosing the model it should be borne in mind that paramagnetic substances with large relaxation times were used in masers and rather often, the relation $\gamma_\perp \gg \kappa$ is not satisfied. Therefore, we should use Eqs. (3.81) as the reference point. Analysis is simplified by the fact that the emission stage takes a time much shorter than T_1, thus making it possible to neglect the population relaxation of the working levels. For the same reason, it is convenient to renormalize the field and polarization (magnetization) variables as $F_m = \tilde{\gamma}^{1/2} F$ and $P_m = \tilde{\gamma}^{1/2} P$. Thus, Eqs. (3.81) transform to

$$\frac{dF_m}{d\tau} = \tilde{\kappa}(P_m \cos\Phi - F_m), \tag{6.79a}$$

$$\frac{dP_m}{d\tau} = nF_m \cos\Phi - P_m, \tag{6.79b}$$

$$\frac{dn}{d\tau} = -P_m F_m \cos\Phi, \tag{6.79c}$$

$$\frac{d\Phi}{d\tau} = \Delta_{0c}(\tau) - (nF_m / P_m + \tilde{\kappa} P_m / F_m)\sin\Phi. \tag{6.79d}$$

Figure 6.11 shows the results of numerical simulations of the processes in such a maser for linear sweeping of the detuning $\Delta_{0c} = \Delta(0) - u_{sw}\tau$, where $u_{sw} = \gamma_\perp^{-2} d\omega_0/d\tau$. The solution for the field has the form of the succession of isolate spikes. Each spike is accompanied by a relatively large reduction in inversion. Meanwhile, each new spike is emitted of conditions of noticeable excess above the laser threshold. The sweep rates adopted in these calculations and those corresponding to the experimental values are almost the same as the critical rate of the adiabatic retuning $u_{sw}^{cr} = \tilde{\kappa}\tilde{L}/\tilde{L}'$, which is of order of unity in this particular case.

In the experiment the formation of isolated spikes by two-level paramagnetic masers was observed only once [487]. In other cases the signal had the form of single pulse with oscillating tail. This form of emission can be explained by the inhomogeneous line broadening of paramagnetic resonance under the real experimental conditions. Only those paramagnetic ions participate in generation at a particular moment, which have their transition frequencies close to ω_c at that time. They give way to other ions as sweeping proceeds. In this respect, the analogy between sweeping and pumping is apparent.

Under the condition of inhomogeneous broadening the population difference is a function of the variables τ and Δ_0, owing to which

$$\frac{dn}{d\tau} = \frac{\partial n}{\partial \tau} + u_{sw}\frac{\partial n}{\partial \Delta_0}. \tag{6.80}$$

Substituting Eq. (6.80) into Eq. (6.79) we notice that the term $u_{sw}\partial n/\partial\Delta_0$ is similar to $A-n$ in Eq. (6.1b). In the most crude approximation

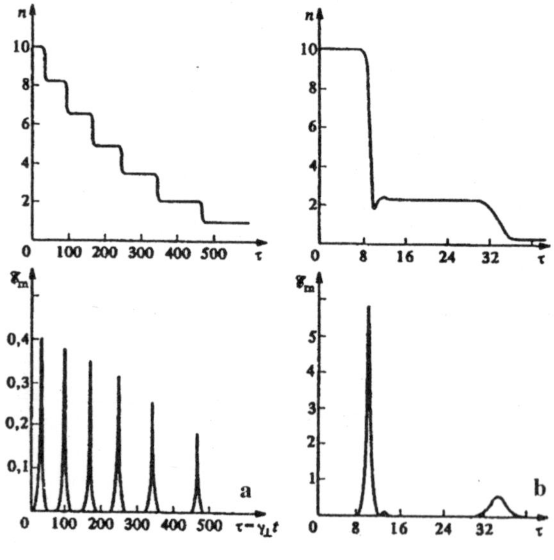

Fig. 6.11. Examples of the numerical solutions of the equations of a two-level paramagnetic maser (6.79). The initial inversion $n(0) = 10$; the sweeping rates $u_{sw} = 0.01$ (*a*), 0.25 (*b*).

$\partial n / \partial \Delta_0 \approx n_0 - n$, where n_0 is initial value of n. We, therefore, have, instead of (6.79c)

$$\frac{\partial n}{\partial \tau} = -\tilde{\gamma}_{\text{eff}}(n - n_0) - F_m P_m \cos\Phi \, .$$

The additional term takes into account the 'pumping' due to the sweeping and $\tilde{\gamma}_{\text{eff}} = \gamma_\perp^{-2} \mathrm{d}\omega_0/\mathrm{d}t = u_{\text{sw}}$.

The problem is reduced, as in the case of monotonically varied parameter, to analysis of the transient processes in the equivalent generator with continuous pumping. For the case $\gamma_\perp >> \kappa$ such an analysis was performed in Section 6.2. The properties of the emission process are in qualitative agreement with the experimental data. All features of the paramagnetic masers mentioned above find a natural explanation. The damping rate and frequency of pulsations are defined by the sweeping rate.

Nowadays two-level paramagnetic masers are of purely historical interest. Nevertheless, the search for understanding of the origin of their pulsations was an essential motive in the development of the theory of laser dynamics. It is both of fundamental importance and rather great irony that the maser pulsations were of technical rather than a natural autonomous origin.

6.3.2 Laser Parameter Oscillations Under Free-Running Conditions

We have established that periodic modulation of the cavity is an effective way to excite laser pulsations, especially by modulation on frequencies near the relaxation oscillation frequencies. We now demonstrate that similar modulations leading to intensity pulsations can be caused by uncontrolled monotonic variations of the cavity geometry or of the laser rod temperature.

Consider first a laser rod, which has its end surfaces parallel to the cavity mirrors. The collection of a several parallel partially reflecting surfaces can be treated as a set of coupled cavities. Displacement of one boundary with respect to the others (Fig. 6.12) modifies the system and influences, in particular, its losses. The position of the ends of the laser rod is

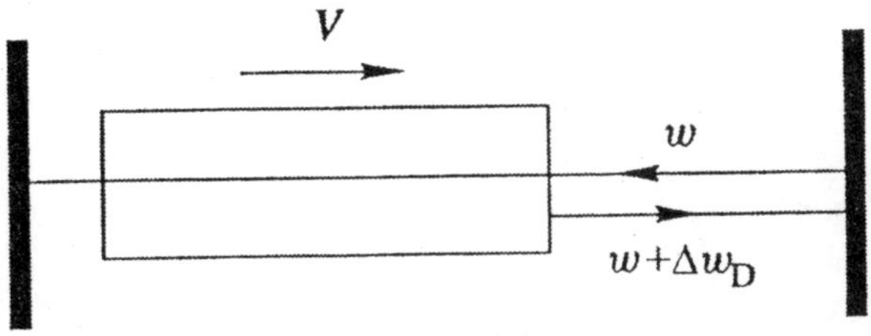

Fig. 6.12. Scheme of composite laser cavity with a moving interface.

also important to the electromagnetic field energy distribution among the sections of the cavity occupied by the laser medium and those, which is vacant [495]. Since all positions of the rod ends, differing by integral number of half wavelengths, are equivalent, uniform translational motion of the rod along the cavity axis produces periodic oscillations of the cavity Q and filling factors with a period $T = \lambda/2V_{refl}$, where V_{refl} is the velocity of the reflecting boundary.

Loss modulation from translational motion of a boundary in a composite cavity can also be treated as the result of the Doppler frequency shift as the wave is reflected at this boundary. If the laser rod moves and the mirrors are fixed, then the cavity eigenfrequencies can be assumed to be constant. The wave reflected from the end of the moving rod acquires the Doppler shift $\Delta\omega_D = 2\omega V_{refl}/c$ with respect to the transmitted wave. Thus, besides the wave of frequency ω, there are also waves with shifted frequencies $\omega \pm \Delta\omega_D$ in the cavity. The amplitudes of the Doppler shifted components are defined by the reflections of the inner boundaries and by reflected wave interference conditions if there are several moving boundaries. With the Doppler components taken into account, the laser equation reduce to the form (6.2), where $\Omega = \Delta\omega_D/\gamma_\parallel$, and β_{loss} can exceed the Fresnel reflectivity.

Doppler modulation also occurs if the laser rod is motionless and the cavity mirror is moved. This version features time-dependent cavity eigenfrequencies, which follow the mirror motion.

The loss modulation frequency (Doppler shift) becomes resonant for the translational motion of one reflecting boundary, $\Delta\omega_D/\gamma_\parallel = \Omega_1$, provided

$$V_{refl} = V_1 = (\lambda/4\pi)\sqrt{2\gamma_\parallel\kappa(A-1)}\,. \tag{6.81}$$

Example 6.1
Ruby laser

$$\left.\begin{aligned}\lambda &= 7\cdot10^{-5}\ \text{cm},\ \text{A}=10,\\ \gamma_\parallel &= 10^3\ \text{s}^{-1},\ \kappa = 10^7\ \text{s}^{-1},\end{aligned}\right|\ V_1 = 0.5\ \text{cm/s}.$$

Example 6.2
Nd:YAG laser

$$\left.\begin{aligned}\lambda &= 10^{-4}\ \text{cm},\ \text{A}=2,\\ \gamma_\parallel &= 2\cdot10^3\ \text{s}^{-1},\ \kappa = 10^7\ \text{s}^{-1},\end{aligned}\right|\ V_1 = 1\ \text{cm/s}.$$

In ordinary laser operation the motion of the reflecting surfaces is purely accidental. The external kicks of the laser components produce microvibrations. The velocity of any vibrating element surface does not remain constant, of course, but if it is close to resonance velocity for a short time, then intense pulsations may hold up.

According to measurements reported in [496], he instantaneous velocity of a vibrating mirror reaches 0.2 cm/s. This value is somewhat lower than the estimates of resonance velocities, which require higher amplitude vibrations. The experiment with Nd:YAG laser has shown that a several fold increase in the vibration amplitude leads to a sharp enhancement of the pulsations.

Periodic modulation of the cavity Q-factor also occurs during monotonic variation of the temperature of the laser rod, since both the refractive index η and the optical length $L_{\text{eff}} = L_{\text{a}}\eta$ are temperature dependent. Equating the effective velocity of cavity lengthening $V_{\text{eff}} = \mathrm{d}L_{\text{eff}}/\mathrm{d}t = L_{\text{eff}}\mathrm{d}\eta/\mathrm{d}t$ to the value V_1 introduced above we find the heating rate, which cause the resonance modulation of cavity losses:

$$\left(\frac{\mathrm{d}T}{\mathrm{d}t}\right)_1 = V_1\left(L_{\text{a}}\frac{\mathrm{d}\eta}{\mathrm{d}T}\right)^{-1}. \tag{6.82}$$

Using in (6.82) the values $L_{\text{a}} = 10\,\text{cm}, V_1 = 1\,\text{cm/s}$ and $\mathrm{d}\eta/\mathrm{d}T = 1.5\cdot 10^{-5}\ \text{K}^{-1}$, which correspond to ruby [497], we find $(\mathrm{d}T/\mathrm{d}t)_1 = 10^4\ \text{K/s}$. Heating rates such as this are quite realistic in a ruby laser pumped by a flash lamp.

The build-up of large-amplitude pulsations from the resonant perturbations requires that the loss modulation index satisfy the inequality $\beta_{\text{loss}} \geq \beta_{\text{loss}}^{\text{cr}}$, which is the inverse of Eq. (6.9). In the absence of an antireflection coating, the reflectivity of the ends of the laser rod is few percent. The kinematic or temperature loss modulation index has the same order of magnitude and is therefore well over $\beta_{\text{loss}}^{\text{cr}}$.

The simplest and most effective method of weakening the parasitic modulation of the cavity Q-factor is to tilt the laser rod ends in combination with applying antireflection coatings. At the angles of incidence exceeding the beam divergence, the waves reflected at the rod ends are completely removed from the cavity and the lost power is independent of the laser rod position. If the Gaussian beam of radius a_0 is generated, then this criterion can be written as [1]

$$\vartheta_{\text{incl}} > \vartheta_1 = \lambda/(2\pi a_0). \tag{6.83}$$

6.3.3 Instabilities Due to Sweeping of the Cavity Eigenfrequency

A tilted laser rod may function as an interferometric longitudinal mode selector, the properties of which depend on ϑ_{incl} and L_{eff}. The variation of the refractive index with heating causes the transmission maxima to move with respect to the cavity eigenfrequencies. Mirror vibration sweeps the cavity mode spectrum across the selector passband. However, the tilted laser rod ceases to operate as an interferometer if

$$\vartheta_{\text{incl}} > \vartheta_2 = a_0 / L_{\text{eff}} . \tag{6.84}$$

As a rule $a_0^2 / \lambda L_a >> 1$, owing to which $\vartheta_2 >> \vartheta_1$.

When inequality (6.84) is satisfied, neither mechanical vibrations nor heating of the laser elements leads to Q-modulation. Q-modulation is also absent in a laser with mirrors at the rod ends. In all these cases only the mechanisms, which sweep the mode spectrum with respect to the gain line continue to perturb the laser action. Let us make estimates of them using the expression for the longitudinal mode frequency $v_q = q/(2L_{\text{eff}})$, where $q = 2L_{\text{eff}} / \lambda$ is an integer and $L_{\text{eff}} = L + (\eta - 1)L_a$. The cavity tuning rate is given by

$$U_{\text{sw}} = \frac{\text{d}v_q}{\text{d}t} = -\frac{1}{\lambda L_{\text{eff}}} \frac{\text{d}L_{\text{eff}}}{\text{d}t} . \tag{6.85}$$

$$U_{\text{sw}} = \frac{\text{d}v_q}{\text{d}t} = -\frac{1}{\lambda L_{\text{eff}}} \frac{\text{d}L_{\text{eff}}}{\text{d}t} . \tag{6.85}$$

If the cavity is tuned by moving the mirror, then $\text{d}L_{\text{eff}}/\text{d}t = V_{\text{refl}}$ and Eq. (6.85) can be rewritten as

$$V_{\text{refl}} = \lambda L_{\text{eff}} U_{\text{sw}} . \tag{6.86}$$

The condition of excitation of intense pulsations is represented by inequality (6.78). As applied to a ruby laser ($\lambda = 7 \cdot 10^{-5}$ cm, $L_{\text{eff}} = 100$ cm, $U_{\text{sw}} > 10^{-2}$ cm$^{-1}/\mu$s) Eq. (6.86) yields $V_{\text{refl}} > 100\,\text{cm/s}$. This value greatly exceeds the resonance velocity V_1 found previously and, of course, it is not achieved in a free-running operation.

Let us estimate the rate of heating of the laser rod, which corresponds to the critical tuning rate. Using the Eq. (6.82), which relates the rod heating rate to the tuning rate,

$$\frac{\text{d}T}{\text{d}t} = \frac{V}{L_a \text{d}\eta/\text{d}T} = \frac{\lambda L_{\text{eff}}}{L_a} \frac{U_{\text{sw}}}{\text{d}\eta/\text{d}T} , \tag{6.87}$$

and inequality (6.78), we arrive at the condition for excitation of intense pulsations in a ruby laser: $\text{d}T/\text{d}t > 10^5$ K/s, assuming $L_{\text{eff}} = L_a \eta = 10$ cm. This value slightly exceeds the ruby heating rate under the flash lamp pumping conditions.

The temperature of the rod influences both the refractive index of its host and the position of spectral lines of impurity ions. The temperature line drift velocity

$$\frac{d\nu_0}{dt} = \frac{dT/dt}{dT/d\nu_0}, \tag{6.88}$$

depends on the parameter $dT/d\nu_0$, which is equal to 8 K/cm^{-1} for ruby at room temperature [497]. According to Eq. (6.78) the critical drift velocity corresponds to a heating rate of $dT/dt \approx 10^5$ K/s. Hence, it can be concluded that temperature drift should not lead to intense pulsations of solid-state lasers. However, more moderate amplitude pulsations can be excited in this manner. This conclusion agrees with the results of numerical investigations of a two-mode model, given in Ref. [498], where such a drift mechanism of spiking was proposed.

In a multimode laser, the change of modes can occur against the background of uninterrupted lasing. Therefore, the transient process, which appear each time a new mode begins to oscillate, are superimposed on each other. On the whole, the time-dependent laser action is seen as undamped although it consists of a series of damped transient pulsations. Numerical simulations that confirm these premises were performed in [496, 499] on the basis of the multimode rate equations.

6.3.4 Role of Spatial Effects

In Section 4.1.3 we gave the proof of the theorem that regardless of the number of modes the set of rate equations has a unique steady-state solution, which is globally stable. This means that spatial inhomogeneity of the inversion by itself cannot be the reason for undamped laser pulsations. However, small-scale structure on the order of the wavelength, which is due to nonuniform saturation of the laser medium by the radiated field, does increase the sensitivity of the laser to perturbations. Using enormous experimental data, partly given in Chapter 1, it has been shown that spiking regimes, especially irregular ones, are associated, as a rule, with a small number of modes excited simultaneously. Any measures, which reduce the nonuniform dynamical saturation of the laser medium, contribute to stabilization of the laser operation. Using these premises we can assert that criteria like Eq. (6.78), which take into account mode discrimination based on mode detuning from the line centre, give only an upper limit on the critical rates of laser tuning that lead to instabilities. Under real conditions, spatial inhomogeneities can noticeably decrease the critical velocity.

These ideas on the origin of spiking are in full accord with the experimental results for Nd:YAG lasers. Measures such as the increase of the construction rigidity, stabilization of the temperature, inclination of the ends of the laser rod and giving them an antireflection coating completely eliminate spiking. In ruby lasers measures aimed at smoothing out the longitudinal inhomogeneity of the inversion have also been added and contribute to added stability. experimental details, which demonstrate these results, where given in Section 1.2.3.

However, the reasons for the pulsations of a solid-state laser are not limited to parameter instabilities. Another cause of instabilities can be the nonlinearity of the laser rod host, which is discussed in the next Chapter. The reasons for spiking in semiconductor lasers are similar.

It should be noted that the acuteness of the problem of spiking regimes has become a thing of the past, along with the lasers of first generation. Modern sources of optical pumping for solid-state lasers, in contrast to flash lamps, have narrow radiation spectra matched to the absorption lines of active elements. Therefore, they do not create excessive thermal and mechanical tensions, which is peculiar to the flash lamps. Thus, the reasons of intense technical fluctuations of parameters, which lead to spiking, are eliminated.

Chapter 7

Lasers with Nonlinear Parameters

In addition to active parameter modulation discussed in the previous chapter there is another effective method of influencing the laser dynamics. This method is based on the use of elements with the optical properties, which change depending on intensity of the light illuminating them. Such elements can be responsible for a dependence of the cavity Q-factor on he laser output or, equivalently, they execute what is called 'passive Q-switching'. The decrease of losses with increasing power reduces the laser stability threshold and enhanced pulsations. On the contrary, the increase of losses with increasing power leads to more stable laser behaviour. For example, the oscillatory transient can be transformed to aperiodic one.

Sometimes, the nonlinearity of the media filling the laser cavity has greater influence on the laser dynamics than the time dependence of the parameters. This has been used to explain, for example, the undamped spiking of a neodymium glass laser, which depends on the spectral composition of pumping. In semiconductor lasers thermal and mechanical instabilities of parameters have little influence on the laser dynamics, so that material nonlinearities are the only possible cause of the frequently observed fast pulsations.

7.1 Laser with an Opto-Electronic Feedback

Automatic control of cavity losses was first used to remove spikes in a ruby laser [500–503]. An electrooptic Kerr cell in combination with a polarizer served as the control element. The control voltage applied to the cell electrodes was the amplified signal from a photocell that intercepted part of the laser output power. The control voltage was added to a D.C.

bias that was used to define the position of the operation point on the modulator characteristic.

The chain 'photocell–amplifier–device controlling the cavity Q-factor or pumping power' has been given the name 'opto-electronic feedback' (earlier the term 'external additional feedback' was used). Depending on the sign of the control voltage the additional feedback can be either positive (the cavity losses decrease with an increase in the output power). For negative feedback without delay the emitted power is stabilized. However, if the negative feedback is delayed this situation can lead to enhancement of the pulsations. Undamped pulsations are also enhanced in a laser with positive feedback [502].

Opto-electronic feedback (by control of the injection current) is also used in semiconductor lasers to influence the laser behaviour [504].

7.1.1 Single-Mode Laser with Self-Controlling of the Cavity Losses

To model these phenomena we turn again to the rate equations for class B lasers. The dependence of the cavity losses on laser power can be written conveniently as $\kappa = \kappa_0[1+\beta(m)]$ assuming $\beta(m) \geq 0$. Generalization of the rate equations (12.1) to the case considered [451, 502] yields

$$\frac{dm}{d\tau} = Gm[n-1-\beta(m)], \tag{7.1a}$$

$$\frac{dn}{d\tau} = A - n(m+1)]. \tag{7.1b}$$

The total number of fixed points of Eqs. (7.1) depends on the function $\beta(m)$. One of these points ($\bar{n} = A, \bar{m} = 0$) is trivial. Its instability occurs when the self-excitation condition

$$A > 1 + \beta(0) \tag{7.2}$$

is met. The coordinates of the remaining fixed points ($\bar{m}, \bar{n}$) can be found by solving the set of algebraic equations

$$\bar{n} = 1 + \beta(\bar{m}), \tag{7.3a}$$

$$\bar{n} = A/(\bar{m}+1). \tag{7.3b}$$

When Eqs. (7.1) are linearized in the vicinity of a nontrivial fixed point they take the form

$$\frac{d(\delta m)}{d\tau} = G\bar{m}(\delta n - \beta'\delta m), \tag{7.4a}$$

$$\frac{d(\delta n)}{d\tau} = -n(\bar{m}+1)\delta n), \tag{7.4b}$$

where $\delta m = m - \overline{m}$, $\delta n = n - \overline{n}$, $\beta' = d\beta/dm$ and the derivative is taken in the point $m = \overline{m}$. Eqs. (7.4) have characteristic roots

$$\lambda_{1,2} = -\frac{1}{2}(G\overline{m}\beta' + \overline{m} + 1) \pm \sqrt{\frac{1}{4}(G\overline{m}\beta' + \overline{m} + 1)^2 - G\overline{m}[(\overline{m}+1)\beta' + \overline{n}]}. \tag{7.5}$$

Nonoscillatory transients take place if the fixed point is a stable node (i.e., the characteristic roots are real and negative). Since $G \gg 1$, the condition for the real form of roots can be reduced to $(G\overline{m}\beta')^2 > 4G\overline{m}\overline{n}$. The roots will be negative if $\beta' > 0$. Combining the two inequalities yields

$$\beta' > 2\sqrt{\frac{\overline{n}}{G\overline{m}}}, \tag{7.6}$$

on condition that the fixed point under investigation is a stable node.

In a laser, in which condition (7.6) is met, the losses increase with increasing power (i.e., the additional feedback is negative). The phase portrait of such a laser system is presented in Fig. 7.1.

To find the steady-state values of the laser intensity and inversion one should know the form of the function $\beta(m)$. If we do not go beyond the limits of inequality $|\beta'| \ll 1$, which does not contradict Eq. (7.6), then its difference from Eq. (12.2) is expressed by a small term of order β'. In this case

$$\overline{m} \approx A - 1; \quad \overline{n} \approx 1$$

and the condition for the absence of pulsing (7.6) transforms to

$$\beta' > \beta'_{cr} = 2[G(A-1)]^{-1}. \tag{7.7}$$

In lasers with positive external feedback the forms of the behaviour are more diversified. If there is no delay in the reduction of cavity losses with

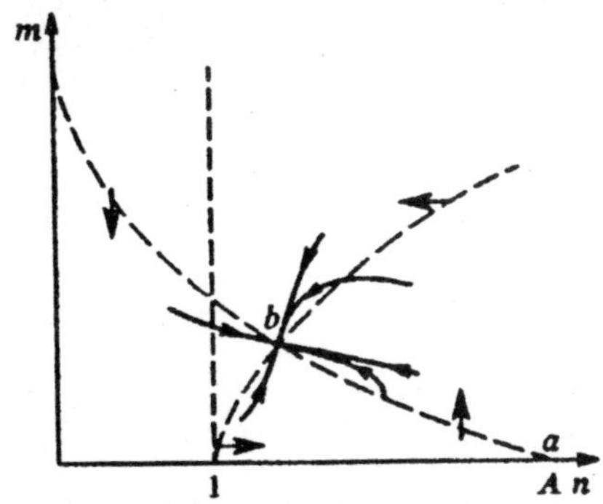

Fig. 7.1. Phase portrait of a laser with additional negative feedback. The dashed lines are the isoclinic ones $n = 1 + \beta(m)$ with horizontal and $m = A/n - 1$ with vertical tangents.

the growth of intensity, the stability of the fixed points decreases. If this solution is stable it corresponds to a fixed point of a focus type. To make the focus unstable one has to make the feedback stronger by requiring

$$G\overline{m}\beta' + \overline{m} + 1 < 0. \tag{7.8}$$

This implies the necessary condition $\beta' < 0$. The unstable focus is enclosed be a stable limit cycle (Fig. 7.2a).

It should be noted that the self-excitation condition (7.2) is fulfilled if there is one or any arbitrary odd number of fixed points in the upper half-plane. Things are different if the number of fixed points in the upper half-plane is even. Then the inequality (7.2) remains unsatisfied and lasing can be initiated only with a strong perturbation (hard excitation), by external signal injection, for example. We restrict ourselves to discussing the properties of a system with two nontrivial equilibrium positions.

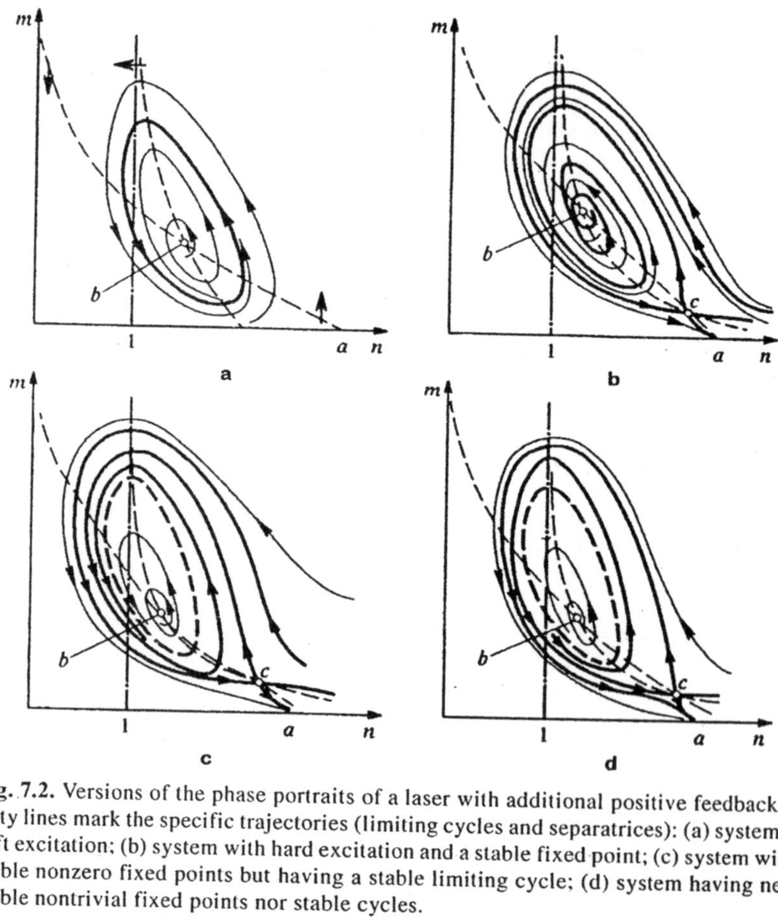

Fig. 7.2. Versions of the phase portraits of a laser with additional positive feedback. The fatty lines mark the specific trajectories (limiting cycles and separatrices): (a) system with soft excitation; (b) system with hard excitation and a stable fixed point; (c) system without stable nonzero fixed points but having a stable limiting cycle; (d) system having neither stable nontrivial fixed points nor stable cycles.

It is easy to ascertain the type of each fixed point. Point *a* is a stable node. At point *c* the characteristic roots (7.5) are real and opposite in sign, which is indicative of a saddle. All that was said above about the type of fixed point *b* and its stability remains valid. The phase portrait for the case where *b* is a stable focus is presented in Fig. 7.2b. One separatrix leaves the saddle and tends to point and the other tends to point *a*. When point *b* is unstable the system has a limit cycle. The stability of the cycle is defined by the sign of the expression

$$\sigma_0 = P'_m(\overline{m}_c, \overline{n}_c) + Q'_n(\overline{m}_c, \overline{n}_c), \qquad (7.9)$$

in which $P(\overline{m}_c, \overline{n}_c)$ and $Q(\overline{m}_c, \overline{n}_c)$ stand for the right-hand sides of Eqs. (7.1). The prime means the differentiation with respect to the variable indicated by the subscript. According to [211] the limit cycle is stable if $\sigma_0 < 0$.

Substituting the right-hand side of Eqs. (7.1) into Eq. (7.9) we find

$$\sigma_0 = -[G\overline{m}_c\beta'(\overline{m}_c) + \overline{m}_c + 1].$$

The stability of the cycle depends on the values that $\overline{m}_c$ and $\beta'(\overline{m}_c)$ assume. The only indisputable fact is that the limit cycle is stable for small $\overline{m}_c$. The phase portrait of a laser with hard excitation and a stable limit cycle is given in Fig. 7.2c. The separatrix, which tends toward point *b* in Fig. 7.2b, has the cycle as a limit. The phase portrait of a system without a stable limit cycle and without a stable fixed point in the upper half-plane is shown in Fig. 7.2d. All the nonfixed phase space trajectories end at point *a*. This implies that sustained laser operation is impossible, since even if lasing is initiating by an external signal, the laser will emit a powerful pulse and return permanently to the state of no output.

7.1.2 Multimode Laser with Selective and Combined Feedback

The method of the opto-electronic feedback has found further development in the systems with the selective and combined derivative feedback. The fact that the control of laser parameters is achieved using the derivative of the output signal makes it easier to perform analysis because, changing the topology of the phase space in the vicinity of the fixed points, such a feedback does not change the position of these points. But what is most interesting is the possibility of selective influence on the laser behaviour using the selected mode. The selective feedback can be combined with the feedback proportional to the total output intensity. The advantages of such a system are illustrated by an example of combining the negative feedback through the total intensity with the positive selective feedback, considered below [167].

The idea of a laser with the combined feedback can be seen in the simplified scheme presented in Fig. 7.3. A Nd:YAG crystal is pumped along

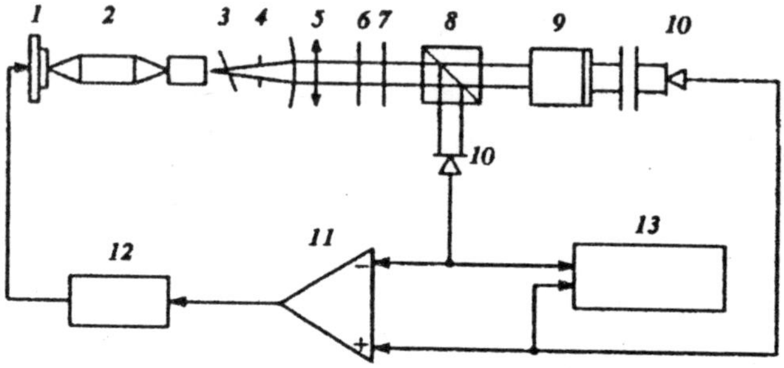

Fig. 7.3. Experimental setup showing a multimode laser with combined feedback: 1 – laser diode; 2 – shaping optics; 3 – polarizer (Brewster plate); 4 – diaphragm; 5 – collimating lens; 6 – filter at 1.06 mm; 7 – half wave plate; 8 – beamsplitter; 9 – Fabry–Perot interferometer; 10 – photodetector; 11 – derivative amplifier; 12 – power supply of laser diode; 13 – spectrum analyzer [167].

its axis with radiation of a diode laser. The output radiation is split into two beams; either of them is directed on its own photodiode. However, one of those beams has been passed previously through a Fabry–Perot interferometer with the aim to separate the chosen longitudinal mode. In the electric part of the feedback net the signals taken with a certain weight are subtracted, with differentiation and amplification of the signal, which control the function of the laser diode supply unit and then the pumping power of the solid-state laser.

The equations of a laser with combined feedback,

$$\frac{dm_k}{dt} = Gm_k\left[\tilde{L}_k(n_0 + n_k) - 1 - \beta_k\right],$$

$$\frac{dn_0}{d\tau} = A_0\left[1 + \beta_{\text{pump}}\cos(\Omega\tau)\right] + \Upsilon_{\text{total}}\sum_{j=1}^{N}\frac{dm_j}{d\tau} + \Upsilon_i\frac{dm_i}{d\tau} - n_0\left(1 + \sum_{j=1}^{N}\tilde{L}_j m_j\right) - \sum_{j=1}^{N}\tilde{L}_j m_j n_j \tag{7.10}$$

$$\frac{dn_k}{d\tau} = -n_k\left(1 + \sum_{j=1}^{N}\tilde{L}_j m_j\right) - \frac{1}{2}\tilde{L}_k m_k n_0$$

differ from Eqs. (12.36) in the pumping parameter which looks here as

$$A_0[1 + \beta_{\text{pump}}\cos(\Omega\tau)] + \Upsilon_{\text{total}}\sum_{j=1}^{N}\frac{dm_j}{d\tau} + \Upsilon_i\frac{dm_i}{d\tau},$$

where Υ_{total} is the feedback coefficient taken on the total intensity, Υ_i is the feedback coefficient taken on the i-th intensity. The weak sinusoidal pump modulation is introduced in the last expression for obtaining the transfer functions.

The set of equations (7.10) has been investigated by numeral methods. The presented results are obtained neglecting the gain dispersion ($\widetilde{L}_j = 1$) while all the mode discrimination refers to losses. It is assumed that extra losses β_j linearly increase with the mode number:

$$\beta_j = \beta(j-1).$$

If the mode discrimination is weak, this simplification does not change qualitatively the results.

The dependence of the decrements of relaxation oscillations on one of the feedback coefficients (in this case this is Y_4) when a negative value of Y_{total} is fixed is illustrated in Fig. 7.4. The horizontal axis is divided into three domains by two bifurcation points.

The steady state is stable in domain I. As the system approaches the right boundary of the domain I the resonance peak at the frequency Ω_4 becomes narrower, as seen in Fig. 7.5a. All the transfer functions are normalized to their maximal values and, therefore, we do not see the growth of the peak. It should be noted that the existence of a selective positive feedback makes the low-frequency relaxation oscillations uncompensated and it is more convenient to detect them in the total intensity where they are absent without the selective feedback. Effects of the peak narrowing and increasing of its amplitude after switching the selective feedback are observed also in the spectrum of fluctuations of the total intensity of a Nd:YAG laser [167].

Transition in domain II is accompanied by changing of the sign of the decrement θ_4, which means loss of stability and rise of the regime of periodic intensity oscillations at a frequency close to Ω_4. The modulation depth increases with increasing Y_4 and reaches 100% close to the right boundary of domain II.

In domain III the in-phase relaxation oscillations at frequency Ω_1 are

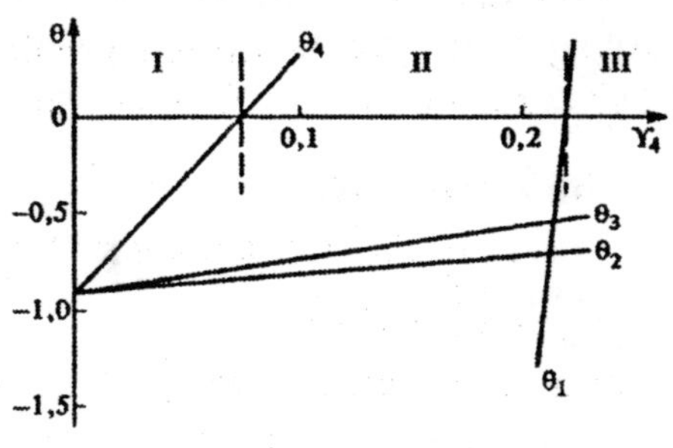

Fig. 7.4. Phase diagram of the diode-pumped laser with combined derivative feedback. $G = 7 \cdot 10^3$; $A = 2$; $\beta = 0.03$; $Y_{total} = -0.005$ [167].

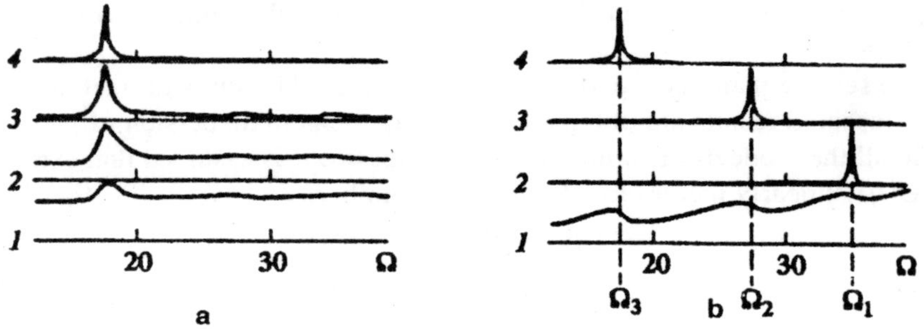

Fig.7.5. Transfer function $K_{total}(\Omega)$ at $G = 7 \cdot 10^3$; A = 2; β = 0.03 for different values of Y_{total} and Y_j parameters: (a) $Y_{total} = -0.015$, $Y_1 = Y_2 = Y_3 = 0$, $Y_4 = 0.010$ (1); 0.030 (2); 0.060 (3); 0.070 (4); (b) $Y_{total} = 0$; $Y_j = 0$ (1); $Y_{total} = -0.015$; $Y_1 = Y_3 = Y_4 = 0$, $Y_2 = 0.039$ (2); $Y_{total} = -0.015$; $Y_1 = Y_2 = Y_4 = 0$, $Y_3 = 0.041$ (3); $Y_{total} = -0.015$; $Y_1 = Y_2 = Y_3 = 0$, $Y_4 = 0.075$ (4) [167].

also undamped and in the heart of this domain oscillations become chaotic. So, with increasing of the coefficient of the selective feedback Y_4 the following scenario is realized in the system: subcritical Hopf bifurcation – regular oscillations – transition to chaos through quasiperiodicity.

The position of the bifurcation points depends on the chosen value of the feedback parameter Y_{total}. If $Y_{total} = 0$, then domain II is absent; the growth and narrowing of the resonance peak takes place only when Y_4 is small. Further increasing of Y_4 leads to a simultaneous excitation of undamped pulsations at the frequencies Ω_4 and Ω_1. In this case a different scenario is realized, according to which the chaotic pulsations at the frequency Ω_1 are established just after passing the threshold of the Hopf bifurcation, which is now a supercritical one.

In the considered example the weakest of the excited laser modes was chosen to realize the positive feedback. The intensity of this mode is denoted as m_4. The most flexible in this case is the relaxation oscillation at the lowest frequency Ω_4. Switching the input of the feedback net to any other mode m_j except m_1, we pass from the relaxation oscillation Ω_4 to Ω_j (Fig. 7.5.b). Thus, we confirm the existence of the definite connection between the chosen cavity mode and the concrete relaxation oscillation.

7.2 Laser with a Nonlinear Absorber

On a level with opto-electronic feedback some other methods of influencing dynamics of an autonomous laser, especially those, which use the nonlinear properties of intracavity materials, have become widespread. Nonlinearities of different types are good for these purposes but saturable absorbers are used most frequently.

The scheme of the laser with an extra nonlinear element is rather simple.

Such an element (nonlinear filter) is placed coaxially with the rod inside the laser cavity. This can reduce the laser stability, and produce nonlinear pulsations if the density of the absorber is high enough. This effect occurs in solid-state lasers with intracavity nonlinear filters of organic dyes [505] and in molecular gas lasers with nonlinear absorbing gas cells [506–509]. The latter is most suitable for CW operation and, therefore, CO_2 lasers with various nonlinear filters have been used in a majority of experimental investigations of complicated dynamical processes in lasers of this design [510–521].

Stabilization of the laser power requires the use of a nonlinear element, in which the losses grow with an increase in the laser power [505, 522, 523]. An example of such a medium is a zinc phtalocyanin solution, the nonlinearity of which is due to the absorption from the excited state of phtalocyanin molecules. Introducing a cell with this solution into the cavity of a ruby laser leads to noticeable smoothing of the pulsations. Simultaneously the divergence of the laser beam is decreased and the field distribution over the beam becomes more uniform.

Nonlinear growth of the losses can also be ensured by stimulated scattering or by intracavity second harmonic generation.

7.2.1 Two-Level Rate Equation Model of Laser with a Nonlinear Filter; Steady States and Their Stability

Consider the simplest model of laser with a nonlinear filter [524,525]. Assume that the laser field is uniform both in the laser rod and in the nonlinear cell although the field intensities in these two volumes are not necessarily the same. The absorbing medium may have the same or the different chemical composition as the laser medium. The only important thing is that the transition frequencies in two media are matched and that they can be regarded as two-level media. Fast relaxation of the polarization permit the adiabatic elimination of the corresponding variables and use of a set of rate equations

$$\frac{\mathrm{d}m}{\mathrm{d}\tau} = Gm(n+n_{\mathrm{a}}-1)\,, \tag{7.11a}$$

$$\frac{\mathrm{d}n}{\mathrm{d}\tau} = A-n(m+1)\,, \tag{7.11b}$$

$$\frac{\mathrm{d}n_{\mathrm{a}}}{\mathrm{d}\tau} = \delta A-n_{\mathrm{a}}(\rho m+\delta)\,. \tag{7.11c}$$

The quantities that refer to the nonlinear filter are marked by subscript *a*. Equations (7.11) are a generalization of Eqs. (4.9). We have introduced dimensionless variables

$$\tau=\gamma_{\|}, \quad m=\frac{\beta_a B}{\gamma_{\|}}M, \quad n=\frac{B}{\kappa}N, \quad \delta=\frac{\gamma_{\|a}}{\gamma_{\|}}, \quad \rho=\frac{B_a\,|\psi(\mathbf{r}_a)|^2}{B\,|\psi(\mathbf{r})|^2}. \tag{7.12}$$

In what follows we assume that A is positive while A_a may have either sign in principle.

Equations (7.11) are valid for media with either simple or complex energy structure as long as the rate at which quasiequilibrium is established between sublevels within the energy level is high enough. This means that rate equations such as (7.11) are useful for the analysis of the dynamical processes in molecular gas lasers with gaseous nonlinear filters.

Equations (7.11) have three branches of steady-state solutions: the trivial solution

$$\overline{m}_0=0, \quad \overline{n}_0=A, \quad \overline{n}_{a0}=A_a \tag{7.13}$$

and two nonzero-intensity branches

$$\overline{m}_{\pm}=\frac{1}{2}\left\{\left[A-1+\frac{\delta}{\rho}(A_a-1)\right]\pm\sqrt{\left[A-1+\frac{\delta}{\rho}(A_a-1)\right]^2+4\frac{\delta}{\rho}(A+A_a-1)}\right\}, \tag{7.14a}$$

$$\overline{n}=\frac{A}{1+\overline{m}}, \tag{7.14b}$$

$$\overline{n}_a=\frac{A_a}{1+\overline{m}\rho/\delta}. \tag{7.14c}$$

The branches that correspond to the real and positive values of $\overline{m}$ are physically relevant. If the self-excitation condition

$$A+A_a-1>0 \tag{7.15}$$

is met, i.e., the laser is subject to soft excitation, then there are two steady states: $\overline{m}_0$ and $\overline{m}_+$. The first is unstable while the stability of the second requires investigation. If the inequality (7.15) is reverted then only hard excitation is possible and there are three steady states. The domain of hard excitation in the plane (A, A_a) is separated from the domain in which lasing is impossible, by the boundary curve

$$\left[A-1+\frac{\delta}{\rho}(A_a-1)\right]^2+4\frac{\delta}{\rho}(A+A_a-1)=0, \tag{7.16}$$

and from the domain of soft excitation by the threshold line

$$A + A_a - 1 = 0 \tag{7.17}$$

Both of these curves are shown in Fig. 7.6a. The point of their intersection has the coordinates

$$A_1 = \frac{1}{1-\delta/\rho}, \quad A_{a1} = -\frac{\delta/\rho}{1-\delta/\rho}. \tag{7.18}$$

Since $A_1 > 0$, from Eq. (7.18) we get a necessary condition for the existence of a hard excitation domain

$$\rho > \delta. \tag{7.19}$$

Sometimes it is more convenient to use the steady-state solutions $\overline{m}, \overline{n}$ rather than A, A_a as the parameters. In the plane ($\overline{m}, \overline{n}$) the threshold line (7.17) is given by

$$\overline{n} = \overline{n}_2 = \frac{1}{1-\delta/\rho}, \tag{7.20}$$

and the boundary curve (7.16) is

$$\overline{m} = m_1 = \overline{n}(1-\delta/\rho) - 1. \tag{7.21}$$

Partition of the parameter plane ($\overline{m}, \overline{n}$) into domains with different types of laser excitation is shown in Fig. 7.6b. The branch of steady-state solutions $\overline{m}_+$ is located above and the solutions $\overline{m}_-$ below the line (7.21). Lasing is impossible for $\overline{n} < 1$.

Linearization of Eqs. (7.11) near any of the nontrivial fixed points leads to a cubic characteristic equation $\lambda^3 + a_1\lambda^2 + a_2\lambda + a_3 = 0$ with coefficients

$$\begin{aligned} a_1 &= C_1 + C_2, \\ a_2 &= C_1C_2 + G(D_1 + D_2), \\ a_3 &= G(C_1D_2 + C_2D_1), \end{aligned} \tag{7.22}$$

where

$$C_1 = \overline{m} + 1, \quad C_2 = \rho\overline{m} + \delta, \quad D_1 = \overline{m}\overline{n}, \qquad D_1 = \rho\overline{m}\overline{n}_a. \tag{7.23}$$

According to the Routh–Hurwitz criterion the instability of the fixed point occurs when either of the two inequalities $a_3 < 0, \quad a_1a_2 - a_3 < 0$ is satisfied. Writing them in the form

$$C_1D_2 + C_2D_1 < 0, \tag{7.24}$$

$$C_1C_2(C_1 + C_2) + G(C_1D_1 + C_2D_2) < 0 \tag{7.25}$$

and remembering that $C_1 > 0, C_2 > 0, D_1 > 0$, we find the necessary instability condition

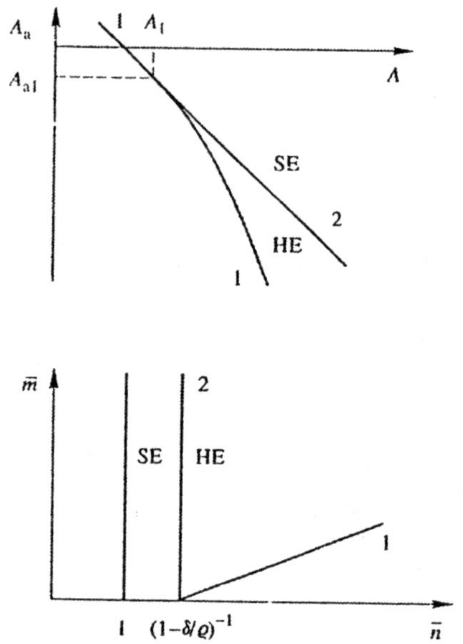

Fig. 7.6. Diagram of steady states of a laser with a saturable absorber (a) in the parameter plane (A, A_a) and (b) in the plane $(\bar{n}, \bar{m})$: 1 – boundary line; 2 – threshold line that separates the regions of soft (SE) and hard (HE) excitation of laser.

$$A_a < 0. \tag{7.26}$$

This means that only the absorbing medium can make the steady state unstable.

Using (7.23) and

$$\bar{n} + \bar{n}_a = 1 \tag{7.27}$$

inequality (7.24) is transformed to

$$\bar{m} < \bar{m}_1. \tag{7.28}$$

This means that the branch $\bar{m}_-$ is absolutely unstable.

Consider now inequality (7.25). One necessary condition, under which it is satisfied, is (7.26). The other follows from the requirement that the sum $C_1D_1 + C_2D_2$ should be negative. In expanded form this requirement reads

$$\bar{m}[\bar{n}(1-\rho^2)+\rho^2]+[\bar{n}(1-\rho\delta)+\rho\delta]<0. \tag{7.29}$$

The curve

$$\bar{m} = \frac{\bar{n}(\rho\delta-1)-\rho\delta}{\bar{n}(1-\rho^2)+\rho^2} \tag{7.30}$$

isolates the domain of absolutely stable lasing from the plane $(\bar{m}, \bar{n})$. For $\rho < \delta$, when only soft excitation is possible, this domain is located above the curve (7.30). If, in addition, $\rho\delta < 1$, then the steady state is stable for any arbitrary parameter values. In the opposite case, for $\rho > \delta$, the domain of absolute stability is situated below the curve (7.30) but above $\overline{m} = \overline{m}_1$.

In the expanded form, the sufficient instability condition (7.25) looks like

$$(\overline{m}+1)(\rho\overline{m}+\delta)(\overline{m}+1+\rho\overline{m}+\delta)+G\overline{m}[\overline{n}(\overline{m}+1)+\rho(1-\overline{n})(\rho\overline{m}+\delta)]<0. \tag{7.31}$$

The curve that limits the unstable domain

$$\overline{n} = \frac{G\rho\overline{m}(\rho\overline{m}+\delta)+(\overline{m}+1)(\rho\overline{m}+\delta)(\overline{m}+1+\rho\overline{m}+\delta)}{G\overline{m}[\overline{m}(\rho^2-1)+\rho\delta-1]}, \tag{7.32}$$

is rather complicated. However, the asymptotic behaviour for $\overline{n} \to \infty$ is easy to ascertain. For $\rho > 1$ the asymptotes are the straight lines

$$\overline{m} = 0, \quad \overline{m} = G\rho^{-1}(\rho-1)\overline{n}, \tag{7.33a}$$

and for $\rho < 1$ the asymptotes are

$$\overline{m} = 0, \quad \overline{m} = (\rho\delta-1)/(1-\rho^2), \tag{7.33b}$$

When the plain ($\overline{m}, \overline{n}$) is mapped onto the plain (A, A_a), the curve (7.30) transforms onto straight line

$$A + \rho\delta A_a = 0. \tag{7.34}$$

Different variants of the stability diagrams in both representations are given in Fig. 7.7.

In the limiting cases of practical interest the general criteria of stability can be noticeably simplified.

1. $\rho/\delta >> 1$. For the steady state the approximate expressions

$$\overline{m} = A-1, \quad \overline{n} = 1, \quad \overline{n}_a = \frac{\delta}{\rho}\frac{A_a}{A\text{-}1} \tag{7.35}$$

are valid, which follows directly from Eq. (7.14). Inequality (7.31) with the additional condition $\rho\delta|A_a| >> A$ takes the form

$$|A_a| > \frac{\rho}{G\delta}A(A-1). \tag{7.36}$$

If, for example, we consider a ruby laser with a methanol solution of vanadyl phtalocyanin as a nonlinear filter, then we have the estimates [526, 527]; $\delta = 10^3$ ($\gamma_\parallel \approx 10^3\ s^{-1}$, $\gamma_{\parallel a} \approx 10^6\ s^{-1}$), $\rho \approx 10^4$, i.e., $\rho/\delta \approx 10$. Assuming $G = 10^5$ and $A = 10$ we obtain the instability condition $|A_a| < 10^{-2}$.

2. $\rho/\delta << 1$. The steady-state solution is

$$\overline{m} = \frac{A-1-|A_a|}{1+|A_a|}, \quad \overline{n} = 1+|A_a|, \quad \overline{n}_a = A_a\left(1-\frac{\rho}{\delta}\frac{A-1-|A_a|}{1+|A_a|}\right). \tag{7.37}$$

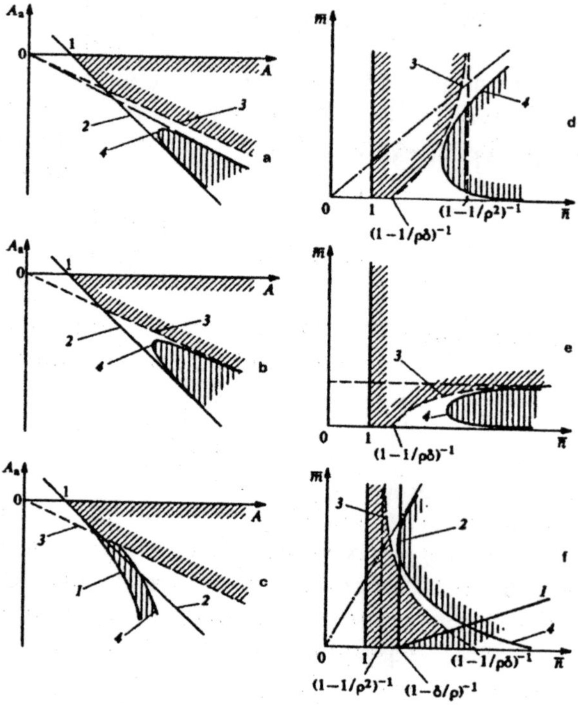

Fig. 7.7. Diagrams of stability of the steady-state solutions to Eqs. (7.10): (a,d) $1<\rho<\delta$; (b,e) $\rho<1$, $\rho\delta>1$; (c,f) $\rho>\delta$ (1 – boundary line; 2 – threshold line; 3 – boundary of the domain of absolute stability (skew hatching); 4 – boundary of the unstable region (vertical hatching)).

The steady state becomes unstable under the condition

$$|A_a|>\frac{\delta}{G\rho}\frac{A}{A-1}. \tag{7.38}$$

Using the values $G=10^5, \rho=10^4, \delta=10^5$, which correspond to $\gamma_{\parallel a}\approx 10^8\ \mathrm{s}^{-1}$ [528, 529], we find the density of the nonlinear filter for which the steady state is unstable to be given by $|A_a|>10^{-4}$.

3. $\rho/\delta\approx 1$. Here

$$\bar{m}=A-1-|A_a|,\quad \bar{n}=\frac{A}{A-|A_a|},\quad \bar{n}=\frac{A_a}{A-|A_a|} \tag{7.39}$$

and, since the first term in (7.31) can be neglected, the instability criterion is written as

$$\rho\delta\,|A_a|>A. \tag{7.40}$$

The approximate values of the roots of the characteristic equation are easily found if two of them are complex with predominantly imaginary parts. In this case

$$\lambda_{1,2} \approx -(a_1 a_2 - a_3)/(2a_2) \pm i\sqrt{a_2}, \quad \lambda_3 \approx a_3 / a_2. \qquad (7.41)$$

Thus, crossing the boundary $a_1 a_2 = a_3$ means a Hopf bifurcation (of a supercritical type).

7.2.2 Nonlinear Pulsations in a Laser with Saturable Absorber

For a single-component laser medium, the fixed points determine unambiguously the global structure of the phase space of the laser. Rather often things are not so simple if a nonlinear filter is used. This can be demonstrated using a laser system with $\rho / \delta >> 1$ as an example. The presence of small parameters δ / ρ and $1/G$ make it possible to divide the solution of Eq. (7.11) into several stages [289, 530].

The pumping stage is characterized by such low laser intensity ($m < \delta / \rho$) that stimulated transitions do not influence the populations of either the laser medium or the nonlinear filter. Thus, the equations

$$\frac{dn}{d\tau} = A - n, \quad \frac{dn_a}{d\tau} = \delta(A_a - n_a) \qquad (7.42)$$

are valid, which follows from Eqs. (7.11) for $m = 0$.

The segments of phase space trajectories that correspond to the pumping stage are located in the plane (n, n_a). Their analytical expression is the first integral of Eqs. (7.42)

$$\frac{A_a - n_a}{A_a - n_{a1}} = \left(\frac{A - n}{A - n_1} \right)^{\delta}, \qquad (7.43)$$

where n_1 and n_{a1} are the initial values. Above the self-excitation boundary (the line n+n_a = 1) the motion along the trajectories (7.43) is unstable. Therefore, the segments of the curves in this domain are shown by dash lines in Fig. 7.8.

Beyond the self-excitation boundary the laser fluctuations cause the representative point to move to one of the trajectories going away from the plane (n, n_a). Nevertheless, Eqs. (7.42) and, therefore, (7.43) remain valid until the laser intensity reaches the value $m_2 \approx \delta / \rho$. This time is sufficient for the inversion to grow to the value n_2, which can be estimated using Eq. (3.29)

$$\tilde{\eta} = n_2 - 1 - n_a \approx \left[\frac{2}{G} (A - 1 - |A_a|) \ln \frac{m_2}{m_{\min}} \right]^{1/2}. \qquad (7.44)$$

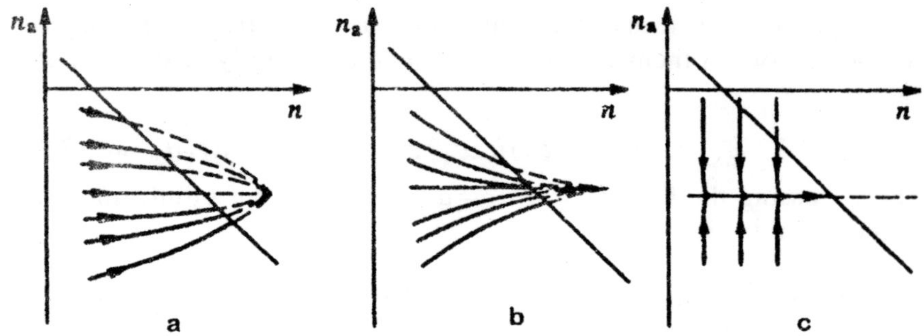

Fig. 7.8. Varieties of phase trajectories at the pumping stage in a laser with saturable absorber at (a) $\delta < 1$; (b) $\delta > 1$; (c) $\delta \gg 1$.

The saturation stage of nonlinear filter occurs for intensities given by $\delta/\rho < m < A-1$. As before, the saturation of laser medium is small, such we can assume $n = n_2$, since the filter saturation stage is rather short. This stage is described by

$$\frac{\mathrm{d}m}{\mathrm{d}\tau} = Gm(n_2 + n_a - 1), \quad \frac{\mathrm{d}n_a}{\mathrm{d}\tau} = -\rho n_a m .$$

Integrating these equations within the limits

$$\delta / \rho \le m \le A - 1, \quad n_{a2} \le n_a \le \bar{n}_a$$

we find

$$\ln \frac{\bar{n}_a}{n_{a2}} = \frac{\bar{n}_a}{n_{a2}} - 1 + \frac{(A-1)\rho}{Gn_{a2}}, \tag{7.45}$$

bearing in mind that $\rho / \delta >> 1$ and $n_2 - 1 = -n_{a2}$. The filter saturation will be complete ($\bar{n}_a / n_{a2} << 1$) if $(A-1)\rho / G >> |A_a|$. This condition follows from Eq. (7.45) and is assumed to be satisfied elsewhere below.

The high field intensity ($m > A-1$) and the complete saturation of the nonlinear filter is characteristic of the *emission stage*. This stage is described by a set of equations

$$\frac{\mathrm{d}m}{\mathrm{d}\tau} = Gm(n-1), \quad \frac{\mathrm{d}n}{\mathrm{d}\tau} = -mn ,$$

which have the first integral

$$m - m_2 = G\left(n_2 - n - \ln \frac{n_2}{n}\right). \tag{7.46}$$

By the end of this stage the inversion falls off to the value n_3, which can be found by putting $m_3 = m_2$ in Eq. (7.46):

$$n_2 - n_3 = \ln\frac{n_2}{n_3}. \tag{7.47}$$

The value $n = n_3$ is the initial inversion for the next pumping stage. Matching these segments of the phase space trajectory we find the whole trajectory. Projections of these trajectories onto the plane (n, n_a) are presented in Fig. 7.9.

We now specify the conditions for existence of trajectories that do not tend to the time-independent solutions. Assume for simplicity that the representative point moves along the curve (7.43) until the self-excitation boundary is crossed. Bearing in mind that $n_{a2} = 1 - n_2, n_{a1} = 0$, and introducing the notation

$$x_1 = 1 - n_1, \quad y_2 = n_2 - 1$$

we write (7.43) as

$$y_2 = |A_a|\left[1 - \left(\frac{A - 1 - y_2}{A - 1 + x_1}\right)^{\delta}\right]. \tag{7.48}$$

This equation transforms the points of set x_1 to set y_2.

Taking into account that points n_3 and n_1 belong to one set we write (7.47) as

$$\ln(1 - x_1) + x_1 = \ln(1 + y_2) - y_2. \tag{7.49}$$

This equation transforms the set y_2 to set x_1.

Eq. (7.49) is satisfied by the function $x_1 = \Phi(y_2)$. Near $y_2 = 0$ this function reduces to

$$x_1 \approx y_2 - \frac{2}{3}y_2^2,$$

and in the limit of $y_2 \to \infty$ we have $\Phi(y_2) \to 1$. In Fig. 7.10 the function is represented by curve 1. The function $x_1 = F(y_2)$, which satisfies (7.48), lies in the range $0 \le y_2 \le |A_a|$ and $F(y_2) \to \infty$ for $y_2 \to |A_a|$. The function $F(y_2)$ is monotonic under the condition $\delta |A_a| < A - 1$ (curve 2 in Fig. 7.10) and it has the form of curve 3 with a minimum in the opposite case.

The curves $x_1 = \Phi(y_2)$ and $x_1 = F(y_2)$ will intersect if their derivatives at the origin of coordinates satisfy the inequality $\partial\Phi/\partial y_2 > \partial F/\partial y_2$, which is equivalent to

$$2\delta |A_a| > A - 1. \tag{7.50}$$

This inequality represents the desired criterion for existence of trajecto-

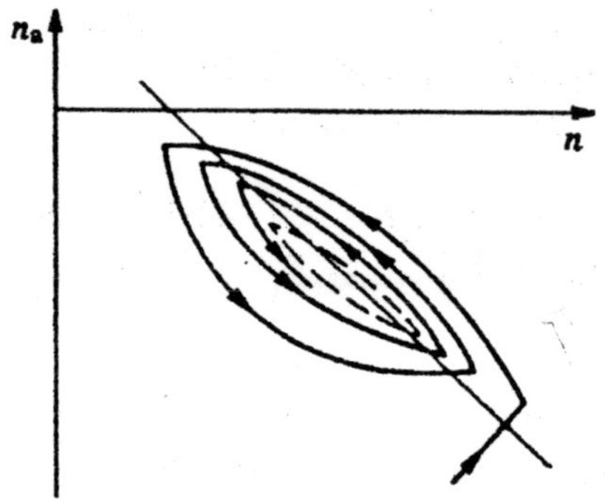

Fig. 7.9. The phase space trajectory projection onto the plane (n, n_a). The fatty line shows the limit cycle.

ries that do not tend to the steady states. The intersection of the curves corresponds to a closed trajectory, or a limit cycle. The fact that the sequence of points of set x_1 converges to the point of intersection of the sequence functions indicates that the limit cycle is stable. This conclusion is illustrated by a broken dash line in Fig. 7.10, which follows the point-to-point transformations.

An important result is that inequality (7.50) does not coincide with (7.36). For $2\rho A>G$ the inequality (7.36) is stronger. In the domain

$$\frac{\rho}{G}A(A-1)>\delta\,|\,A_a\,|>\frac{1}{2}(A-1) \tag{7.51}$$

there is a stable limit cycle in spite of the fact that the fixed point is stable as well. Therefore, either CW lasing or emission of isolated pulses is possible depending on the initial conditions.

In order to find the energy and temporal characteristics of the laser pulses, one needs to solve the set of Eqs. (7.48)–(7.49). If the equilibrium populations in the nonlinear filter are completely restored during the period between pulses, then the approximate solution is apparent: $y_2=A_a$. The constraint $A_2 \ll 1$ permits the use of Eqs. (3.33)–(3.35).

Since the limit cycle intersects the self-excitation boundary, it can be inferred that powerful undamped pulsations are impossible in the case of hard excitation. After the first pulse is over, the system will be below threshold and return to the zero fixed point. Thus, the instability of the steady-state solution for hard excitation (bistability) means, as a rule, that long-term laser action is not possible at all.

Let us consider one more important situation. Assume that the nonlinear filter is inertialess in the sense that $\gamma_{\|a} \gg \kappa, \gamma_{\|}$. We thus can eliminate adiabatically the variable n_a from the equations putting $dn_a/d\tau=0$. Eqs. (7.11) are most greatly simplified if field intensities noticeably saturate the filter are not achieved. The maximum field intensity is

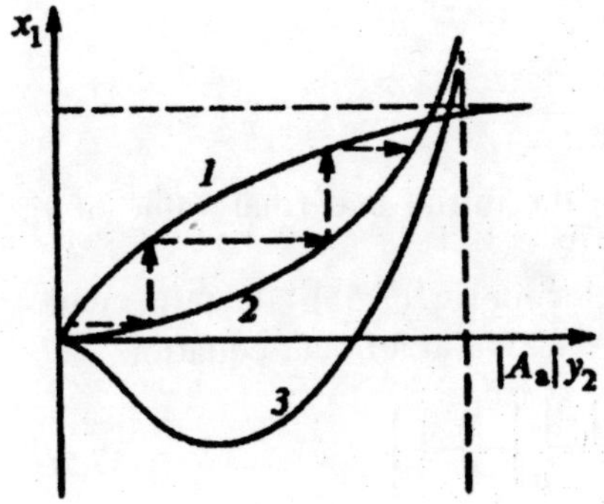

Fig. 7.10. Diagram explaining the conditions for existence of a limit cycle and its finding by the point-to-point transformations.

$m_{max} \approx G(|A_a| + \tilde{\eta})^2/2$. Its smallness, compared to δ/ρ, guarantees the filter saturation. The necessary condition is

$$\tilde{\eta}^2 << \frac{2\delta}{G\rho}, \tag{7.52}$$

or, taking account of relation (3.29),

$$\frac{\delta}{\rho} >> (A-1-|A_a|)\ln\frac{A-1-|A_a|}{m_{min}}. \tag{7.53}$$

Numerically, this inequality means $\delta/\rho > 25$ if we use the solid-state laser as an example. The nonlinear filter density has an upper limit due to the condition

$$|A_a| << \sqrt{\frac{2\delta}{G\rho}}, \tag{7.54}$$

which is similar to Eq. (7.52).

The smallness of the derivative and the assumption of weak saturation of the filer make it possible to reduce (7.11) to

$$\frac{dm}{d\tau} = Gm\left[n-1+A_a\left(1-\frac{\rho}{\delta}m\right)\right], \tag{7.55a}$$

$$\frac{dn}{d\tau} = A - n(m+1). \tag{7.55b}$$

These equations represent the particular case of (7.1), where $\beta = 1 - A_a(1 - m\rho/\delta)$. The criterion of instability of the steady-state solution coincides both with Eq. (7.38) and Eq. (7.8).

Partition of the phase space trajectories into segments with fast and slow motion is the same as for the case of free oscillations. In the pumping stage we assume $m = 0$ and find

$$\eta_2 = -\eta_1 - \frac{2}{3}\frac{\eta^2}{A-1},$$

where $\eta = n - 1 + A_a$, and η_1 and η_2 are the initial and final value of η, respectively, during the pumping stage.

During the emission stage we can neglect in Eq. (7.55b) the terms without m and dividing (7.54a) by (7.54b) we arrive at a linear equation

$$\frac{dm}{dn} = \frac{G\rho}{\delta}\frac{m}{n} + G\left(\frac{1+|A_a|}{n} - 1\right). \tag{7.56}$$

It has the first integral

$$m_2 n_2^S - mn^S = G\left[\frac{1}{S+1}(n^{S+1} - n_2^{S+1}) - \frac{1+|A_a|}{S}(n^S - n_2^S)\right], \tag{7.57}$$

where $S = (G\rho/\delta)|A_a|$. During the emission stage, the initial values are n_2 and $m_2 = m_b$, which are achieved by the end of the pumping stage.

Let us find the inversion n_3 by the end of the emission stage when $m_3 = m_2$. To do this we expand n^S in a Taylor series:

$$n^S = (1+|A_a|)^S + S(1+|A_a|)^{S-1}\eta + \frac{1}{2}S(S-1)(1+|A_a|)^{S-2}\eta^2 + \frac{1}{6}S(S-1)(S-2)(1+|A_a|)^{S-3}\eta^3 \tag{7.58}$$

Higher order terms can be neglected by virtue of the condition

$$S\eta = \frac{G\rho}{\delta}|A_a\eta| << 1,$$

which is due to inequalities (7.52) and (7.54). Substituting (7.58) into (7.57) we find, to accuracy within terms of order η^3:

$$\eta_3 = -\eta_2 - \frac{2}{3}(S-1)\frac{\eta^2}{1+|A_a|}.$$

The necessary condition for existence of a limit cycle $|\eta_3 - \eta_1| \geq 1$ is given by

$$|A_a| \geq \frac{\delta}{G\rho}\frac{A}{A-1},$$

which fully coincides with (7.38).

7.2.3 Experiments with a CO_2 Laser. The Rate Equation Model of a Three-Level Laser with a Two-Level Nonlinear Filter

The most systematic experimental investigations of the influence of a saturable absorber on laser dynamics were carried out with CO_2 lasers [510-

521]. Besides the fact of the steady-state instability, these experiments have revealed the features of laser behavior, which do not find explanation in terms of the simplest two-level model described above. The most significant of these features is that two types of time-dependent processes are observed. The first one begins to exhibit sinusoidal oscillations just above the instability threshold. The amplitude increases and the oscillation transform to a sequence of smooth single pulses as the laser is pushed inside the unstable region (Fig. 7.11*a,b*). The process of other type exhibits quite a different pulse shape where the leading intense peak is followed by a tail of diverging oscillations and then the laser cutoff occurs. After a time of the order of microseconds the process restart (Fig. 7.11*c,d*) the pause can be shorter than the emission stage of the process.

The scenarios, by which the dynamic regimes change when the control perimeters (the pumping level, the cavity tuning with respect to the transition frequencies in both media, and the density and relaxation rates of saturable absorber) are varied, can also be different. The existence of chaotic regimes has been revealed experimentally in Refs. [519,520]. The route to chaos may contain a cascade of period doubling. The matter is so when smooth pulses are generated. In the second case the process of changing the number of oscillation in the pulse tail is also takes place.

The abundance of control parameters and the diversity of experimental investigations by different authors greatly impede a systematic description of the existing possibilities. We, therefore, restrict ourselves to the experimental phase diagram in the parameter plane (cavity tuning, absorbing cell pressure) taken from [519,520]. The diagram given in Fig. 7.12 clearly demonstrates the rich dynamic potential of the system.

The existence of regimes with complex form of signals is associated with the topological situation in the phase space described by Shil'nikov [531]. According to his theory, the representative point, having escaped along the separatrix from the vicinity of a saddle-focus fixed point, then returns to it moving along a spiral trajectory. Therefore, a train of damped oscillations should follow the first intense peak. However, the inverse sequence takes place in CO_2 lasers, i.e., the oscillations in the pulse tail grow instead of being damped. Such a sequence corresponds to the fixed point of an unstable focus – stable node type. The pause with a zero field value between pulses means that the system spends some time near the other, trivial fixed point.

Numerical simulation shows that solutions of the two types mentioned above can be obtained in the laser model assuming a three-level approximation of the laser medium [532, 533]. Using the energy diagram shown in Fig. 7.13 we write the set of equations in dimensionless form consistent with Eqs. (7.11)

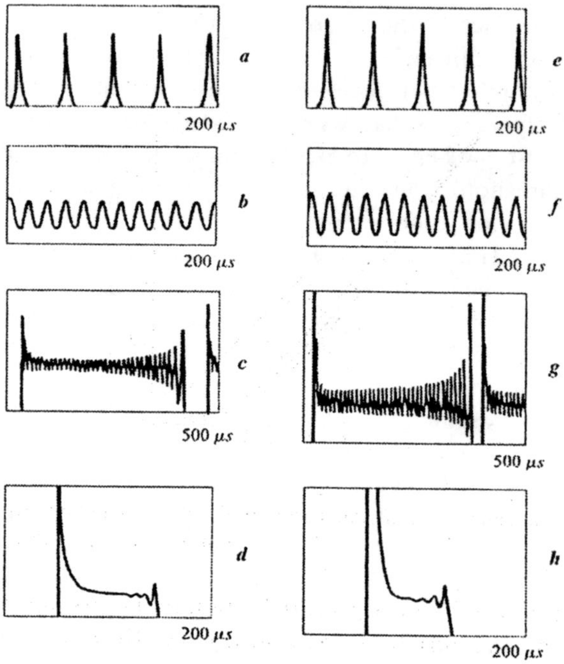

Fig. 7.11. Observed (a–d) and calculated (e–h) using Eqs. (7.59) shapes of the output intensity of a CO_2 laser with a saturable absorber [513]: a – CH_3OH pressure is 325 mTorr, discharge current 6 mA; b – CH_3OH pressure is 325 mTorr, discharge current 8 mA; c – HCOOH pressure is 77 mTorr; d – SF_6 pressure is 23 mTorr.

$$\frac{dm}{d\tau} = Gm(n + n_a - 1),$$

$$\frac{dn}{d\tau} = A - n(m+1) + \left(\tilde{w}_{21} - \tilde{w}_{31} - \frac{3A}{R}\right)n_2, \tag{7.59}$$

$$\frac{dn_2}{d\tau} = -\tilde{w}_{21}n_2 + \tilde{w}_{32}n + \frac{1}{2}mn,$$

$$\frac{dn_a}{d\tau} = \delta A_a - n_a(\rho m + \delta).$$

These equations (excluding the last one) have been obtained from Eqs. (3.113) by adiabatic elimination of the polarization $\tilde{P}_{32}$ and time renormalization from $\tau = \gamma_{32}t$ to $\tau = (w_{31} + 2w_{32})t$. The parameters marked by a dash on top are also normalized to $w_{31} + 2w_{32}$. As an additional vari-

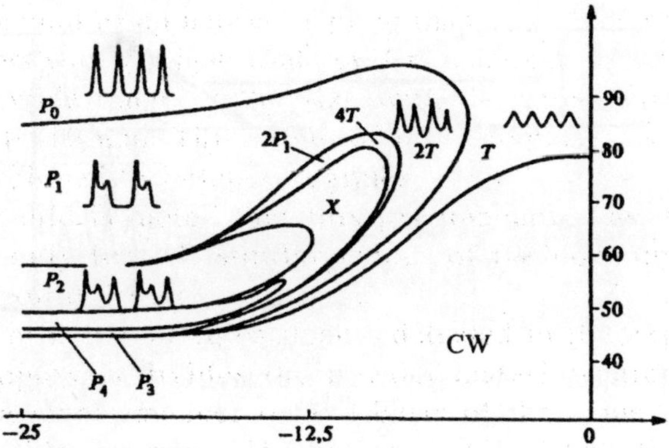

Fig. 7.12. Phase diagram of a CO_2–laser with a CH_3I absorber in the control parameter plane (cavity detuning, pressure in the absorption cell): P_n stands for the domains of existence of regimes with n oscillations at the pulse tail, C is the region of chaotic behaviour [519].

able with respect to (7.11) we have chosen the population of the lower laser level n_2 as in Section 3.5.

Examples of numerical solutions of the set of rate equations similar to (7.59) are given in Fig. 7.11*e–h*. The similarity of these solutions to the oscillograms of the real process in a CO_2 laser is notable in almost all details. An important role is played by the absorber saturation parameter ρ/δ. This parameter is less than unity in the upper two Figs. (7.11*e* and 7.11*f*) and more than unity in the lower two Figs. (7.11*g* and 7.11*h*). The pause between pulses decreases with decreasing ρ/δ and when it disappears the regime of the second type is replaced by the first one.

The similarity of the numerical simulations in Fig. 7.11 and the experimental oscillograms is, of course, very impressive. This alone, however, does not allow to think that the problem of interpretation of the experimental results is fully exhausted. Obviously, the resemblance between the solutions and experimental data can also be achieved by parameter fitting in the model of laser with a two-level active medium without adiabatic elimination of the polarization of both (amplifying and absorbing) media [534]. These models have been intensely investigated by many authors but are not considered in this book.

7.3 Laser with a Nonlinear Dielectric

In the previous section we have demonstrated the connection between the loss nonlinearity in the medium filling the cavity and the laser dynamics. Besides the bulk losses, the laser cavity has other forms of losses, which are sensitive to the laser field. For example, the reflectivity at the inter-

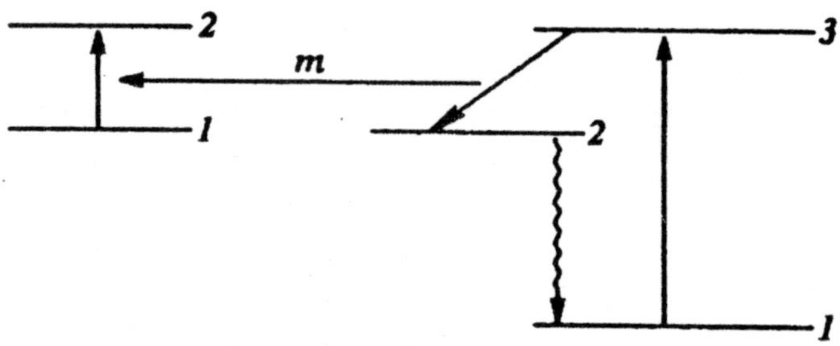

Fig. 7.13. Energy-level diagram of a three-level laser with a two-level absorber.

faces can be changed. In semiconductors the reflectivity depends on the electron and hole densities in the surface layer. The carrier density increases with increasing intensity of light incident onto the semiconductors surface, thus increasing the reflectivity. Therefore, the cavity Q-factor is an increasing function of laser power if one of the cavity mirrors is a semiconductor crystal. A laser with such a cavity can generate giant pulses [535–537].

Another possibility is passive modulation of diffraction losses. This requires the use of a medium for which the refractive index depends on the field intensity. A layer of such a medium acts on a beam that passes through it as a lens with an intensity-controlled focal length. Such nonlinear properties in the optical range are exhibited by many dielectrics. They are strongest in organic solvents [538] and weaker in the laser crystal hosts [539, 540].

The presence of a nonlinear dielectric in the cavity leads to dynamic deformation of the modes, which causes a dependence on the laser power of both the diffraction losses and the effective filling factor.

7.3.1 Examples Showing the Influence of the Refractive Index Nonlinearity on Laser Dynamics

The influence of a nonlinear dielectric on solid-state laser dynamics has been observed experimentally. When an organic solvent is placed in a plane-parallel laser cavity, the spikes are shortened and their amplitude and repetition time increase [541]. The changes in laser behaviour are markedly increased if one of the cavity mirrors is tilted [542–544].

There are some facts, however, which do not permit an unambiguous interpretation, although superficially they resemble those discussed above. In a ruby laser with a misaligned (or, more generally speaking, unstable) cavity, intense spikes arise without adding nonlinear elements [89–91,103, 545]. Possibly we should add to this list the result of Ref. [546] concerning a ruby laser with longitudinal pumping by a narrow beam of an argon ion laser. Sharp enhancement of the spike amplitudes was provided by slight tilting of the pumping beam with respect to the cavity axis.

While for solid-state and most of other types of lasers the nonlinear dielectric is modeled as a lens, one must consider a nonlinear dielectric act as a waveguide when dealing with semiconductor devices [547], since the active region has the form of a layer only of few microns thick. The localization of the field in the active layer is due to the dielectric constant gradient associated with the *p–n* junction (injection laser) or the increased electron density (semiconductor laser excited by an electron beam). Owing to the strong absorption of the wave outside the active layer, the laser is very sensitive to the degree of field localization. The nonlineariry of the refractive index of the active layer has greater influence on laser dynamics when the waveguide properties are weaker. This assumption has been substantiated experimentally: doping of the surface layer of a GaAs crystal with Zn to produce a dielectric waveguide weakens the pulsations in an electronically pumped laser [548].

7.3.2 Passive Modulation of Diffraction Losses by a Dielectric with Cubic Nonlinearity (Kerr Medium Inside a Cavity with Aperture)

It is the goal of this section to demonstrate that the steady state can, in principle, become unstable owing to the nonlinear refractive index of the medium inside the laser cavity. Consider a laser operated on the fundamental Gaussian mode [549]. For a rigorous analysis one should use a spatial-extended model, since the effects in question are due to the dynamical deformation of the field profile. The situation can be simplified by disregarding the influence of the transverse dynamics effects in the laser medium and pay attention only to the dependence of diffraction losses on the laser intensity on the beam axis.

In order to estimate the diffraction losses we assume that the energy of that part of light beam, which is outside an aperture, is lost in each trip round the cavity. The influence of the finite sizes of the mirrors on the transverse field structure is assumed to be insignificant. Such an approximation is appropriate for a laser with a stable cavity, which is taken to be far from concentric so that we may ignore the dependence of the beam cross-section on the longitudinal coordinate. We, therefore, have the following definition of the diffraction loss coefficient:

$$\Pi_{\text{diffr}} = 1 - \frac{\int_{V_c} \psi^2 \mathrm{d}V}{\int \psi^2 \mathrm{d}V}.$$

The integral in the denominator is taken over the entire space, whereas the cavity aperture over the volume limits the integral in the numerator. By the normalization condition the second is $V_c = \pi b^2 L$, where b is the radius of the mirror while the first should be calculated assuming the Gaussian function of the transverse intensity distribution $\psi = \exp(-r^2/a^2)$, where a is the effective radius of the light beam. It is easy to see that

$$\Pi_{\text{diffr}} = \exp\left(-\frac{b^2}{a^2}\right). \tag{7.60}$$

The damping rate of the 'empty' cavity and the loss coefficient are related by

$$\kappa = \frac{c}{2L}(\Pi_{\text{loss}} + \Pi_{\text{diffr}}), \tag{7.61}$$

where $\Pi_{\text{diffr}} << \Pi_{\text{loss}}$.

In the aberrationless approximation for the light beam of finite size, a layer of dielectric with a nonlinear refractive index acts as a lens with variable focal length F_{nl}. Bearing in mind that the laser rod also has lens properties because of nonuniform heating, the equivalent lens inside the cavity is described by

$$\frac{1}{F} = \frac{1}{F_0} + \frac{1}{F_{\text{nl}}}. \tag{7.62}$$

a plane-parallel cavity with the lens at its centre is equivalent, provided $F \gg L$, to a spherical cavity of the same length with focal F of its mirrors [550]. In this case

$$a = \left(\frac{4LF}{k^2}\right)^{1/4} \tag{7.63}$$

Suppose that the refractive index depends on the photon density

$$\widetilde{M} = \frac{M}{\pi a^2 L}\exp\left(-\frac{r^2}{a^2}\right) \tag{7.64}$$

as

$$\eta = \eta_0 + 4\pi\eta_2\widetilde{M}. \tag{7.65}$$

The nonlinear part of susceptibility is proportional to the square of the field intensity and the polarization is proportional to the cube of the field intensity, owing to which this nonlinearity is called cubic. A Gaussian beam with a plane phase front

$$E_{\text{in}} = E_0 \exp\left(-\frac{r^2}{2a^2}\right),$$

after passing through a dielectric layer, transforms to

$$E_{\text{out}} = E_0 \exp\left[-\frac{r^2}{2a^2} + i\Psi \exp\left(-\frac{r^2}{a^2} \right) \right], \tag{7.66}$$

where

$$\Psi = \frac{4kL_{\text{nl}}\eta_2}{\eta_0 La^2} M \, .$$

L_{nl} denotes the thickness of the layer of nonlinear medium and M the total number of photons inside the cavity.

In the aberrationless approximation we replace Eq. (7.66) by

$$\tilde{E}_{\text{out}} = E_0 \exp\left(-\frac{r^2}{2a^2} + i\tilde{\Psi} \frac{r^2}{a^2} \right). \tag{7.67}$$

In doing so, we expand Eq. (7.66) in Hermitian functions, the first of which is Eq. (7.67). The quantity $\tilde{\Psi}$ is chosen such that the coefficient of the first term of the expansion, i.e.,

$$\int E^*_{\text{out}} \tilde{E}_{\text{out}} \mathrm{d}S \, ,$$

is as close to

$$\int | E_{\text{out}} |^2 \, \mathrm{d}S$$

as possible. Under this condition, neglect of the higher order terms is the most correct. Correspondingly, $\tilde{\Psi} = -\Psi/4$ and the focal length of the aberrationless lens is given by

$$F_{\text{nl}} = \frac{\eta_0 La^4}{2L_{\text{nl}}\eta_2 M} \, . \tag{7.68}$$

Using Eqs. ((7.62), (7.63) and (7.68) we find the radius of the Gaussian beam. For $F_{\text{nl}} \gg F_0$

$$a^2 = \left(\frac{4F_0 L}{k^2} \right)^{1/2} \left(1 - \frac{k^2 L_{\text{nl}}\eta_2}{\eta_0 L^2} M \right) \tag{7.69}$$

and, according to (7.60),

$$\Pi_{\text{diffr}} = \exp\left(-\frac{b^2}{a_0^2} \right)\left(1 - \frac{b^2 k^2 L_{\text{nl}}\eta_2}{a_o^2 \eta_0 L^2} M \right). \tag{7.70}$$

Inserting Eq. (7.70) into Eq. (7.61) we arrive at

$$\kappa = \kappa_0 (1 - gM) \, , \tag{7.71}$$

where

$$a_0 = \left(\frac{4F_0L}{k^2}\right)^{1/2}, \quad \kappa_0 = \frac{c\Pi_{\rm loss}}{2L}, \quad g = \frac{b^2k^2L_{\rm nl}\eta_2}{a_0^2\eta_0L^2}\frac{\exp(-b^2/a_0^2)}{\Pi_{\rm loss}}. \quad (7.72)$$

The dynamical change of the beam radius influences both the diffraction losses and the gain $\int N\psi^2 dV$. The role of the latter effect is enhanced if the radius of the light beam exceeds the radius of the active medium. This version of the problem will be considered later. So far we shall focus here on the opposite case, disregarding the gain modulation. This makes it possible to assume that the laser medium is not space-extendent and, hence, use the conclusion drawn in Section 7.1. The derivative β' in Eq. 7.8 is related to coefficient g defined in Eq. (7.72) by

$$\beta' = -\frac{\pi La^2\gamma_\parallel}{c\sigma_{\rm tr}}g. \quad (7.73)$$

Substituting Eq. (7.73), as well as $\overline{m} = A-1$ into Eq. (7.8) we find the instability condition for a laser with a Kerr medium:

$$\left(\frac{\Pi^0_{\rm diffr}}{\Pi_{\rm loss}}b^2\frac{\pi k^2L_{\rm nl}\eta_2}{c\eta_0L}\right)\left(\frac{2\kappa}{\sigma_{\rm tr}}\frac{A-1}{A}\right) > 1. \quad (7.74)$$

The quantities in the left-hand side of the inequality (7.74) are collected into two factors. It is convenient to compare Eq. (7.74) with the instability condition for a laser with saturable absorber (7.38), which can be rewritten in similar form

$$\left(A_{\rm a}\frac{\sigma_{\rm tr}}{\gamma_{\parallel\rm a}}\right)\left(\frac{2\kappa}{\sigma_{\rm tr}}\frac{A-1}{A}\right) > 1. \quad (7.75)$$

The second brackets of both inequalities are completely coincident. Although the first brackets are different, their physical meaning is similar. Each of them characterizes the efficiency of the respective nonlinear medium as a passive loss modulator. It is possible to combine both inequalities, (7.74) and (7.75),

$$K_{\rm nl}\frac{2\kappa}{\sigma_{\rm tr}}\frac{A-1}{A} > 1 \quad (7.76)$$

using the generalized coefficient

$$K_{\text{nl}} = \begin{cases} A_{\text{a}} \dfrac{\sigma_{\text{tr}}}{\gamma_{\|\text{a}}} & \text{for saturable absorber,} \\ \dfrac{\Pi^{(0)}_{\text{diffr}}}{\Pi_{\text{loss}}} b^2 \dfrac{\pi k^2 L_{\text{nl}} \eta_2}{c \eta_0 L} & \text{for Kerr medium.} \end{cases} \tag{7.77}$$

In order to estimate whether the instability threshold can be reached using a spherical cavity we transform inequality (7.74) with the help of Eq. (7.72) to

$$F_0 > \frac{k^2 b}{4L} \left[\ln \frac{\pi b^2 k^2 L_{\text{nl}} \eta_2 (A-1)}{\eta_0 L^2 \sigma_{\text{tr}} A} \right]^{-1} . \tag{7.78}$$

Estimating the argument of the logarithmic function to unity we find the minimum value $L_{\text{nl}}\eta_2$, for which instability might occur in principle

$$(L_{\text{nl}}\eta_2)_{\min} = \frac{A}{A-1} \frac{\eta_0 \sigma_{\text{tr}} L^2}{\pi k^2 b^2} . \tag{7.79}$$

Assuming $\sigma_{\text{tr}} = 10^{-20}$ cm^2, $k = 9 \cdot 10^4$ cm^{-1}, $L = 10^2$ cm, $b = 0.5$ cm, $A \gg 1$, we obtain $(L_{\text{nl}}\eta_2)_{\min} = 4.10^{-26}$ cm^4. Introducing a nonlinear medium equivalent to a centimetre length carbon bisulfate layer ($\eta_2 = 2 \cdot 10^{-23}$ cm^3 [551]) into a ruby laser cavity destabilizes the lasing for $F_0 > 10^5$ cm. Using the nonlineariry of the ruby crystal host, which, according to [552], is characterized by $\eta_2 = 3.8 \cdot 10^{-25}$ cm^3, we find that the instability criterion (7.75) can also be satisfied, since $L_{\text{nl}}\eta_2 \approx 4 \cdot 10^{-24}$ cm$^4 > (L_{\text{nl}}\eta_2)_{\min}$ for a rod length $L_{\text{a}} = L_{\text{nl}} = 10$ cm.

Our treatment above refers rigorously to a parallel arrangement of the cavity mirrors. However, even a small angle between them, comparable to the available accuracy of alignment, can markedly change the result. Mirror tilt to an angle ϑ causes the beam to shift toward the mirror side by $\Delta x = F\vartheta$ on single pass. Misalignment plays a determining role when the beam shift Δx is comparable to the mirror radius (cavity aperture). Using this simple consideration we can reliable estimate the critical value of the focal length in the misalignment:

$$F_0^{\text{cr}} \approx b / \vartheta .$$

Assuming $\vartheta = 10''$ and $b = 0.5$ cm we obtain $F_0^{\text{cr}} \approx 10^4$ cm.

Experimental observation of the influence of nonlinear dielectrics on laser dynamics is impeded by optical inhomogeneities of the laser rods. Owing to nonuniform heating the dielectric behaves as an excessively strong (according to the estimates proposed) positive lens with $F_0 \le 10^3$ cm. Approaching the instability threshold requires cavity misalignment or positive lens compensation, which was observed in the Refs. [541–545].

7.3.3 Passive Modulation of the Mode Filling factor (Kerr Medium Inside a Cavity Without Aperture)

The uniform distribution of inversion across the cavity cross-section is never achieved. Hence, the dynamic deformation of the beam profile should be accompanied with the cavity filling factor modulation or, in other words, passive gain modulation. In studying this effect it is impossible to neglect the dependence of N on the transverse coordinate. However, the dependence of N on the longitudinal coordinate may be considered as previously, to be in an inconsequential range. Thus, by adiabatic elimination of the polarization (the off-diagonal elements of the density matrix) we transform Eqs. (2.76) to

$$\frac{\mathrm{d}|F|^2}{\mathrm{d}t}+2\kappa|F|^2=\frac{d\omega L_a}{\hbar\gamma_\perp V_c}|F|^2\int N\psi^2\mathrm{d}S\,, \tag{7.80a}$$

$$\frac{\mathrm{d}N}{\mathrm{d}t}+\gamma_\parallel(N-N^{(0)})=-\frac{\beta_a d^2}{2\hbar^2\gamma_\perp\gamma_\parallel}|F|^2\,N\psi^2 \tag{7.80b}$$

A nontrivial question arises: What the cavity volume should mean V_c? Since the field profile is given by eigenfunction $\psi=\exp(-r^2/2a^2)$, the cavity volume is calculated as

$$V_c=\int\psi^2\mathrm{d}V=\pi a^2 L\,.$$

The dependence on laser power is also contained in this factor owing a^2.

Introducing the dimensionless notation

$$\tau=\gamma_\parallel t,\quad \rho=\frac{r}{a_0},\quad m=-\frac{\beta_a d^2}{2\hbar^2\gamma_\parallel\gamma_\perp}|F|^2,\quad n=\frac{2\pi\omega d^2 L_a}{\hbar\gamma_\perp\kappa L}N\,, \tag{7.81}$$

$$\alpha^2=\left(\frac{a}{a_0}\right)^2=1-g^*m,\quad g^*=\frac{\pi k^2 L_{nl}\eta_2 a_0^2\gamma_\parallel}{cL\eta_0\sigma_{tr}}\,,$$

we rewrite Eqs. (7.80) as

$$\frac{dm}{d\tau}=Gm\left[\alpha^{-2}\int_0^{\rho_a} n exp\left(-\frac{\rho^2}{a^2}\right)d(\rho^2)-1\right], \tag{7.82a}$$

$$\frac{\mathrm{d}n}{\mathrm{d}\tau}=A-n\left[1+m\exp\left(-\frac{\rho^2}{a^2}\right)\right]. \tag{7.82b}$$

impossible to neglect the dependence of N on the transverse coordinate

However, the dependence of on the longitudinal coordinate may be considered as previously, to be in an inconsequential range. Thus, by adiabatic elimination of the polarization (the off-diagonal elements of the density matrix) we transform Eqs. (2.76) to

$$\frac{\mathrm{d}|F|^2}{\mathrm{d}t}+2\kappa|F|^2=\frac{d\omega L_{\mathrm{a}}}{\hbar\gamma_\perp V_{\mathrm{c}}}|F|^2\int N\psi^2\mathrm{d}S \tag{7.80a}$$

$$\frac{\mathrm{d}N}{\mathrm{d}t}+\gamma_\parallel(N-N^{(0)})=-\frac{\beta_a d^2}{2\hbar^2\gamma_\perp\gamma_\parallel}|F|^2 N\psi \tag{7.80b}$$

A nontrivial question arises: What the cavity volume should mean ? Since the field profile is given by eigenfunction $\psi=\exp(-r^2/2a^2)$ the cavity volume is calculated as

$$V_{\mathrm{c}}=\int\psi^2\mathrm{d}V=\pi a^2 L$$

The dependence on laser power is also contained in this factor owing to a^2.

Introducing the dimensionless notation

$$\tau=\gamma_\parallel t,\quad \rho=\frac{r}{a_0},\quad m=-\frac{\beta_a d^2}{2\hbar^2\gamma_\parallel\gamma_\perp}|F|^2,\quad n=\frac{2\pi\omega d^2 L_{\mathrm{a}}}{\hbar\gamma_\perp\kappa L}N, \tag{7.81}$$

$$\alpha^2=\left(\frac{a}{a_0}\right)^2=1-g^*m,\quad g^*=\frac{\pi k^2 L_{\mathrm{nl}}\eta_2 a_0^2\gamma_\parallel}{cL\eta_0\sigma_{\mathrm{tr}}},$$

we rewrite Eqs. (7.80) as

$$\frac{dm}{d\tau}=Gm\left[\alpha^{-2}\int_0^{\rho_a} n\exp\left(-\frac{\rho^2}{a^2}\right)d(\rho^2)-1\right], \tag{7.82a}$$

$$\frac{\mathrm{d}n}{\mathrm{d}\tau}=A-n\left[1+m\exp\left(-\frac{\rho^2}{a^2}\right)\right]. \tag{7.82b}$$

By ρ_{a} we mean the radius of the laser rod while the beam radius is defined by Eq. (7.69). The quantity α differs little from unity in the whole range of m, thus making it possible to put $\overline{\alpha}=1$. Linearizing Eqs. (7.82) and assuming solutions in the form $\exp\lambda\tau$, we find, in a standard manner of the linear stability analysis, the approximate damping rate of small oscillations:

$$2\theta_0 = G\overline{m}\left[-\Omega_0^{-2}\int_0^{\rho_a} A\exp(-2\rho^2)d(\rho^2) + K_{nl}\int_0^{\rho_a}\overline{n}(1-\rho^2)\exp(-\rho^2)d(\rho^2)\right].$$

The oscillation frequency is given by

$$\Omega_0^2 = G\overline{m}\left[\int_0^{\rho_a}\overline{n}\exp(-2\rho^2)d(\rho^2)\right]. \tag{7.83}$$

The steady state becomes unstable ($\theta_0 > 0$) if two conditions,

$$\int_0^{\rho_a}\overline{n}(1-\rho^2)\exp(-\rho^2)d(\rho^2) > 0, \tag{7.84}$$

and

$$g^* > \frac{\int_0^{\rho_a} A\exp(-2\rho^2)d(\rho^2)/\Omega_0^2}{\int_0^{\rho_a}\overline{n}(1-\rho^2)\exp(-\rho^2)d(\rho^2)} > 0, \tag{7.85}$$

are satisfied.

The factor $1-\rho^2$ indicates two opposite tendencies. A decrease in the mode radius improves the lasing conditions in a laser medium of finite size. On the one hand, this leads to the development of instability because of the increase of the mode-filling factor. On the other hand, these are a tendency towards stability via the increase in the pumping excess over the laser threshold.

The last inequalities are greatly simplified if the dependence of n on ρ within the active region can be neglected. The approximation is justified either when the active zone radius is small or the laser threshold is slightly exceeded. This is easily seen from

$$\overline{n} = \frac{A}{1+\overline{m}\exp(-\rho^2)}.$$

Assuming $\overline{n} = \text{const}$ we transform (7.84) to

$$\overline{n}\rho_a^2\exp(-\rho_a^2) > 0 \tag{7.86}$$

and see that this inequality is satisfied within the framework of the approximations we have adopted.

The steady-state solution of Eqs. (7.82) is given by

$$\overline{n} \approx 1/\rho_a^2, \quad \overline{m} \approx A\rho_a^2 - 1. \tag{7.87}$$

Integrating Eq. (7.85) and using Eqs. (7.81), (7.83) and (7.87), we arrive at

$$\left(a_0^2\frac{\pi k^2 L_{nl}\eta_2}{cL\eta_0}\right)\left(\frac{2\kappa}{\sigma_{tr}}\frac{A\rho_a^2-1}{A\rho_a^2}\right) > 1, \tag{7.88}$$

which is the steady state instability condition for a laser with Kerr medium in a cavity without aperture. The inequality (7.88) is similar to (7.74). Only in place of the aperture parameter $b^2\Pi^{(0)}_{\text{diffr}}/\Pi_{\text{loss}}$ we have the beam parameter a_0^2. Thus we come to instability criteria like Eq. (7.76), but it should borne in mind that

$$K_{\text{nl}} = a_0^2 \frac{\pi k^2 L_{\text{nl}} \eta_2}{cL\eta_0}. \tag{7.89}$$

If the laser rod host acts as a nonlinear dielectric, then the greatest influence of self-focusing on laser dynamics is achieved in a cavity with the mirrors at the rod ends ($L_{\text{nl}} = L$). From Eq. (7.88) it follows that instability threshold is independent of the cavity length in this case. Using typical parameter values for a ruby laser ($\sigma_{\text{tr}} = 10^{-20}$ cm^2, $k = 10^5$ cm^{-1}, $\eta_0 = 1.5$, $\kappa = 10^8$ s^{-1}) for $A\rho_{\text{a}} >> 1$, $a_0 = 0.05$ cm leads to the necessary condition for instability $\eta_2 > 0.5 \cdot 10^{-25}$ cm^3. For ruby $\eta_2 = 3.8 \cdot 10^{-25}$ cm^3, and this estimate indicates that such instability is quite real.

The instability of this type should be most readily exhibited in a laser with a very thin active element immersed in an index-matched medium. This might occur, for example, if a sapphire shell surrounds a ruby rod. Index-matching is needed to avoid reflections at the sides of the active medium, which can localize the light field inside. For a four-level laser medium, then the same result is obtained by pumping with a narrow beam propagated along the cavity axis. However, in a three-level medium with a narrow pumping beam, outside the inversion region the absorption is strong and the situation nearer to that of passive loss modulation.

In essence, we have considered two special cases in this section. We have demonstrated that an instability can arise from self-focusing in a nonlinear medium placed inside the cavity of a class *B* laser. This is a key to understanding the experimental results described in Section 7.3.1 and earlier in Section 1.2.3. However, this should not be treated as a quantitative theory suitable for detailed comparison with the experiment.

7.4 Passive Mode Locking in Lasers

Various aspects of multimode operation were discussed in the previous chapters. We considered relatively slow processes, with the time scales much larger that the cavity round trip time T_0. In these cases the presence of many modes and their interaction exert noticeable influence. However, the most obvious consequence of multimode lasing is the presence of fast oscillations of the laser intensity due to temporal mode interference. These fast oscillations may vary, depending on the amplitudes, frequencies and phases of the laser modes. Of all these quantities, the most important role is played by the mode frequencies. If they form a spectrum with equal spacing, then the output intensity is periodically modulated. Otherwise the output has the form of irregular oscillations.

Equality of the spacing in the mode spectrum is a necessary condition for mode locking, which is very important for practical purposes. In the broad meaning of mode locking, equidistancy of the spectrum is a sufficient condition. In quantum electronics, however, the term 'mode locking' usually has a narrower meaning, i.e., this should not be simply periodic, and the pulse duration should be minimum possible for a given spectral width. The last requirement imposes limits on the form of the spectral envelope, which, roughly speaking, should be smooth, and on the mode phases, the frequency dependence of which should be linear. Generally, the regime that satisfies all these requirements is referred to as 'phased mode locking'.

Nonlinear mode-mode coupling in the media inside the laser cavity is a key factor in mode locking. Indeed, nonlinear mode-mode coupling in absorbing media is used to achieve passive mode locking [553–568]. Mode-mode coupling in a laser medium can force equidistancy in the spectrum [559–563] as mentioned in Section 4.5.1, and it can determine some phase relations between the modes. Similarly, phase relations among the modes are the reference point in the theory of passive mode locking.

7.4.1 Role of the Laser Medium Nonlinearity in Mode Locking

It is a priori clear that owing to its nonlinear properties, the active medium should influence the mode locking process in lasers. The first study of the nature of this influence was made in Refs. [559–563]. Postulating a 'maximum emission principle' the authors infer that the combination tone mode-mode coupling in a laser medium can lead to phase locking of the modes. It appeared, however, that the 'maximum emission principle' contradicted the conclusion that follows directly from the laser equations and that the phase relations between modes found in such manner are erroneous [564].

In order to study the problem of phase relations controlled by the combination tone mode-mode coupling we consider Eqs. (4.79). Assuming the inequality (4.85) is satisfied we suppose that the mode frequencies are equally spaced and neglect the influence of dispersion on the mode frequencies. Thus, the steady-state phases can be defined by a set of equations

$$\mathrm{Re}(\widetilde{G}_{k+q} - \widetilde{G}_k - \widetilde{G}_{\mu+q} + \widetilde{G}_\mu)^{\mathrm{comb}} = 0, \tag{7.90}$$

in which, according to Eq. (4.77),

$$\tilde{G}_k^{\text{comb}} = \frac{i}{4}\tilde{F}_k \sum_{\mu,q} \frac{\tilde{F}_\mu(n_0+n_\mu)+\tilde{F}^*_{\mu+q}(n_0+n_{\mu+q})}{1+iq\Delta/\tilde{\gamma}} \frac{f_\mu f^*_{\mu+q} f_{k+q}}{f_k}. \tag{7.91}$$

Assuming that the medium is sufficiently slow in its relaxation ($\Delta/\tilde{\gamma} = \Delta\omega/\gamma_\parallel >> 1$) and passing over to the real amplitudes and phases ($f_k = F_k \exp i\varphi_k$) we write instead of Eq. (7.91)

$$\text{Re}\,\tilde{G}_k^{\text{comb}} \approx \frac{\tilde{\gamma}}{4\Delta} \sum_{\mu,q} \frac{2n_0+n_\mu+n_{\mu+q}}{q} \frac{F_{k+q}F_\mu F_{\mu+q}}{F_k} \cos\Phi_{kq\mu}, \tag{7.92}$$

where

$$\Phi_{kq\mu} = \varphi_{k+q} - \varphi_k - \varphi_{\mu+q} + \varphi_\mu. \tag{7.93}$$

Based on the assumption that phased mode locking occurs, i.e., $\Phi_{kq\mu} = 0$, the laser spectrum has a rectangular form, $F_k \equiv F$, and, in addition, $n_{kq\mu} \equiv \bar{n}$, we have

$$\text{Re}\,\tilde{G}_k^{\text{comb}} \approx \frac{\bar{n}(A-1)\tilde{\gamma}}{2\Delta}\left(\ln\frac{1-2k/N}{1+2k/N} + \frac{2k}{N}\right). \tag{7.94}$$

The enumeration of the modes is such that the spectrum boundaries correspond to the indices $k = \pm(N-1)/2$. The function $F(k/N)$ in the brackets is different from a linear one and, therefore, the expression (7.94) does not satisfy the phase equations (7.90). Thus, we infer that mode locking due to the combination tone interaction in a slowly relaxing laser medium is not possible.

The assumption that all the quantities $n_{kq\mu}$ are equal is reasonable only in the case of uniform saturation of the laser medium. Undoubtedly, such an assumption is valid for an unidirectional ring laser. In a Fabry–Perot cavity, a situation similar to that occurs when many modes are excited or when the laser threshold is only slightly exceeded. Moreover, the laser medium should uniformly fill the cavity or be concentrated in a relatively thin layer near the mirror. Any other arrangement of the laser rod changes the situation with n_μ, but this fact is not essential to the conclusion that the phase self-mode-locking is not possible.

In steady state the form of the spectrum is typically parabolic rather than rectangular. However, this is not very important in this context. Only the finite width of the spectrum matters. If the laser frequency band is infinite, then the condition $\Phi_{kq\mu} = 0$ leads to $\text{Re}\,\tilde{G}_k^{\text{comb}} \equiv 0$ making the self-mode-locking (spontaneous formation of δ-pulses) feasible.

Assuming that the laser medium is fast, i.e., $\Delta/\tilde{\gamma} << 1$ Eq. (7.91) transform to

$$\tilde{G}_k^{\text{comb}} = \frac{i}{4}\tilde{F}_k \sum_{\mu,q}(2n_0 + n_\mu + n_{\mu+q})\frac{f_{k+q}f_\mu f_{\mu+q}}{f_k}. \qquad (7.95)$$

Disregarding the nonequidistancy of the cavity eigenfrequences Eqs. (4.78) are written as

$$\frac{d\varphi_k}{d\tau} = \frac{1}{4}G\sum_{\mu,q}(2n_0 + n_\mu + n_{\mu+q})\frac{F_{k+q}F_\mu F_{\mu+q}}{F_k}\sin\Phi_{kq\mu}. \qquad (7.96)$$

These equations are satisfied by the steady states $\overline{\Phi}_{kq\mu} = 0$, which correspond to phased mode locking, and the problem is reduced to studying the stability of such a solution.

Complete analysis of the laser stability of the mode-locked state is rather time consuming. However, it is relatively easy to investigate the laser stability with respect to a particular form of perturbation involving only the mode phases. Assume that the phase deviation of the k – th mode, $\delta\varphi_k = \varphi_k - \overline{\varphi}_k$, is small, and linearize Eqs. (7.96) near $\overline{\Phi}_{kq\mu} = 0$. The linearized equations written as

$$\frac{1}{\delta\varphi_k}\frac{d(\delta\varphi_k)}{d\tau} = \frac{1}{4}G\sum_{\mu,q}\frac{F_{k+q}F_\mu F_{\mu+q}}{F_k}(2n_0 + n_\mu + n_{\mu+q}), \qquad (7.97)$$

have a positive right-hand side. This is an indication that the steady-state solution is unstable.

This result means that combination tone mode-mode coupling in an inertialess laser medium does not lead to phase locking either. Meanwhile, it is seen from Eq. (7.97) that a sign reversal of the population difference changes the situation. Thus, we have arrived at the idea of passive mode locking in a laser with a saturable absorber.

7.4.2 Threshold Conditions of Passive Mode Locking

In the presence of a saturable absorber the phase equations (4.78) transform to

$$\frac{d\varphi_k}{d\tau} = \frac{1}{2}G\Delta_{ck} - \frac{1}{\tilde{\gamma}}\Delta_k + G(\mathrm{Re}\,\tilde{G}_k + \mathrm{Re}\,G_{ak}). \qquad (7.98)$$

Consider the most important practical case where the transition frequencies and the active and the absorption media coincide ($\omega_0 = \omega_a$) and the absorber is fast ($\Delta\omega/\gamma_{\|a} << 1$). If the lasing modes occupy the frequency band ($N\Delta\omega << \gamma_{\|a}$), then for the saturable absorber the expression

$$\mathrm{Re}\,\tilde{G}_k^{\text{comb}} = \frac{G\rho}{4\delta}\sum_{\mu,q}(2n_0 + n_\mu + n_{\mu+q})\frac{F_{k+q}F_\mu F_{\mu+q}}{F_k}\sin\Phi_{kq\mu} \qquad (7.99)$$

is valid, which is similar to Eq. (7.92). As before, ρ is the ratio of the cross-sections of the transitions in absorbing and amplifying media and $\delta = \gamma_{\parallel a} / \gamma_{\parallel}$.

Complete phased mode locking corresponds to $\operatorname{Re}\tilde{G}_{ak}^{\text{comb}} = 0$. As shown above, the condition $\sin\Phi_{kq\mu} = 0$ does not satisfy the equations of laser without a nonlinear filter even if all the frequencies in the laser spectrum are equally spaced. However, the use of a saturable absorber makes it possible to come close to this condition. The requirement is that the nonlinear parameter of the absorber $(\rho/\delta)n_a$, be larger than that of the laser medium, $(\gamma_{\parallel}/\Delta\omega)n$. For $\rho/\delta \ll 1$ Eqs. (7.37) are valid and yield

$$n \approx 1 + A_a, \quad n_a \approx A_a$$

and make it possible to write this criterion in the form

$$A_a > A_a^{\text{lock}} = \left(\frac{\rho \Delta\omega}{\gamma_{\parallel a}} - 1 \right)^{-1}. \tag{7.100}$$

We therefore get the necessary condition

$$\frac{\rho \Delta\omega}{\gamma_{\parallel a}} > 1. \tag{7.101}$$

Substituting into (7.100) typical values $\rho = 10^4, \Delta\omega = 10^9\ \text{s}^{-1}$ and $\gamma_{\parallel a} = 10^{11}\ \text{s}^{-1}$, which certainly satisfy the inequality (7.101), yields $A_a^{\text{lock}} = 10^{-2}$. Hence, in a steady state mode locking is achieved if the unsaturated losses in the nonlinear filter exceed 1% of the total cavity losses.

With the chosen parameters of the nonlinear filter, the inequality (7.101) can be treated as an upper limit on cavity length

$$L < \frac{\pi c \rho}{\gamma_{\parallel a}}. \tag{7.102}$$

This conclusion has been confirmed experimentally. In Ref. [565] it is found that in a neodymium glass laser passive mode locking is achieved only for $L < 60\ \mu\text{m}$. This fact is in quantitative agreement with Eq. (7.102) provided $\gamma_{\parallel a} = 2 \cdot 10^{11}\ \text{s}^{-1}$.

The threshold condition of passive mode locking is weaker than the low-frequency instability criterion of a class B laser if

$$\Delta\omega A > \kappa(A - 1),$$

i.e., if the resonance curves of the cavity modes do not overlap. This shows the feasibility of passive mode locking of a solid-state laser without spikes.

The threshold condition that coincides with Eq. (7.100) can be obtained by both taking the modal approach discussed above and the spatio-temporal approach used in [566]. A common feature is the necessity of taking into account the population difference oscillations in the laser medium. This validates the assumption of mode decoupling by the nonlinearity of the active medium.

In the previous section we mentioned that there is close analogy between the influences of absorptive and refractive (Kerr-like) nonlinearities on the laser stability. The steady state instability criteria have been written in a generalized form introducing the parameter K_{nl}, which was defined by Eq. (7.77). This analogy is useful in an analysis of passive mode locking conditions as well.

Let us write the inequality (7.100) in the same generalized form:

$$K_{nl}\frac{\Delta\omega}{\sigma_{tr}}>1 .$$

Correspondingly, for a laser with the Kerr medium we have

$$\left(\frac{\Pi^{(0)}_{diffr}}{\Pi_{loss}}b^2\frac{\pi k^2 L_{nl}\eta_2}{\sigma_{tr}\eta_0 L}\right)\left(\frac{\Delta\omega}{\sigma_{tr}}\right)>1, \tag{7.103}$$

which is the criterion for passive mode locking via self-focusing in a cavity with aperture.

For a ruby laser (L_{nl}= 15 cm, L = 100 cm, b = 0.5 cm, η_0 = 1.76, η_2 = $3.6\cdot10^{-25}$ cm^3, $\sigma_{tr}=10^{-20}$ cm^2) mode locking is possible for $\Pi_{diffr}/\Pi_{loss}>10^{-3}$. The requirements slightly more stringent in the case of Nd:YAG laser ($\eta_0=1.82, \eta_2=7\cdot10^{-25}$ cm^3, $\sigma_{tr}=4.5\cdot10^{-19}$ cm^2). These estimates provide sufficient ground for attributing the early experimental results on self-mode-locking in ruby [567] and Nd:YAG [568] lasers to self-focusing in the laser crystal host. There have recently been a great number of experimental publications devoted to self-mode-locking in lasers using Ti:sapphire and other crystals with a broad gain line as laser media. Generation of femtosecond pulses is achieved both in diaphragmed laser cavities [569–571] and under the condition when the aperture effects cannot play an essential role [572]. In the latter case it is reasonable to assume the action of the mode-locked mechanism through passive modulation of the filling factor. The threshold condition

$$\left(a_0^2\frac{\pi k^2 L_{nl}\eta_2}{\sigma_{tr}\eta_0 L}\right)\left(\frac{\Delta\omega}{\sigma_{tr}}\right)>1 \tag{7.104}$$

can be obtained by substituting the "diffraction" factor $b^2\Pi^{(0)}_{diffr}/\Pi_{loss}$ in Eq. (7.103) for beam parameter a_0^2 by analogy with Eqs. (7.77) and (7.89).

7.5 Processes in a Traveling Wave Laser with Saturable Absorber

The occurrence of mode locking (ultrashort pulse generation) depends on two conditions. First, simultaneous excitation of a large number of modes is required. Second, definite phase and frequency relations between these modes should be provided. These two conditions have a threshold nature and can relate to each other in different ways, thus leading to a variety of physical situations. When the multifrequency threshold is above the mode-locking threshold, bistability and hysteresis occur, and the mode-locking regime can be achieved through hard excitation.

In order to investigate the situation arising here one should consider the problem of the multimode laser threshold, i.e., stability of single mode oscillation in the presence of a saturable absorber. If the cavity modes are standing waves and the length of laser medium is comparable to the cavity length, the multimode regime will be established for a small excess of the pumping above the lasing threshold, owing to the weak spatial mode competition (see Section 4.1.2).

The problem of stability of single-frequency operation of a travelling wave laser was discussed in Section 4.6. Coherent interaction of laser radiation with the active medium may destabilize a single-frequency regime, but considerable excess pumping above the laser threshold is required. This requirement is not so rigorous if the laser cavity contains a saturable absorber. Of course, one is tempted to simplify the problem by making use of the rate equation approach. It should be borne in mind, however, that adiabatic elimination of the laser medium polarization greatly changes the dispersive properties of the model, influencing the result inevitably. There has recently been an increase of interest to this fact in connection with the spatio-temporal structures in the laser radiation field [573]. Much earlier this problem was considered by Gurevich [574], and we use his method in what follows.

7.5.1 Rate Equation Model of a Traveling Wave Laser Without Bulk Losses

Unlike the problem formulation in Section 4.6, we now assume that the linear losses are attributed to the cavity mirrors and that the laser medium occupies part of the cavity perimeter (Fig. 7.14). Wave propagation in the laser medium is described by the equations

$$\frac{\partial M}{\partial t}+\frac{\partial M}{\partial z}=BMN\,, \qquad (7.105a)$$

$$\frac{\partial N}{\partial t} + \gamma_{\parallel}(N - N^{(0)}) = -BMN\,, \tag{7.105b}$$

that represents an apparent generalization of Eqs. (3.5) to the case of a travelling wave without losses. These equations are used to define the relationship between the input and output wave intensities. For this, one must turn briefly to the moving frame of reference, $\vartheta = t - z/c, z = z$, by transforming Eqs. (7.105) to

$$c\frac{\partial M}{\partial z} = BMN\,, \tag{7.106a}$$

$$\frac{\partial N}{\partial \vartheta} + \gamma_{\parallel}(N - N^{(0)}) = -BMN\,. \tag{7.106b}$$

The variable N, expressed through M by means of Eq. (7.106a), is substituted into (7.106b), and the resulting equation is integrated with respect to z. This transformation is admissible if $M \neq 0$. In a fixed frame of reference the resulting equality looks like

$$\left(\gamma_{\parallel} + \frac{\partial}{\partial t}\right)\ln[M(t,z)] + BM(t,z) = \frac{\gamma_{\parallel}BN^{(0)}}{c}z + f\left(t - \frac{z}{c}\right). \tag{7.107}$$

The arbitrary function $f(t - z/c)$ can be excluded from our consideration by writing Eq. (7.107) for the initial cross-section $z = L_1$ and time t, as well as for the input cross-section $z = 0$ and the advanced time $t - L_1/c$. Deducing one equation from the other we obtain

$$\left(\gamma_{\parallel} + \frac{\partial}{\partial t}\right)\left[\ln[M_1(t) - \ln M_0\left(t - \frac{z}{c}\right)\right] + B\left[M_1(t) - M_0\left(t - \frac{z}{c}\right)\right] = \frac{\gamma_{\parallel}BN^{(0)}L_1}{c} \tag{7.108}$$

where $M_1(t) = M_1(t, L_1), \quad M_0(t) = M_0(t, L_0)$.

For passing through the empty cavity section the wave requires time $(L - L_1)/c$. The power transfer coefficient for this section is R and, therefore,

$$M_0\left(t - \frac{L_1}{c}\right) = RM_1\left(t - \frac{L}{c}\right). \tag{7.109}$$

Two equations, (7.108) and (7.109), can be combined into one equation. Using the dimensionless variables

$$\tau = \gamma_{\parallel}t, \quad m = B\gamma_{\parallel}^{-1}M, \quad A = BN^{(0)}L_1/(c\ln R^{-1}), \quad \zeta_0 = L\gamma_{\parallel}/c, \tag{7.110}$$

we write this equation as

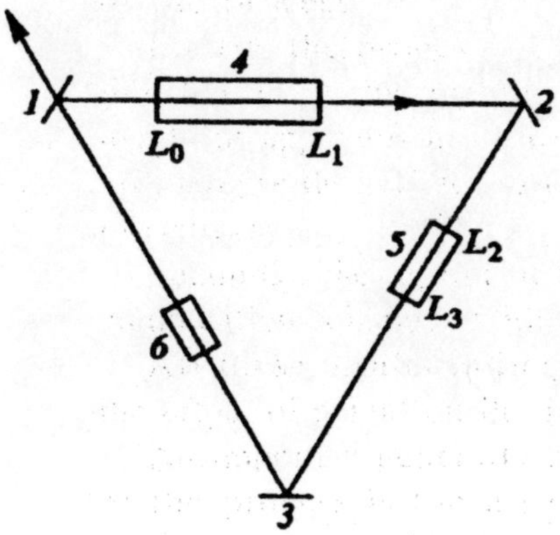

Fig. 7.14. Scheme of a ring laser with nonlinear absorbing cell and a selective element which represents the dispersive properties of the laser medium: 1 – output mirror, 2,3 – total reflecting mirrors, 4 – laser rod, 5 – nonlinear cell, 6 – linear bandpass

$$\left(1+\frac{\partial}{\partial\tau}\right)\ln\frac{m(\tau)}{Rm(\tau-\zeta_0)}+m(\tau)-Rm(\tau-\zeta_0)=A\ln\frac{1}{R}. \quad (7.111)$$

The nonlinear differential-difference Eq. (7.111) has the unique steady-state solution

$$\overline{m}=\frac{\ln R^{-1}}{1-R}(A-1). \quad (7.112)$$

If the losses are small, such that $1-R<<1$, then Eq. (7.112) is slightly simplified:

$$\overline{m}=A-1.$$

Note that Eq. (7.111) does not have a trivial solution, since it was assumed that $M\neq 0$ when this equation was derived.

In the vicinity of the steady state Eq. (7.111) can be linearized with respect to the variable $\delta m=m-\overline{m}$. The solutions of the linearized equation have the form $\delta m_0\exp(\lambda\tau)$, where l is the root of the characteristic equation

$$\frac{\lambda+1+R\overline{m}}{\lambda+1+\overline{m}}=\exp(\zeta_0\lambda). \quad (7.113)$$

First we will find the roots, which satisfy the inequality $\zeta_0|\lambda|<<1$ and admit an expansion $\exp(\zeta_0\lambda)=1+\zeta_0\lambda$. Eq. (7.113) transforms to

$$\lambda^2+(\overline{m}+1)\lambda+(1-R)\zeta_0^{-1}\overline{m}=0. \quad (7.114)$$

Note that

$$\frac{1-R}{\zeta_0}=\frac{c}{L\gamma_\parallel}(1-R)\approx\frac{2\kappa}{\gamma_\parallel}=G \quad (7.115)$$

and, therefore, for $1-R\ll 1$ Eq. (7.114) coincides with Eq. (3.19).

We now consider the roots of Eq. (7.113), which satisfy the condition $|\lambda|>>1$. When one of these roots is substituted into Eq. (7.113), the left-hand side of the equality is close to unity. This means that

$$\lambda = i\Omega + \theta = 2\pi i q \zeta_0^{-1} + i\tilde{\Delta} + \theta ,$$

where $|i\tilde{\Delta}+\theta| << \zeta_0^{-1}$. Multiplying Eq. (7.113) by the complex conjugate equation we find:

$$\frac{\Omega^2 + (\theta + 1 + R\overline{m})^2}{\Omega^2 + (\theta + 1 + \overline{m})^2} = \exp(2\zeta_0\theta) \approx 1 + 2\zeta_0\theta ,$$

hence

$$2\zeta_0\theta \approx \frac{\overline{m}(1-R)[2+\overline{m}(1-R)]}{(1+\overline{m})^2 + \Omega^2} . \tag{7.116}$$

For $q = 1, 2, ...$ the first term in the denominator can be neglected so that

$$\theta \approx \frac{1}{2\zeta_0\Omega^2}\overline{m}(1-R)[2+\overline{m}(1-R)] , \tag{7.117}$$

and in the case $R \approx 1$, we find, in view of (7.115)

$$\theta = -G\Omega^{-2}A(A-1) . \tag{7.118}$$

This equation yields a monotonic decrease of the perturbation damping rate as the perturbation frequency grows. This result is true as long as the rate equation approach, which ignores the laser medium dispersion, is valid. Taking the dispersion into account is critical exactly in the domain of high-frequency perturbation. To see this we assume, following [574], that the laser cavity contains, besides the nondispersive laser medium, a hypothetical frequency filter with the passband $\gamma_\perp$ to transform the laser intensity by the law

$$\left(1 + \frac{1}{\gamma_\perp}\frac{\partial}{\partial\tau}\right) M_{\text{out}}(t) = M_{\text{in}}\left(t - \frac{L_{\text{filt}}}{c}\right) . \tag{7.119}$$

Using this equation instead of Eq. (7.109) we obtain the set of equations

$$\left(1 + \frac{\partial}{\partial\tau}\right)\ln\frac{m_1(\tau+\zeta_1)}{m_0(\tau)} + m_1(\tau+\zeta_1) - m_0(\tau) = A\ln\frac{1}{R} , \tag{7.120a}$$

$$\left(1 + \tilde{\gamma}\frac{\partial}{\partial\tau}\right) m_0(\tau+\zeta_0) = Rm_1(\tau+\zeta_1) \tag{7.120b}$$

instead of (7.111). Here, $\zeta_i = \gamma_\parallel(L_i - L_0)$ is the dimensionless coordinate of each cross-section shown in Fig. 7.14 and m_i is the field intensity in the corresponding cross-section. Linearization of Eqs. (7.120) yields a characteristic equation

$$\frac{\lambda+1+m_0}{(\lambda+1+\overline{m}_1)(\tilde{\gamma}\lambda+1)}=\exp(\zeta_0\lambda)\,. \tag{7.121}$$

Multiplying (7.121) by the complex-conjugate equation and neglecting the small terms we find the perturbation decay rate as a function of the perturbation frequency:

$$\theta=\frac{1}{2\zeta_0}\left\{\frac{\Omega^2+(1+\overline{m}_0)^2}{[\Omega^2+(1+\overline{m}_1)^2](\tilde{\gamma}\Omega^2+1)}-1\right\}. \tag{7.122}$$

This dependence has an extremum for the frequency

$$\Omega_{\max}=\tilde{\Omega}^{-1/2}[(2+\overline{m}_1+\overline{m}_0)(\overline{m}_1-\overline{m}_0)^{1/4}\,. \tag{7.123}$$

In a wide range of frequencies, including $\Omega=\Omega_{\max}$, Eq. (7.122) is approximated by an expression

$$\theta=-\frac{\tilde{\gamma}}{2\zeta_0}\frac{2A(A-1)(1-R)+(\tilde{\gamma}\Omega^2)^2}{\tilde{\gamma}\Omega^2}\,. \tag{7.124}$$

The parameter $\tilde{\gamma}/2\zeta_0$ is nearly the same for different types of lasers. We assume for estimates that $\gamma_\perp=10^{12}$ s^{-1} and $L=1.5$ m, which corresponds to $\tilde{\gamma}/2\zeta_0=10^{-4}$.

The diagram of function (7.124) is given in Fig. 7.15. The frequency of the perturbation assumes the discrete set of values

$$\Omega_q=2\pi q/\zeta_0\ (q=1,2,3,\dots). \tag{7.125}$$

The frequency $\Omega_{\min}$ corresponding to the minimum damping of the perturbation enters this set if $\Omega_1<\Omega_{\min}$, which is equivalent to

$$L>L_{\min}\approx 2\pi c(\gamma_\|\gamma_\perp)^{-1/2}[2A(A-1)(1-R)]^{-1/4}\,. \tag{7.126}$$

Using numerical values $\gamma_\|=5\cdot10^3$ s^{-1}, $\gamma_\perp=10^{12}$ s^{-1}, $A=2$, $R=0.9$ yields $L_1\approx30$ m.

For solid-state lasers with cavity length of order 1 m each perturbation frequency is larger than $\Omega_{\min}$. Thus, for these parameter values and $L=$ 1.5 m we have $\tilde{\gamma}^{1/2}\Omega_1\approx20$ and $\tilde{\gamma}^{1/2}\Omega_{\min}\approx0.65$. For a dye laser with $\gamma_\|=2\cdot10^8$ s^{-1}, from Eq. (7.126) we have: L_m =1 7 cm, and from Eq. (7.125) it follows $\tilde{\gamma}^{1/2}\Omega_1\approx0.1$.

In Fig. 7.15 the dashed line shows the function $\theta(\Omega)$ in the ordinary rate-equation approximation, which ignores the gain dispersion. It is clearly seen that this approximation is of course inapplicable for study of the processes with frequencies $\Omega_1\geq\Omega_{\min}$. Therefore, it is complete inapplicable in a solid-state laser, since its whole spectrum of admissible perturbation frequencies is larger than Ω_{min}. However, this approximation is quite acceptable for a dye laser, and it is not surprising that a rate equation approach [575] yields about the same results as the analysis based on a more general semiclassical model [576].

7.5.2 Instability Threshold of Single-Frequency Operation of a Traveling-Wave Laser with Nonlinear Filter

Let us return to the laser scheme shown in Fig. 7.14. The change of the intensity of the light wave as it passes through the laser medium is described by Eq. (7.108). This equation is also useful for describing interaction with a nonlinear absorber. Adding Eq. (7.119), which adequately takes into account the dispersion of the laser system we arrive at the desired laser model:

$$\left(1+\frac{\partial}{\partial\tau}\right)\ln\frac{m_1(\tau+\zeta_1)}{m_0(\tau)}+m_1(\tau+\zeta_1)-m_0(\tau)=A\ln\frac{1}{R}, \tag{7.127a}$$

$$\left(1+\frac{1}{\delta}\frac{\partial}{\partial\tau}\right)\ln\frac{m_3(\tau+\zeta_3)}{m_1(\tau+\zeta_1)}+\frac{\rho}{\delta}[m_3(\tau+\zeta_3)-m_1(\tau+\zeta_1)]=-A_a\ln\frac{1}{R}, \tag{7.127b}$$

$$\left(1+\frac{\partial}{\partial\tau}\right)m_0(\tau+\zeta_0)=Rm_3(\tau+\zeta_3). \tag{7.127c}$$

All these symbols take on the meanings that have been used previously in this chapter.

Equations (7.127) satisfy the following relations between the stationary laser intensities in specified cross-sections over the cavity perimeter:

$$\overline{m}_1-\overline{m}_0=\frac{(A-A_a-1)\ln R^{-1}-(\rho/\delta)\overline{m}_0(R-1)}{1-\rho/\delta}, \tag{7.128a}$$

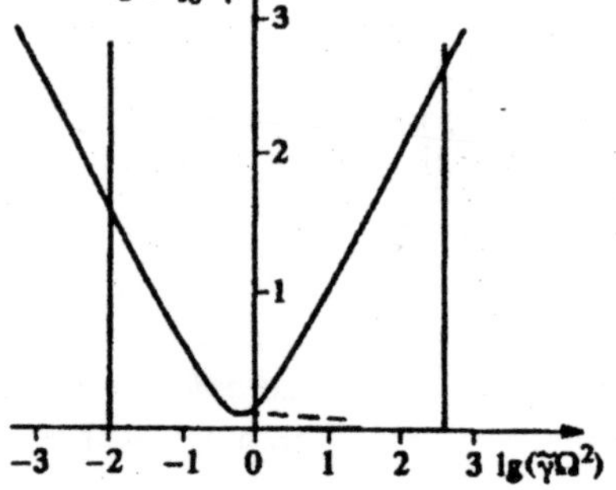

Fig. 7.15. Decay rate of the disturbance as a function of its frequency. The left vertical line indicates the low-frequency limit of the disturbance spectrum which corresponds to the minimum intermode spacing for a dye laser. The right vertical line has the same sense but for a solid-state laser. This diagram corresponds to the following values of parameters: $A=2;\ 2\zeta_0/\tilde{\gamma}=10^4;\ R=0{,}9$.

$$\overline{m}_1-\overline{m}_3=\frac{(A-A_a-1)\ln R^{-1}-\overline{m}_0(R-1)}{1-\rho/\delta}, \tag{7.128b}$$

The values $\overline{m}_i$ are obtained in explicit form only in the case $1-R\ll 1$ and $A_a\ll 1$ where taking into account the first two terms of the expansion of $\ln(\overline{m}_1/\overline{m}_0)$ in a Taylor series leads to the expression for $\overline{m}_0$ coincident with Eq. (7.14). Assuming below $\rho/\delta\ll 1$ and using Eq. (7.37),

$$\overline{m}_0=\frac{A-A_a-1}{A_a+1},$$

which is valid in this limit, we find from Eqs. (7.128)

$$\overline{m}_1-\overline{m}_0\approx(A-1)\ln R^{-1},\quad \overline{m}_1-\overline{m}_3\approx A_a(A-1)\ln R^{-1}. \tag{7.129}$$

Linearization of Eq. (7.127) in the vicinity of the steady state (7.128) leads to a characteristic equation

$$\frac{(\lambda+\delta+\rho\overline{m}_1)(\lambda+1+\overline{m}_0)}{(\lambda+\delta+\rho\overline{m}_3)(\lambda+1+\overline{m}_1)(\tilde{\gamma}\lambda+1)}=\exp(\zeta_0\lambda), \tag{7.130}$$

which transforms under the condition $|\theta|\ll\Omega$ to

$$2\zeta_0\theta=\frac{(\Omega^2+D_1^2)(\Omega^2+C_0^2)-(\Omega^2+D_3^2)(\Omega^2+C_1^2)(\tilde{\gamma}\Omega^2+1)}{(\Omega^2+D_3^2)(\Omega^2+C_1^2)(\tilde{\gamma}\Omega^2+1)}. \tag{7.131}$$

We have used the notations

$$C_0=1+\overline{m}_0,\quad C_1=1+\overline{m}_1,\quad D_1=\delta+\rho\overline{m}_1,\quad D_3=\delta+\rho\overline{m}_3.$$

The instability condition of the single-frequency regime of a travelling wave laser can be written in the form of an inequality

$$D_1^2-D_3^2>\frac{\Omega^2+D_3^2}{\Omega^2+C_0^2}[\tilde{\gamma}^2\Omega^2(\Omega^2+C_1^2)+C_1^2-C_0^2], \tag{7.132}$$

which follows directly from (7.131). In all real situations $\Omega^2\gg C_0^2, C_1^2$, which makes it possible to slightly simplify Eq. (7.132):

$$\overline{m}_1-\overline{m}_3>\frac{\Omega^2+(\delta+\rho\overline{m}_3)^2}{\rho\Omega^2[2\delta+\rho(\overline{m}_1+\overline{m}_3)]}\left[\tilde{\gamma}^2\Omega^4+(2+\overline{m}_0+\overline{m}_1)(\overline{m}_1-\overline{m}_0)\right]. \tag{7.133}$$

For fast-relaxing nonlinear filters $\delta\gg\rho,\Omega$, and the instability condition is further simplified reducing to

$$\overline{m}_1-\overline{m}_3>\frac{\delta}{2\rho\Omega^2}\left[\tilde{\gamma}^2\Omega^4+(2+\overline{m}_0+\overline{m}_1)(\overline{m}_1-\overline{m}_0)\right]. \tag{7.134}$$

Let us consider several important cases.

1. $\Omega_l >> \Omega_{min}$. This relation is characteristic of a solid-state laser with $L << L_{min}$. This means that the first term on the right-hand side of Eq. (7.134) dominates. Inserting the steady-state value (7.129) into (7.134) we find the threshold condition of instability

$$A_a > A_a^{cr} = \frac{\delta\tilde{\gamma}^2\Omega^2}{2\rho(A-1)\ln R^{-1}}. \tag{7.135}$$

Example 7.1

Nd:YAG laser with a fast nonlinear filter

$$\left.\begin{array}{ll} \rho = 10^4, & \Omega_l = 4\cdot 10^5, \\ \delta = 2\cdot 10^6, & A = 2, \\ \tilde{\gamma} = 5\cdot 10^{-9}, & r = 0.9, \end{array}\right| A_a^{cr} \approx 4\cdot 10^{-3}.$$

2. $\Omega_l << \Omega_{min}$. This condition is characteristic of all lasers with $L >> L_{min}$ including dye lasers. The inequality (7.134) transforms to

$$\frac{\overline{m}_1 - \overline{m}_3}{\overline{m}_1 - \overline{m}_0} > \frac{\delta}{2\rho\Omega^2}(2 + \overline{m}_0 + \overline{m}_1). \tag{7.136}$$

Further simplification is possible provided $A_a^{cr} << 1$ and Eqs. (7.129) are valid. Using these equations in Eq. (7.136) we find the instability condition

$$A_a > A_a^{cr} = \frac{\delta A}{\rho\Omega^2}. \tag{7.137}$$

Example 7.2

Dye laser with a fast nonlinear filter

$$\left.\begin{array}{l} \gamma_{\parallel} = 2\cdot 10^8\ \mathrm{s}^{-1}, \rho = 5, \\ \gamma_{\parallel a} = 2\cdot 10^{10}\ \mathrm{s}^{-1}, \delta = 100, \\ L = 100\ \mathrm{cm}, \Omega = 10,\ \mathrm{A} = 2, \end{array}\right| A_a^{cr} = 0.4.$$

This example shows that in ring dye lasers the destabilization of the single-frequency travelling wave regime occurs only when an optically dense nonlinear filter is put into the cavity and if the initial transmission of such a filter is comparable to the mirror reflectivity. Therefore, Eq. (7.137) can be treated as very approximate. One should also remember the role played by the combination tone mode-mode coupling in dye lasers, which was discussed in Chapter 4 but which was not taken into account in this model.

For completeness, we consider another case that corresponds to a solid-state laser with a superlong cavity.

3. $\Omega_l \approx \Omega_{min}$. The term in the right-hand side of Eq. (7.134) is of the same order, so that the instability criterion can be written approximately as

$$\frac{\overline{m}_1 - \overline{m}_3}{\overline{m}_1 - \overline{m}_0} > A_a^{cr} = \frac{3A\delta\tilde{\gamma}}{\rho}. \tag{7.138}$$

Example 7.3

Nd:YAG laser with a fast nonlinear filter

$$\left.\begin{aligned} \gamma_{\|} &= 5\cdot 10^3\ \mathrm{s}^{-1},\ \rho = 10^4, \\ \gamma_{\|a} &= 10^{10}\ \mathrm{s}^{-1},\ \delta = 2\cdot 10^6, \\ \gamma_{\perp} &= 10^{12}\ \mathrm{s}^{-1},\ \tilde{\gamma} = 5\cdot 10^{-9},\quad A = 2, \end{aligned}\right| A_a^{cr} = 6\cdot 10^{-6}.$$

The smallness of A_a^{cr} is worth noting. Nonlinear losses such as these may come from impurities uncontrollably entering the laser medium as it is prepared or due to contaminants in the atmospheric air. Meanwhile, the time of onset of the multifrequency regime in the presence of such a weak nonlinear filter is a few seconds. Hence, such an effect can be seen only in a CW laser.

7.5.3 Soft and Hard Regimes of Ultrashort Pulse Formation in Lasers with Nonlinear Filters

Two important facts have been established in the previous sections of this chapter: (a) the conditions of multimode laser action and of passive mode locking are both of threshold nature; (b) threshold values of the control parameters such as the nonlinear filter optical density for these two effects are different in general. Things are simple when the multimode laser threshold is below the mode-locking threshold. Soft mode locking is sufficient in this case. Otherwise, when the phased locking threshold is exceeded but the threshold of a multimode laser with a smooth spectrum containing all modes is not yet achieved, hard mode locking is possible. The effect is that two stable regimes occur simultaneously: CW and generation of a periodic train of pulses. Either regime can be achieved depending on the initial conditions. Stable generation of a train of pulses occurs, for example, when an initiating pulse is injected into the laser cavity but it may not occur spontaneously when the laser is switched on.

Let us illustrate what has been said using a solid-state laser as an example. In such a laser, the phased mode locking condition is expressed by inequality (7.100) and multimode unidirectional laser operation condition is described by Eq. (7.135). Bistability and, therefore, hard mode locking, occur in the range of values $A_a^{lock} < A_a < A_a^{cr}$. Assuming $\rho\Delta\omega >> \gamma_\parallel$ we simplify inequality (7.116) and reduce the necessary hard mode locking condition $A_a^{lock} < A_a^{cr}$ to

$$\zeta_0 < \pi\left[\frac{4\tilde{\gamma}^2}{(A-1)\ln R^{-1}}\right]^{1/3} \quad \text{i.e.,} \quad L < \frac{\pi c}{\gamma_\parallel}\left[\frac{2\tilde{\gamma}^2}{(A-1)\ln R^{-1}}\right]^{1/3}. \tag{7.139}$$

For a Nd:YAG laser ($\gamma_\parallel = 5\cdot 10^3\ \mathrm{s}^{-1}$, $\tilde{\gamma} = 5\cdot 10^{-9}$, $A = 2$, $R = 0.9$) the estimate using Eqs. (7.139) yields $L < 150\ \mathrm{cm}$. Consequently, a solid-state ring-cavity laser of ordinary size admits hard mode locking.

Things are somewhat different in a dye laser. Here, unlike the solid-state laser, the damping rate of the perturbation decreases with an increase in frequency (Fig. 7.15). This means that for $\theta(\Omega_1) < 0$ we may have $\theta(\Omega_q) > 0$ for $q > 1$, and the bifurcation will lead to an excitation of longitudinal modes some distance from the central mode rather than the neighboring ones. The process will proceed like this provided an inequality

$$\frac{\overline{m}_1 - \overline{m}_3}{\overline{m}_1 - \overline{m}_0} < \frac{\delta}{2\rho\Omega_1^2}(2 + \overline{m}_0 + \overline{m}_1), \tag{7.140}$$

the opposite of Eq. (7.136), is satisfied. The left-hand side of Eq. (7.140) is less than unity whereas $2 + \overline{m}_0 + \overline{m}_1 > 2$. Therefore, the inequality

$$\delta > \delta_{cr} = \rho\Omega_1^2 = 4\pi^2\rho/\zeta_0^2 \tag{7.141}$$

of course guarantees that inequality (7.140) is met. In the limiting case $A_a << 1$ and $1 - R << 1$ one may use the steady-state relations (7.129) and obtain a more exact inequality

$$\delta > \frac{4\pi^2\rho}{\zeta_0^2}\frac{A_a}{A} \qquad \left(\gamma_{\parallel a} > \frac{4\pi^2 c^2\rho}{L^2\gamma_\parallel}\frac{A_a}{A}\right). \tag{7.142}$$

In combination with the passive mode locking condition (7.100), inequality (7.141) or (7.142) make up the necessary condition for hard mode locking in a dye laser.

Example 7.4

$$\begin{array}{l|l} \rho = 5,\ A = 2, & \delta_{cr} = 250. \\ \zeta_0 = 0.4,\ A_a = 0.4, & \end{array}$$

Of course, this analysis is not exhaustive, since the rate-equation model has been used, whereas it was shown in Chapter 4 that the combination tone mode-mode coupling considerable changes the multimode laser dynamics. Therefore, the model of a travelling wave laser with a saturable absorber was numerically investigated without adiabatic elimination of the polarization of both media [577]. Laser parameters were assigned in accordance with recommendations given in Ref. [578] to provide a stable oscillation regime with one pulse in the cavity. If the population relaxation time of a saturable absorber was taken large enough ($\delta < \delta_{cr}$), then the process with one pulse, corresponding to soft mode locking, was established under any arbitrary initial conditions.

A decrease of the relaxation time down to values corresponding to $\delta \geq \delta_c$ leads to crucial changes in laser behaviour. Thus, for $\delta / \delta_{cr} = 2.5$ the initial uniform field first evolves to the field profile with four or five maxima per period with shallow modulation. At this rather long stage of the transient process the absorber saturation is weak. The next step of the transient process has a shorter duration. Owing to the diminishing of the field profile minima the modulation depth starts to increase and only one pulse will survive among the competing ones.

It should be noted that a tendency towards a longer transient process and its variation with decreasing relaxation time of the absorber was also mentioned in [579]. As the inequality $\delta > \delta_{cr}$ increases, the stage of laser action with shallow random-form modulation becomes longer. At $\delta = 20\delta_{cr}$ this stage was still in progress when the extensive computation was over.

Meanwhile, the process is quite different after a solitary pulse is injected into the laser. The transient process excited this way ends very soon if the pulse is injected during the initial stage and if the initiating pulse far exceed the noise level.

Numerical simulations of the dynamical processes in a dye ring laser with saturable absorber have confirmed the conclusions that follow from these analytical considerations. Passive mode locking was difficult when an absorber with a very short population relaxation time was used. Such a regime can be achieved in a pulsed laser only with hard pulse initiation because of the long transient process. Owing to limits in computing time, numerical simulations may not answer whether soft mode locking is feasible in a CW dye laser with a fast saturable absorber.

The numerical solution of the Maxwell–Bloch-type equations has confirmed the onset of bistability, which was revealed using the rate-equation model of a travelling wave dye laser with a saturable absorber. However, the alternative mode-locking regime is obviously the regime with a small number of excited modes and with irregular change of the field profile rather than the single-mode operation.

Many issues concerning the dynamics of lasers with nonlinear intracavity elements have not been tackled in this chapter. The list includes the influence of self-focusing on processes in the laser. In particular, there is no systematic analysis of the stability of single mode lasers in the presence of self-focusing due to the nonuniform saturation of the active medium itself by the laser field.

Also beyond the scope of this chapter is a wide range of issues concerning the nonadiabatic models of lasers with a saturable absorber. This is due to the multivariable nature of this problem, which would require too much space to be presented clearly in sufficient detail. Our neglect of this topic is justified by the facts that the theory is not convincingly tied to experiments and mathematical subtleties seem to outweigh the physical results at present.

Chapter 8

Giant Pulse Regime (*Q*-Switching)

The peak output power of a pulsed laser increases, the more the initial inversion exceeds the threshold level. In free running operation the deviations from the threshold level do not go beyond several per cent and the peak power of a solid-state laser is limited to tens of kilowatts. Considerably more powerful pulses, called giant pulses, are obtained when lasing is delayed for a time needed for a high inversion level to be reached [580]. The goal can be achieved by Q-switching of the laser cavity. In one method the laser threshold is held high, while the population difference increases, and the threshold is then reduced rapidly to the minimum possible level once the required inversion is reached and an output pulse is desired. An alternative method uses powerful pumping to ensure a sufficiently fast growth of inversion. But the latter did not find application.

8.1 Active $Q-$switching

8.1.1. Active methods of generating giant pulses

Methods of the cavity Q-switching are divided into active and passive groups. Active methods use modulation devices that change the cavity losses by a given law or in accordance with an external control signal. Passive modulating elements are those controlled directly by the radiation field in the laser cavity.

Active modulation devices are also divided into two groups: opto-mechanical and electrooptical. The simplest modulator of the first type is a punched disk of an opaque material [581]. Rotated around an axis parallel to the cavity axis, the disk blocks and opens the path between the mirrors periodically. The Q-switching time provided by such a device cannot be less than a millisecond. This value is considerable larger than the rise time of a giant pulse, which is the main drawback of a disk modulator.

Much shorter switching times are achieved by rotation of one of the reflectors around an axis perpendicular to the cavity axis [582]. A totally reflecting prism is often used as a rotating reflector and the rib of the prism is normal to the rotation axis. The turning angle at which the Q-factor of a plane-parallel resonator is half the maximum is about 2′ [583]. At the rotation velocity of $6 \cdot 10^4$ rpm the switching time is of order 10^{-7} s, which is sufficient for the single pulse generation. A cavity with spherical mirrors is not so sensitive to misalignment and its switching time is less than the mentioned above. The effective rate of Q-factor variation by a rotating element is reduced by optical imperfections of the laser rod [584].

Modulators of the second types are based on the electrooptical effect, i.e., on the dependence of the normal mode refractive indices on an applied electric field [1,585]. A Kerr cell is an example of such an electrooptical element, in which the effect is proportional to the square of applied voltage. It consists of a cell filled with a liquid with a high Kerr constant, such as nitrobenzene. Two parallel plane electrodes are immersed in the liquid to produce a homogeneous electric field.

Figure 8.1 shows the scheme of a laser with the Kerr shutter. Linearly polarized light is incident on the cell from the laser rod side owing to the polarizer. As the light completes a roundtrip in the cell its polarization changes depending on the electric field in the cell. Hence the intensity of the light transmitted back through the polarizer to the laser rod side also changes. The modulation depth is maximum if the principal directions of the polarizers form the angle of 45^0 with the electric field vector.

The time for Q-switching by an electrooptical Kerr modulator is limited by the rate of voltage variation in the cell, which is mainly restricted by the cell capacitance. The minimum switching time achievable by conventional methods is 10^{-8} s.

The electrooptical effect is linearly proportional to the electric field (the Pockels effect) in some crystals, and this is most pronounced in crys-

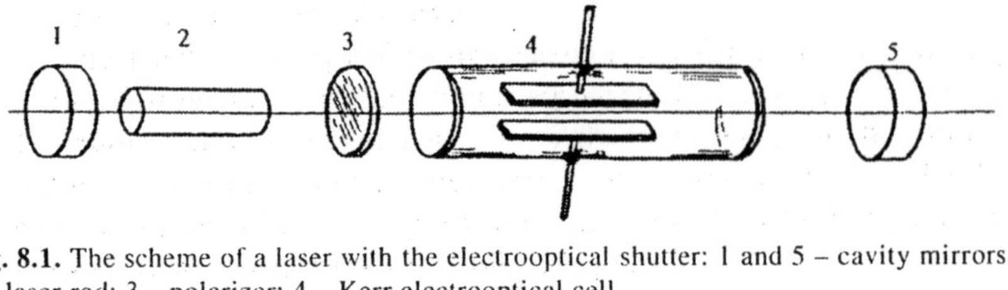

Fig. 8.1. The scheme of a laser with the electrooptical shutter: 1 and 5 – cavity mirrors; 2 – laser rod; 3 – polarizer; 4 – Kerr electrooptical cell.

tals without a centre of symmetry, such as KDP, DKDP, ADP, $LiNbO_3$, and $LiIO_3$. Crystal cells can be smaller in volume and, therefore, have smaller capacitance. In addition, these crystals do not require such high control voltages for reasonable effects. In application using electrooptical crystals, switching times can be reduced to 10^{-9} s.

The *Q*-switching time is the most important characteristic of a loss modulator. For each combination of values of the pumping power, the cavity length, and the mirror losses, there is a critical rate that demarcate the boundary between the domains of single pulse and multipulse lasing. For example, variation of the behaviour of a ruby laser on the response of a Kerr shutter was investigated experimentally in [586]. When the pumping energy exceeded 2.6 times the threshold value, single pulses were generated for *Q*-switching while the duration of the applied voltage transients remained below 100 ns. Above this limit several pulses were produced and the total energy of these pulses were reduced.

The single pulse operation occurred in lasers with flash lamp pumping. Continuous pumped lasers can generate periodic trains of giant pulses if the cavity *Q* is subjected to deep periodic modulation. This has been observed in Nd:YAG lasers [587, 588] and in molecular gas lasers [589–591].

Such active *Q*-switching typically provides more uniform spatial structure of the laser beam and contributes to spectral broadening in comparison with free-running operation [587, 592]. However, the spectral and spatial characteristics of emission may vary as the giant pulse develops. This evolution is readily apparent for lasers with mechanical shutters in which the cavity geometry is continuously varied. In a laser with a fast electrooptical shutter it is not obvious that such evolution would occur but it has been observed experimentally [593–596].The lasing first occurs in the central (axial) part of the rod where the inversion is greater. Then emitting region expands during 5–20 ns and the intensity maximum shifts towards the rod boundary. The pulses emitted by the central and the peripheral parts of the active region are shifted in time and each of them is several times shorter than the total pulse duration. Similar behaviour is observed in *Q*-switched gas lasers [591]. Thus, quite generally, a giant pulse represents the envelope of a succession of shorter bursts emitted by separate parts of the active region.

To observe the fast pattern variations of an optical pulse one can use a streak camera based on the electron-optical image converter (EOIC). The chosen area of the beam cross-section can be electrically swept on the EOIC display; thus, this device is faster than a streak camera with mechanical sweeping, which was mentioned in Section 1.2.3. An EOIC has also been useful to measure the spatial coherence of pulsed laser emission.

Temporal sweeping of the interference fringe pattern produced by two slits shows the presence of short-duration space coherence. This is indicated by the fixed position of the interference fringes during about one third of the pulse length. Stabilization of the interference pattern was achieved by introducing a small diaphragm into the laser cavity [595].

The development of a giant pulse in a ruby laser is accompanied by irreversible drift of the emitted line frequency to the shorter wavelength side. This effect was established by several authors [594,597,598]. The magnitude of the frequency shift depends on the pulse energy and varies from tens to hundreds of megahertz.

The features of giant pulse duration and energies have natural limits. The maximum pulse duration is limited by the growth time of the spike under linear and nonlinear amplification, while the minimum duration is limited by the photon lifetime in the cavity, which governs the pulse decay. The power of such giant pulses in solid-state lasers can reach 10^9 W.

8.1.2 Pulse Delay Time with Instantenous Q-Switching

Theoretical analysis of the emission process in single-mode lasers with fast Q-switching is given in [229,246,599–605]. The factor that makes the treatments much easier is the small, compared to T_1, risė time of giant pulses. Thus, we can neglect the relaxation and pumping processes and make use of the set of equations:

$$\frac{\mathrm{d}m}{\mathrm{d}\tau} - Gm[n - \phi_{\text{loss}}(\tau)] = G\varepsilon(n + n_S), \tag{8.1a}$$

$$\frac{\mathrm{d}n}{\mathrm{d}\tau} = -mn. \tag{8.1b}$$

Equation (8.1a) is an obvious generalization of Eq. (3.51).

The giant pulse formation begins at the moment $\tau = 0$ when the self-excitation condition $n > \phi_{\text{loss}}(0)$ is satisfied. In the initial stage of signal evolution the inversion is almost unchanged and the laser intensity grows in accordance with the linear Eq. (8.1a), in which $n = n_0 = \text{const}$. The solution of this is given by

$$m = \exp\left[G\int_0^\tau (n - \phi_{\text{loss}})\mathrm{d}\tau'\right]\left\{m_0 + G\varepsilon(n_0 + n_S)\int_0^\tau \exp\left[-G\int_0^{\tau'} (n_0 - \phi_{\text{loss}})\mathrm{d}\tau''\right]\mathrm{d}\tau'\right\}, \tag{8.2}$$

where m_0 is the value of the fluctuating field intensity at the moment $\tau = 0$.

We then assume that the Q-switching from one value to another is instantaneous:

$$\phi_{\text{loss}} = \begin{cases} 1 & \text{for } \tau > 0, \\ \phi_0 > n_0 & \text{for } \tau \le 0. \end{cases} \tag{8.3}$$

Putting $dm/d\tau = 0$ in Eq. (8.1*a*) we find

$$m_0 = \frac{\varepsilon(n_0 + n_S)}{\phi_0 - n_0}. \tag{8.4}$$

We substitute relations (8.3) and (8.4) into Eq. (8.2) and obtain, after integration

$$m = \varepsilon \frac{n_0 + n_S}{n_0 - 1} \exp[G(n_0 - 1)\tau] \left\{ 1 - \exp[-G(n_0 - 1)\tau] + \frac{n_0 - 1}{\phi_0 - n_0} \right\}. \tag{8.5}$$

A simplification is achieved at relatively large τ when $G(n-1)\tau >> 1$ and (8.5) transforms to

$$m = C \exp[G(n_0 - 1)\tau]. \tag{8.6}$$

By C we have designated

$$C = \varepsilon \frac{n_0 + n_S}{n_0 - 1} \left(1 + \frac{n_0 - 1}{\phi_0 - n_0} \right) \approx \varepsilon \frac{n_0 + n_S}{n_0 - 1}. \tag{8.7}$$

The second term in the brackets is small, since the losses before Q-switching typically are more than several times larger than the threshold value.

Formula (8.6) is valid until a certain value m_{d} is reached, which can be found by inserting (8.6) into (6.1*a*) and integrating:

$$m_{\text{d}} = G(n_0 - 1) \ln \frac{n_0}{n_{\text{d}}}. \tag{8.8}$$

The duration of this stage, which is the time delay of the pulse after the Q-switching event, is found from the equality of (8.6) and (8.8):

$$G\tau_{\text{d}} = \frac{1}{n_0 - 1} \ln \left[\frac{G(n_0 - 1)}{C} \ln \frac{n_0}{n_{\text{d}}} \right]. \tag{8.9}$$

In calculating τ_{d} there is some arbitrariness in choosing n_{d}, but the dependence of τ_{d} on n_{d} is too weak to noticeably influence the result.

Example 8.1
Ruby laser

$n_S = 100; \varepsilon = 10^{-13};$	$G\tau_d = 15;$
$n_0 = 3.5; C = 4\cdot 10^{-12};$	$t_d = 150$ ns.
$n_0 / n_d = 1.1; G = 10^5;$	
$T_c = 10^{-8}$ s;	

These estimates confirm that the assumptions of $G\tau_d >> 1$ and the transformation from Eq. (8.5) to Eq. (8.6) are reasonable.

We can use the formulas given above if the Q-switching time is small compared to t_d. From the example 8.1 we see that the switching time for the shutter should be of order 10—100 ns.

8.1.3. Energy Characteristics of a Giant Pulse

A giant pulse is emitted at the stage when the field intensity in the cavity exceeds the spontaneous emission level by many orders of magnitude. We therefore can ignore the spontaneous emission and find the first integral of Eqs. (8.1)

$$m_2 - m_1 = G[n_1 - n_2 - \ln(n_1 / n_2)]. \tag{8.10}$$

Putting $n_1 = n_0, m_1 = m_0, n_2 = 1$ and assuming $m_2 = m_{max} >> m_d$ we have:

$$m_{max} = G(n_0 - 1 - \ln n_0). \tag{8.11}$$

In the limiting case $n_0 - 1 << 1$ the logarithm can be expanded in a Taylor series and Eq. (8.11) can be reduced to (3.33).

The maximum power loss including the output radiation is related to the maximum field intensity in the cavity by $P_{output}^{max} = W_{max} / T_c$, or, in dimensionless form,

$$p_{output}^{max} = Gm_{max} = G^2 (n_0 - 1 - \ln n_0). \tag{8.12}$$

The quantity p_{output}^{max} for a given initial value of the inversion, $N(0)$, is not monotonically dependent on the cavity Q-factor. To find the maximum of this dependence we rewrite Eq. (8.12) in a more convenient form, by noting that $G \sim \kappa$, and $n_0 = N(0)/N_{thr} = a\kappa^{-1}$:

$$p_{output}^{max} \sim \kappa^2 [a\kappa^{-1} - 1 - \ln(a\kappa^{-1})]. \tag{8.13}$$

The losses κ_0, which correspond to the value p_{output}^{max}, are found from the condition $dp_{output}^{max} / d\kappa = 0$, which is equivalent to

$$a\kappa^{-1} - 1 - \ln(a\kappa^{-1}) = 0,$$

or

$$n_0 - 1 = 2 \ln n_0 . \tag{8.14}$$

The equality (8.14) corresponds to the optimal initial excess of inversion over the threshold after the switch is $n_0 = 3.5$.

Using these formulas we can estimate the peak output power of a giant pulse:

$$P_{\text{output}}^{\max} = 10^{-7} \frac{2\hbar\omega\kappa^2 V_c}{\beta_a c \sigma_{\text{tr}}} (n_0 - 1 - \ln n_0) \text{ Watt} . \tag{8.15}$$

In passing from the dimensionless form (8.12) to (8.15) we use relations (3.77). From (8.15) it follows that a decrease in the cross-section of the induced transition, for a fixed initial excess of the inversion over the threshold, leads to an increase in the power of the giant pulse.

Example 8.2

Ruby laser

$\omega = 10^{15}$ s^{-1}; $\sigma_{\text{tr}} = 10^{-20}$ cm^2;	$P_{\text{output}}^{\max} = 50$ MW.
$\kappa = 10^8$ s^{-1}; $V_c = 30$ cm^3;	

During giant pulse generation the laser medium gives a total energy

$$W_p = \beta_a^{-1} \hbar\omega (N_0 - N_f) \tag{8.16}$$

to the field. A ruby laser with the parameter values given in the example 8.2 emits an energy of about 3 J at pulse.

8.1.4 Pulse Duration and Shape

Substituting Eq. (8.10) into Eq. (8.1*b*) and integrating the resulting equation, we find the equality

$$G(\tau - \tau_d) = - \int_{n_d}^{n} \frac{dn}{n\{G^{-1} m_d + [n_0 - n - \ln(n_0 / n)]\}} ,$$

which implicitly defines the function $n(\tau)$ and thus $m(\tau)$ via relation (8.10). Computed pulse shapes are shown in Fig. 8.2.

These results are of practical importance when the inversion exceeds by several times the laser threshold immediately after Q-switching. In this case there is an approximate method for calculating the pulse duration.

Let as demonstrate first that the inversion remains almost unchanged until the laser intensity grows to $m_1 = m_{\max} / 2$ and that the laser medium saturates mainly in the subsequent period of time. To do this, we make use of the equality (8.10), in which we put $m_1 << m_2 = m$ and $n_1 = n_0$. Retaining only the linear term of the expansion $\ln(n / n_0)$ in a power series in

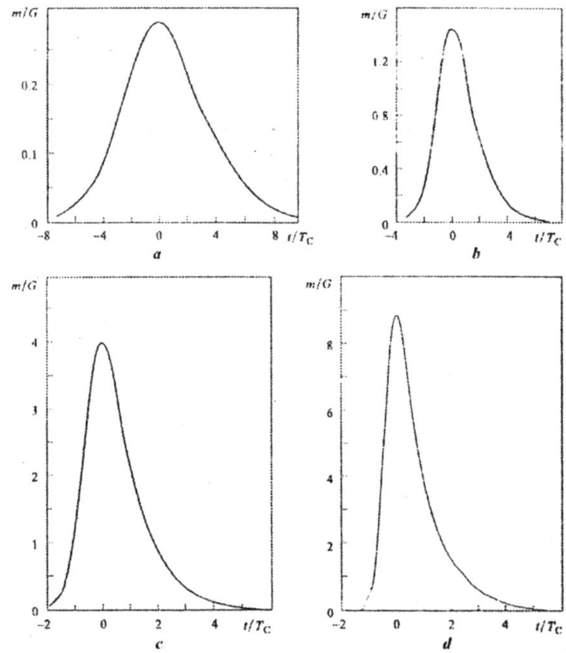

Fig. 8.2. Shape of the giant pulse of different initial exceeding of laser threshold [599]: $n_0 = 1.65$ (*a*); 2.72 (*b*); 4.48 (*c*); 7.40 (*d*).

$(n_0 - n)/n_0$ we find

$$\frac{n_0 - n}{n_0} = \frac{m}{G(n_0 - 1)} \cdot \tag{8.17}$$

The range of application of this formula is limited by the condition $m << G(n_0 - 1)$. Using $m = m_{max}/2 = G(n_0 - 1 - \ln n_0)$ the equality (8.17) takes the form

$$\frac{n_0 - n_i}{n_0} = \frac{n_0 - 1 - \ln n_0}{2(n_0 - 1)} \cdot \tag{8.18}$$

The right-hand of (8.18) tends to zero as $n_0 \to 1$ and to 0.5 as $n_0 \to \infty$. The optimal value $n_0 = 3.5$ corresponds to $(n_0 - n_i)/n_0 = 0.25$. Thus, using Eq. (8.18) we will accurately define the population difference n_i at the moment when m reaches the value $m_{max}/2$.

The time during which the pulse rises from $m_1 = m_{max}/2$ to $m = m_{max}$ is found by integration of (8.1*b*). The $m(\tau)$ variation law is unknown, but the result is little sensitive to this law, since m varies by only a factor of 2 in this range. Assume that the laser intensity obeys the quadratic law

$$m = m_{\max}\left[1 - \frac{(\tau - \tau_{\text{pl}})^2}{2\tau_{\text{pl}}^2}\right], \tag{8.19}$$

where τ_{pl} is the rise time. Then Eq. (8.1*b*) yields the time dependence of the inversion

$$\ln\frac{n_{\text{i}}}{n} = m_{\max}\left[\tau - \frac{(\tau - \tau_{\text{pl}})^3 + \tau_{\text{pl}}^3}{6\tau_{\text{pl}}^2}\right],$$

from which it follows that

$$G\tau_{pl} = \frac{6}{5}\frac{\ln n_{\text{i}}}{n_0 - 1 - \ln n_0}. \tag{8.20}$$

The fall time from $m = m_{\max}$ to $m_{\text{f}} = m_{\max}/2$ is described by the function

$$m = m_{\max}\left[1 - \frac{(\tau - \tau_{\text{p2}})^2}{2\tau_{\text{pl}}^2}\right]. \tag{8.21}$$

In a similar fashion we find

$$G\tau_{p2} = \frac{6}{5}\frac{\ln(1/n_{\text{f}})}{n_0 - 1 - \ln n_0}, \tag{8.22}$$

where τ_{p2} is the fall time, n_{f} is the inversion corresponding to $m = m_{\max}/2$ in the falling stage. The values n_{i} and n_{f} are related by

$$n_{\text{f}} - \ln n_{\text{f}} = n_{\text{i}} - \ln n_{\text{i}}. \tag{8.23}$$

If the fixed value $m = m_{\max}$ is used instead of Eq. (8.19) end Eq. (8.21), then the factor 6/5 is eliminated in (8.20) and (8.22).

Results of the calculation of τ_{pl} and τ_{p2} using Eqs. (8.18)–(8.23) are presented in Table 8.1. Values obtained by numerical integration of Eqs. (8.1) from [599] are given for comparison.

8.1.5. Angular and Frequency Spectra of a Giant Pulse with Instantaneous *Q*-switching

Many modes are excited in a rapidly *Q*-switched laser unless mode selection is provided. Therefore, the question arises: What is the influence of this fact on the temporal and energy characteristics of a giant pulse, as well as on its angular and frequency spectra? In the case of a cavity uniformly filled with the laser medium, the mode composition, which was formed at the linear stage, yields almost the whole information required [605]. The mode composition is not constant, and the problem is much more complicated, if the distribution of the inversion over the cavity cross-

Table 8.1

n_0	n_i	n_f	$G\tau_{p1}$	$G\tau_{p1}$ [599]	$G\tau_{p2}$	$G\tau_{p2}$ [599]
1.65	1.45	0.67	3.04	3.02	3.35	3.48
2.72	2.2	0.36	1.32	1.27	1.75	1.74
4.48	3.1	0.17	0.66	0.7	1.08	1.18
7.39	4.8	0.05	0.47	0.43	0.84	0.94
11.00	6.8	0.007	0.3	0.3	0.8	0.83

section is not uniform [604, 606, 607].

The equations of a Q-switched multimode laser in the case of spatially uniform inversion,

$$\frac{dm_k}{d\tau} - Gm_k\left[\frac{n}{1+\Delta_k^2} - (1+\beta_k)\phi_{\text{loss}}(\tau)\right] = \frac{G\varepsilon}{1+\Delta_k^2}(n+n_s), \quad (8.24a)$$

$$\frac{dn_k}{d\tau} = -n\sum_k \frac{m_k}{1+\Delta_k^2}, \quad (8.24b)$$

follow from Eq. (4.45) under the assumption of a short duration of the giant pulse.

Assumption that the Q-switching is instantaneous we approximate $\phi_{\text{loss}}(\tau)$ by a step-wise function (8.3). The mode intensity growth during the linear stage is described by Eq. (8.24a), the solution of which at $n = n_0$ is given by the function

$$m_k = \exp\left[\left(\frac{n_0}{1+\Delta_k^2} - 1 - \beta_k\right)G\tau\right] \times \left\{m_{k0} + \varepsilon\frac{n_0+n_S}{n_0-(1+\Delta_k^2)(1+\beta_k)}\left[1-\exp\left[-\left(\frac{n_0}{1+\Delta_k^2} - 1 - \beta_k\right)G\tau\right]\right]\right\}. \quad (8.25)$$

This expression can be simplified considerably, since $\Delta_k^2 \ll 1$, $\beta_k \ll 1$ and these terms can be omitted everywhere except in the exponents, in which we use the expansion $(1+\Delta_k^2)^{-1} \approx 1-\Delta_k^2$. In addition, $G\tau_d \gg 1$, which makes the exponential term in the curly brackets negligibly small. Thus, it should be reasonable to replace (8.25) by the simplified expression

$$m_k(\tau) = m_0(\tau)\exp[-(n_0\Delta_k^2 + \beta_k)G\tau], \quad (8.26)$$

where $m_0(\tau)$ is the intensity of the reference mode with $\Delta_k = 0$ and $\beta_k = 0$. The function $m_0(\tau)$ coincides with (8.6).

The calculation of the pulse delay time is essentially the same as that used in the single-mode problem. Introducing Eq. (8.26) into Eq. (8.24*b*) and subsequently integrating leads to an equality that implicitly defines τ_d:

$$\ln\frac{n_0}{n_\mathrm{d}} = \varepsilon\frac{n_0 + n_S}{G(n_0-1)^2}\exp[G(n_0-1)\tau_\mathrm{d}]\sum\exp[-(n_0\Delta_k^2+\beta_k)G\tau_\mathrm{d}]. \tag{8.27}$$

The factor

$$\sum\exp[-(n_0\Delta_k^2+\beta_k)G\tau_\mathrm{d}] = \tilde{N}(G\tau_\mathrm{d})$$

has the meaning of an effective number of excited modes, since from (8.26) it follows that $\tilde{N} = \sum m_k / m_0$.

The expressions for Δ_k and β in the case of a plane-parallel cavity with square mirrors are given in Section 4.1.2. The approximate calculation of $\tilde{N}(G\tau)$ is easy if these expressions are used and the summation is replaced by integration:

$$\tilde{N}(G\tau) = \frac{16b^3\,|\ln R_1R_2|\,\gamma_\perp}{C\lambda}\left(\frac{2}{L\lambda n}\right)^{1/2}(G\tau)^{-3/2} = \tilde{N}_0(G\tau)^{-3/2}. \tag{8.28}$$

We now rewrite Eq. (8.27) as

$$\ln\frac{n_0}{n_\mathrm{d}} = \frac{C\tilde{N}_0}{G(n_0-1)}(G\tau_\mathrm{d})^{-3/2}\exp[G(n_0-1)\tau_\mathrm{d}] \tag{8.29}$$

The coefficient C coincides with (8.7). This equation can be solved by successive approximations. Putting first $(G\tau_\mathrm{d})^{-3/2} = 1$, we find τ_d in the zeroth-order approximation. Then, inserting $(G\tau_\mathrm{d}^{(0)})^{-3/2}$ into (8.29) we find the first-order approximation

$$G\tau_\mathrm{d}^{(1)} = \frac{1}{n_0-1}\ln\left[\tilde{D}\left(\frac{\ln\tilde{D}}{n_0+1}\right)^{3/2}\right]. \tag{8.30}$$

where

$$\tilde{D} = \frac{G(n_0-1)}{C\tilde{N}}\ln\frac{n_0}{n_\mathrm{d}}.$$

Example 8.3

Ruby laser

$b = 0.5$ cm; L = 100 cm;	$\tilde{N}_0 = 10^6$;
$\lambda = 7 \cdot 10^{-5}$ cm; $\gamma_\perp = 10^{12}$ s^{-1};	$G\tau = 11$;
$R_1 R_2 = 0.5$; $n_0 = 3.5$;	$t_d = 110$ ns
$C = 4 \cdot 10^{-12}$; $\ln(n_0 / n_d) = 0.1$;	$\tilde{N}(G\tau_d) = 2 \cdot 10^4$.

It is well known from the single-mode model analysis that the giant pulse duration is much less than the delay time. Thus, it can be asserted that the modes that dominate at the moment τ_d will define the laser characteristics in the pulse maximum as well. These modes ensure uniform saturation of the laser medium. The population difference will be suppressed to the threshold value before the weaker modes have a chance to reach significant strengths. Therefore, Eq. (8.26) with $\tau = \tau_d$ should accurately describe the frequency and angular spectra during the giant pulse.

The width of the frequency spectrum at $1/e$ of its maximum value is determined from the equality $n_0 G\tau_d \Delta_k^2 = 1$, which is equivalent to

$$\Delta\omega = \gamma_\perp (n_0 G\tau_d)^{-1/2} . \tag{8.31}$$

The angular divergence is found from the idea that the mode with the index r has the maximum in the direction $\vartheta = \pm r\lambda / 4b$. The aperture angle, within which the intensity falls to $1/e$ of its value in the forward direction, is determined from the condition

$$\beta_{r1} G\tau_d = 1 . \tag{8.32}$$

This equality together with (4.25), defines

$$r_{max} = \left[\frac{32 b^3 \ln |R_1 R_2|}{G\tau_d \lambda} \left(\frac{2}{\pi\lambda L} \right)^{1/2} \right]^{1/2} . \tag{8.33}$$

We can justify the assumption of smallness $1/G\tau_d$, Δ_k and β_k for the dominant modes. That $G\tau_d >> 1$ is confirmed by our previous numerical examples. From (8.31) and (8.32) it follows that the large value of $G\tau_d$ ensures the smallness of Δ_k and β_k.

Using Eqs. (8.31) and (8.32) we can find the number of longitudinal and transverse modes that participate in lasing. At the same parameter values as those used for estimates in this section, we find $\Delta_k < 0.16$ and $\beta_k < 0.1$, while the maximum index is $r_{max} = 15$, the number of longitudinal modes is 200, and the divergence angle is $4.5'$.

The small difference in relative detuning and losses of the modes that form the main group makes the analysis of the nonlinear stage much easier.

Introducing the total field intensity $m = \sum m_k$, we reduce Eqs. (8.24) to an approximate set of equations, which coincide in form with the single-mode laser equations. The accuracy of the approximation increases with decreasing variation in the number of modes during the pulse. The variation expressed by (8.28) is a limiting one, such that

$$1 > \frac{\tilde{N}(G\tau)}{\tilde{N}(G\tau_d)} > \left(\frac{\tau}{\tau_d}\right)^{-3/2}.$$

The number of modes can be considered constant as long as the pulse duration is small compared to the delay time. This means that the spectrum and the divergence angle remain the same as at the moment τ_d. In this approximation *Q*-switching in a laser with spatially uniform pumping, the initial conditions are the same for all modes. However, the rates of development of these modes do not coincide because of their different losses and their different detunings from the gain line centre. This leads to the significant development of only a limited number of modes. During the nonlinear stage, the modes of this group almost uniformly saturate the laser medium in its entire volume. Thus, the excitation of other modes is excluded.

8.2 Giant Pulse Generation with Passive *Q*-switching

Passive optical shutters use the ability of materials to change their optical properties under the action of incident light. The crudest effect of this type is the destruction of the material. The simplest passive shutter is an absorbing film placard in the laser cavity. When the power density of incident radiation reached the threshold value the film is evaporated thus opening the mirror behind it. The cavity losses are sharply decreased and a giant pulse is generated [608, 609]. There are also materials with increased transparency from photochemical reactions [610]. The drawbacks of such simplest shutters follow from the irreversibility of the processes, which makes multiple uses of the devices impossible.

The idea of using media with saturable reversible absorption in quantum electronics was proposed by Rivlin [611]. They were first used for the giant pulse generation in a ruby laser. Coloured glasses [612] and solution of cyanine dyes such as phtalocyanine [613] and cryptocyanine [614] were employed as the nonlinear filters. Polymethine dye solutions [615] were used for the same purpose in neodymium lasers. The dye should be selected in such a manner that the laser frequency is within the intense absorption band. The flux density of radiation for which saturation is started, range from 10^4–10^7 W/cm^2.

Specific components for passive Q-switching have a stronger effect on the spectral and spatial characteristics and a weaker effect on the energy and shape of the pulses. When a passive shutter in the form of a saturable absorption cell is used, the spatial structure of the light beam exhibits a strong inhomogeneity. In a cavity with spherical mirrors this structure corresponds to an individual transverse mode, most often a higher order [616, 617]. In a plane-parallel cavity the field is divided into filaments [618]. As for the case of an electrooptical cell, the lasing action starts in the central part of the laser rod and then spreads to the periphery [593, 618]. However, it should be noted that field spreading is exhibited only in the case of transversely nonuniform pumping of the laser rod. The use of a pumping device in order to provide an uniform inversion removes the difference in duration of the pulses emitted from the whole of the laser rod cross-section and from any part of it [619]. The radiation spectrum of a laser with saturable absorber is extremely narrow [618, 620].

The model of a multimode laser with passive Q-switching can be easily obtained by generalization of the rate equations (7.11) to this case and adding the terms that take into account the mean spontaneous emission background [353]:

$$\frac{dm_k}{d\tau} - Gm_k\left[\frac{n}{1+\Delta_k^2} - n_a - (1+\beta_k)\right] = G\varepsilon\left(\frac{n+n_S}{1+\Delta_k^2} + A_a - n_a\right),$$

$$\frac{dn}{d\tau} = A - n - n\sum\frac{m_k}{1+\Delta_k^2}, \tag{8.34}$$

$$\frac{dn_a}{d\tau} = \delta(A_a - n_a) - \rho n_a \sum m_k .$$

A wide absorption spectrum of the nonlinear filter makes it possible to neglect the difference of frequency of the laser modes. The variable n_a is defined as $n_{1a}-n_{2a}$ and is, therefore, positive.

During the nonlinear stage the presence of an absorber does not add any other specific features. Thus, the results, obtained in Chapter 4 for free running, are fully applicable here. The long delay time in the pulse leads to a considerable mode discrimination [353, 354]. In a plane-parallel cavity the leading group includes tens of longitudinal and a few transverse modes before the beginning of the nonlinear stage. A single-mode regime occurs when

$$m_k(\tau_d)/m_0(\tau_d) < 1/e$$

and, therefore, the mode losses differ by more than

$$\beta_1 = \left(\frac{2G}{A - A_a - 1} \ln \frac{A - A_a - 1}{m_{\min}} \right)^{-1/2} . \tag{8.35}$$

The parameter values $G = 10^5, m_{\min} = 10^{-10}, A - A_a - 1 \approx 1$ correspond to $\beta_1 = 5 \cdot 10^{-4}$.

Thus, small mode discrimination is sufficient to suppress a particular mode. Hence, random or uncontrolled factors may have a profound effect on the optical spectrum and optical field structure in a giant pulse.

Both the pulse energy and shape can be determined using a single-mode model. The only major deficiency is that it will not take into account the influence of the nonuniform distribution of pumping over the laser rod. The theory of a giant-pulse single-mode laser with saturable absorber was developed in [246, 621–625]. The case of nonlinear filter with relaxation time longer than the pulse duration has been analyzed in the most detail. Under this condition Eqs. (8.34) reduce to

$$\begin{aligned} \frac{dm}{d\tau} &= Gm(n - n_a - 1), \\ \frac{dn}{d\tau} &= -mn, \\ \frac{dn_a}{d\tau} &= -\rho m n_a . \end{aligned} \tag{8.36}$$

From the last two equations of (8.36) we find

$$n_a = \tilde{n}_a (n / \tilde{n})^\rho . \tag{8.37}$$

where $\tilde{n}_a \approx A_a$ and $\tilde{n} \approx 1 + A_a$. Using Eq. (8.37) we eliminate the variable n_a from these equations and reduce (8.35) to a second-order set of equations

$$\begin{aligned} \frac{dm}{d\tau} &= Gm[n - \tilde{n}_a (n / \tilde{n})^\rho - 1], \\ \frac{dn}{d\tau} &= -mn, \end{aligned} \tag{8.38}$$

which have the integral

$$m - \tilde{m} = G\left(n - \tilde{n} - \ln\frac{\tilde{n}}{n} - \frac{\tilde{n}_a}{\rho}\frac{\tilde{n}^\rho - n^\rho}{\tilde{n}^\rho} \right). \tag{8.39}$$

Here $\tilde{m}$ is the value of m which correspond to $n = \tilde{n}$. In contrast to Eqs.

(8.10), which describes the model of laser with instantaneous Q-switching, Eqs. (8.39) has an additional term in the right-hand side. However, for $\rho >> 1$ this addition is inessential and

$$m_{max} \approx G(\tilde{n} - 1 - \ln \tilde{n}) = G[A_a - \ln(1 + A_a)]. \qquad (8.40)$$

The coincidence of Eqs. (8.39) and (8.11) is not accidental. Since the nonlinear filter is saturated much easier than the laser medium ($\rho >> 1$), the giant pulse is developed under the same conditions as with instantaneous Q-switching.

Equations (8.36) are not adequate for describing the processes in the laser with a nonlinear filter having small relaxation time. To be more precise, the last equation of (8.36) is not valid, since the derivative $dn_a/d\tau$ rather than the relaxation term is a small quantity in this case. The last equation of (8.34) should be written as

$$n_a = \frac{A_a}{1 + (\rho / \delta) m},$$

and instead of Eq. (8.36) we should write the equations

$$\frac{dm}{d\tau} = Gm\left[n - 1 - \frac{A_a}{1 + (\rho / \delta) m}\right], \quad \frac{dn}{d\tau} = -mn. \qquad (8.41)$$

Complete saturation of the nonlinear filter is achieved if $m_{max} >> \delta / \rho$. Since m_{max} can never exceed the value defined by Eq. (8.40), the necessary condition for saturation of the filter is given by

$$A_a >> \delta / (G\rho). \qquad (8.42)$$

Typical of solid-state lasers with passive Q-switching are the values $T_1 \approx 10^{-3} - 10^{-4}$ s, $G = 10^5$ $\rho \approx 10^4$, $A_a \approx 1$ for which the inequality (8.42) requires $T_{1a} >> 10^{-12}$ s.

Equations (8.41) have been solved by numerical method [622]. Calculations have shown that under the condition (8.42) the shape and energy of a giant pulse are very weakly dependent on the population relaxation time in the absorber.

SUPPLEMENT

The concept of induced inversion gratings is mostly simply realized in the basic models of multimode lasers if the consideration is limited by the set of longitudinal modes

$$\psi_k = \sqrt{2}\sin(\pi q_k \zeta)\,.$$

Here we must distinguish cases of large ($\Delta_c >> \kappa$) and small ($\Delta_c << \kappa$) intermode frequency spacing.

The starting point for further transformations is the set of equations (4.2) that might be described using the time normalization condition $\tau = \gamma_{\|}t$ in the form

$$\frac{\mathrm{d}f_k}{\mathrm{d}\tau} + i\frac{G}{2}\Delta_{ck} f_k = \frac{G}{2}\left(\int p\psi_k \mathrm{d}\zeta - f_k\right), \tag{S.1a}$$

$$\frac{\partial p}{\partial \tau} - i\frac{\Delta_0}{\tilde{\gamma}} p = \frac{1}{\tilde{\gamma}}\left(n\sum f_j\psi_j - p\right), \tag{S.1b}$$

$$\frac{\partial n}{\partial \tau} = A - n - \frac{1}{2}\sum \psi_j\left(f_j^* p + f_j p^*\right). \tag{S.1c}$$

In the case of large intermode frequency spacing and many modes it is convenient to take the reference frequency ω equal to ω_0, so that $\Delta_0 = 0$. Let us introduce the quantity

$$p_l = \int_0^1 p\psi_l \mathrm{d}\zeta = \sqrt{2}\int_0^1 p\sin(\pi q_l \zeta)\mathrm{d}\zeta\,,$$

that represents the amplitude of the polarization grating. Now we can re-write the equation (S.1a) in the form

$$\frac{\mathrm{d}f_k}{\mathrm{d}\tau} + i\frac{G}{2}\Delta_{ck} f_k = \frac{G}{2}\left(p_l - f_k\right). \tag{S.2}$$

Let us multiply Eq (S.1b) on ψ_j and integrate on the cavity perimeter that leads us to the equation

$$\frac{\mathrm{d}p_l}{\mathrm{d}\tau} = \frac{1}{\tilde{\gamma}}\left(\sum f_j(n_{jl}^- + n_{jl}^+)\right). \tag{S.3}$$

So far as the field and polarization amplitudes have the time dependence like $\exp(-i\Delta_k\tau)$, the procedure of adiabatic elimination of the polarization amplitude leads to the relation

$$p_l = \frac{1}{1 - i\Delta_l} \sum_j f_j (n_{jl}^- + n_{jl}^+), \tag{S.4}$$

where

$$n_{jl}^{\pm} = \mp \int_0^1 n \cos[\pi(q_j \pm q_l)\zeta] \mathrm{d}\zeta \,. \tag{S.5}$$

Note that $n_{kk}^- = n_0$ and $n_{kk}^+ = n_k$. Bearing this in mind we can rewrite (S.4) as

$$p_k = \frac{1}{1 - i\Delta_k} \left[(n_0 + n_k) f_k + \sum_{j \neq k} f_j (n_{jk}^- + n_{jk}^+) \right]. \tag{S.6}$$

Next, using obtained results, we will transform Eq. (S.1c) into equations for the inversion gratings amplitudes

$$\frac{\mathrm{d}n_{\mu\nu}^-}{\mathrm{d}\tau} = -n_{\mu\nu}^- - \frac{\sqrt{2}}{2} \sum_j \int_0^1 (f_j^* p + f_j p^*) \sin(\pi q_j \zeta) \cos[\pi(q_\mu - q_\nu)\zeta] \mathrm{d}\zeta \,. \tag{S.7}$$

As in Eq. (4.1.1), we ignore the combination sums in Eq. (S.6) when we substitute p_j in Eq (S.7) as well as the spatial harmonics that exceed the bounds of the spectrum of lasing modes, and use the condition of harmonic balance for terms with the time dependence like $n_{\mu\nu}^- \sim \exp[-i(\Delta_\mu - \Delta_\nu)/\tilde{\gamma}]$. This results in the expression

$$n_{\mu\nu}^- = -\frac{1}{4} \frac{\tilde{F}_\mu (n_0 + n_\mu) + \tilde{F}_\nu^* (n_0 + n_\nu)}{1 + i(\Delta_\nu - \Delta_\mu)/\tilde{\gamma}} f_\mu f_\nu^* \,. \tag{S.8}$$

Similar expression we find also for $n_{\mu\nu}^+$. Using Eq. (S.8) into Eq. (S.6) bring us to the final formula for gain in presence of four-wave mixing (combination tone mode coupling):

$$p_k = \tilde{F}_k \left[(n_0 + n_k) f_k - \frac{1}{2} \sum_{\mu\nu} \frac{\tilde{F}_\mu (n_0 + n_\mu) + \tilde{F}_\nu^* (n_0 + n_\nu)}{1 + i(\Delta_\nu - \Delta_\mu)/\tilde{\gamma}} f_\mu f_\nu^* f_{k-\mu+\nu} \right]. \tag{S.9}$$

Using Eq. (S.9) in Eq. (S.2) we come to equation

$$\frac{\mathrm{d}f_k}{\mathrm{d}\tau} = \frac{G}{2} \left\{ \tilde{F}_k \left[(n_0 + n_k) f_k - \frac{1}{2} \sum_{\mu\nu} \frac{\tilde{F}_\mu (n_0 + n_\mu) + \tilde{F}_\nu^* (n_0 + n_\nu)}{1 + i(\Delta_\nu - \Delta_\mu)/\tilde{\gamma}} f_\mu f_\nu^* f_{k-\mu+\nu} \right] - 1 \right\}. \tag{S.10a}$$

Together with the material equations

$$\frac{dn_0}{d\tau} = A_0 - n_0\left(1+\sum_{j=1}^{N}\widetilde{L}_j m_j\right) - \sum_{j=1}^{N}\widetilde{L}_j m_j n_j, \tag{S.10b}$$

$$\frac{dn_k}{d\tau} = -n_k\left(1+\sum_{j=1}^{N}\widetilde{L}_j m_j\right) - \frac{1}{2}\widetilde{L}_k m_k n_0, \tag{S.10c}$$

The equation (S.10a) forms a closed set. It reduces to Eq. (4.10) if we ignore the combination sum in Eq. (S.10a). This set of equations is a particular case of Eq. (4.9), but one needs to bear in mind that n_{kk} in Eq. (4.9) corresponds to $n_0 + n_k$ in Eq. (S.10).

Equations of the two-mode laser model with a small frequency spacing between longitudinal modes can be obtained starting from the same set (S.1). However, it is more convenient to use the frequency of one of modes, ω_1, as the reference one. Therefore, instead of (S.4) we have

$$p_k = \frac{1}{1-i\Delta_0}\sum_j f_j(\bar{n}_{jk} + \overset{+}{n}_{jk}). \tag{S.11}$$

The small value of the intermode frequency spacing excludes the possibility to express explicitly $\bar{n}_{\mu\nu}$ and $\overset{+}{n}_{\mu\nu}$ through other variables like (S.8). We can only write differential equations for these variables. this simple procedure leads to the set of equations for the two-mode laser model with the phase-sensitive interaction

$$\frac{df_1}{d\tau} = \frac{G}{2}\left\{-f_1 + \frac{1}{1-i\Delta_0}\left[f_1(n_0+n_1) + f_2(\bar{n}_{12}+\overset{+}{n}_{12})\right]\right\}, \tag{S.12a}$$

$$\frac{df_2}{d\tau} = \frac{G}{2}\left\{-(1+\beta+i\Delta_c)f_2 + \frac{1}{1-i\Delta_0}\left[f_2(n_0+n_2) + f_1(\bar{n}_{12}+\overset{+}{n}_{12})\right]\right\}, \tag{S.12b}$$

$$\frac{dn_0}{d\tau} = A - n_0 - \frac{1}{1-\Delta_0^2}\left[n_0(|f_1|^2+|f_2|^2) + n_1|f_1|^2 + n_2|f_2|^2 + (f_1^* f_2 + f_1 f_2^*)(\bar{n}_{12}+\overset{+}{n}_{12})\right] \tag{S.12.c}$$

$$\frac{dn_1}{d\tau} = -n_1 - \frac{1}{1-\Delta_0^2}\left[n_1(|f_1|^2+|f_2|^2) + \frac{1}{2}n_0|f_1|^2 + \frac{1}{2}(f_1^* f_2 + f_1 f_2^*)(\bar{n}_{12}+\overset{+}{n}_{12})\right] \tag{S.12d}$$

$$\frac{dn_2}{d\tau} = -n_2 - \frac{1}{1-\Delta_0^2}\left[n_2(|f_1|^2+|f_2|^2) + \frac{1}{2}n_0|f_2|^2 + \frac{1}{2}(f_1^* f_2 + f_1 f_2^*)(\bar{n}_{12}+\overset{+}{n}_{12})\right]$$

(S.12*e*)

$$\frac{\mathrm{d}n_{12}^{\pm}}{\mathrm{d}\tau} = -n_{12}^{\pm} - \frac{1}{1-\Delta_0^2}\left[n_{12}^{\pm}(|f_1|^2 + |f_2|^2) + \frac{1}{2}n_{12}^{\mp}(|f_1|^2 + |f_2|^2) + \frac{1}{2}(f_1^* f_2 + f_1 f_2^*)(n_0 + n_1 + n_2)\right] \tag{S.12f}$$

Notations

a	Beam radius
a_0	Unperturbed value of the beam radius
a_j	Coefficient of characteristic equation
A	Pumping parameter
A_0	Normalized initial density of a saturable absorber
A_a^{cr}	Critical value of A_0 corresponding to laser instability threshold
A_a^{lock}	Saturable absorber density corresponding to mode locking threshold
A_{cr}	Pumping parameter at the second laser threshold
b	Mirror radius
$\boldsymbol{B}$	Magnetic induction vector
B	Einstein coefficient
c	Velocity of light in vacuum
c'	Velocity of light in a material
$\mathbf{d}$	Dipole moment vector
d_{mn}	Matrix element of dipole operator
d	Dipole momentum modulus
d_{dif}	Normalized diffusion coefficient
$\boldsymbol{D}$	Electric induction vector
D	Inversion (population difference)
$D^{(0)}$	Unsaturated value of inversion D
D_λ	Spatial harmonic of the inversion
$e_\lambda(t)$	Time-dependent coefficient of the electric field modal decomposition
$\boldsymbol{E}$	Electric field intensity
$\boldsymbol{E}_{damp}$	Fast-damping component of the filed in an open cavity
$\boldsymbol{E}_0$	Amplitude of electric field
$\boldsymbol{E}_\lambda(\mathbf{r})$	Electric eigenfunction of a resonator
F	Complex field amplitude
$F_{\pm 1}$	Amplitudes of counterrunning waves
f	Normalized complex field amplitude
f_λ	Normalized mode amplitude
$f_\pm$	Normalized amplitudes of counterrunning waves
F	Focal length
F_0	Unperturbed value of effective focal length
F_{nl}	Focal length of a nonlinear lens
F_λ	Mode amplitude
F_{sat}	Saturation value of the field amplitude
F_{coh}	Field amplitude corresponding to coherent field-matter interaction
F^{disp}	Function which characterizes a system dispersion

F^{nl}	Function which characterizes a system nonlinearity
$\tilde{F}$	Complex Lorentzian function
G	Big parameter in the class *B* laser theory
G^{net}	Net gain (difference between gain and loss)
$\tilde{G}^{bal}$	Balance (rate equations approach) part of the mode gain
$\tilde{G}^{comb}$	Combination tone part of the mode gain
h	Distribution function
$h_\lambda(t)$	Time-dependent part of the magnetic field modal decomposition
$\boldsymbol{H}$	Magnetic field intensity
$\boldsymbol{H}_\lambda(\boldsymbol{r})$	Magnetic eigenfunction of a resonator
H	Hamiltonian
$\boldsymbol{j}$	Current density vector
k	Wave number
$k_\perp$	Transverse component of the wave number
k_B	Boltzmann constant
k_{gain}	Gain factor
K_{loss}	Amplification factor of loss modulation
K_{pump}	Amplification factor of pump modulation
K_{total}	Transfer function for the total intensity of a multimode laser
K_j	Transfer function for individual mode intensity of a multimode laser
l	Normalized cavity length
L	Cavity length in cm
L_s	Laser medium (sample) length
L_{cr}	Critical value of the cavity length
L_{nl}	Nonlinear medium length
L'	Effective (optical) cavity length
$\tilde{L}$	Lorentzian line shape
m	Normalized number of photons; normalized field intensity
$\overline{m}$	Steady-state value of *m*
$\tilde{m}$	Amplitude of intensity response to parameter modulation; quasi-steady-state value of the intensity in a laser with monotonically varied parameter
$m_0, m_\pm$	Steady states
$\boldsymbol{M}$	Magnetization vector
M	Number of photons
$\boldsymbol{n}$	Unit vector in normal direction to a surface
n	Normalized inversion
$\overline{n}$	Steady-state value of the inversion
$\tilde{n}$	Amplitude of inversion response to parameter modulation; quasi-steady-state value of the inversion in a laser with

	monotonically varied parameter
n_2	Normalized population of one of laser levels
n_r	Amplitudes of inversion gratings
n_s	Normalized molecule density
n_a	Normalized population difference in an absorber
n_d	Inversion arrives at the end of linear stage of pulse formation
n_{jl}^{+}	Amplitudes of small-scale inversion gratings
n_{jl}^{-}	Amplitudes of large-scale inversion gratings
N	Population difference (inversion) per unit of volume
N_s	Total number of molecules per unit of volume
N_{eff}	Effective value of inversion
$N^{(0)}$	Unsaturated value of N
p	Normalized complex amplitude of polarization in an atomic system
$\boldsymbol{P}$	Polarization vector
P	Complex amplitude of polarization of an atomic system
$P_{\pm 1}$	Complex amplitude of counterrunning polarization waves
P_{out}	Output laser power
q	Inhomogeneous broadening parameter; quadratic variable
Q	Quality factor
Q_c	Quality factor corresponding to coupling losses
Q_s	Quality factor corresponding to losses in resonator walls
Q_v	Quality factor corresponding to distributed bulk losses
$\boldsymbol{r}$	Position vector
r	Quadratic variable
r; $r_{\pm}$	Counterrunning wave coupling coefficients
r_{cr}	Critical value of the coupling coefficient
R	Mirror reflectivity; Parameter proportional to the total number of molecules in the theory of three-level laser (Sec. 3.3)
s	Quadratic variable; Parameter in the theory of inhomogeneously broadened laser (Sec. 4.5.)
S	Surface; Sum of energy level populations
t	Time
t_d	Delay time
t_a	Lifetime of a molecule at the energy level
t_{cor}	Time of correlation
t_{trans}	Time constant of transition to aperiodic process in a gaseous laser

T	Temperature
T	Period
T_1	Lifetime of the inversion ("longitudinal" relaxation time)
T_2	Time constant of the dipole-moment decay ("transversal" relaxation time)
T_c	Photon lifetime in the cavity
T_0	Cavity round trip time
T_p	Pulse duration
u	Normalized velocity
U	Velocity (cm per second)
U_0	Most probable value of molecule velocity
v	Normalized volume
v_s	Sample volume
V	Volume
V_{refl}	Velocity of a moving reflective surface
V_1	Resonant value of the velocity of a moving reflective surface inside a laser cavity
V_{eff}	Effective velocity of cavity lengthening
w_{mn}	Rate of the relaxation transition from level m to level n
w_j	Decay rate of j-th atomic level
W_{sp}	Rate of spontaneous emission in a cavity mode
W_{pump}	Rate of the quantum transition induced by pumping
W_j	Rate of the pumping transition at the j-th atomic level
W	Energy
W_0	Band-gap energy
x,y,z	Cartesian coordinates
Z	Impedance
α	Laser line width enhancement factor
β	Normalized losses
β_a	Effective change in the population difference per one event of photon emission
β_{c-s}	Index of the transition cross-section modulation
β^{cr}	Parameter modulation index above which the laser response is considerably nonlinear
β_k	Losses specific for the k-th mode

β_{loss}	Index of the loss modulation
β_{pump}	Index of the pump modulation
β_{NR}	Amplitude nonreciprocity of a ring cavity
$\beta_{\pm}$	Deviations of counterrunning wave losses in a ring laser from the mean
$\gamma_{\parallel}$	Inversion relaxation rate
$\gamma_{\perp}$	Dipole moment polarization relaxation rate
$\hat{\gamma}$	Dimensionless relaxation rate
$\tilde{\gamma}$	Dimensionless cross-relaxation rate
$\gamma_{1,2}$	Complex parameters
Γ	Cross-relaxation rate
$\tilde{\Gamma}$	Normalized cross-relaxation rate
δ	Parameter in the theory of lasers with a saturable absorber
$\delta_{\mu\nu}$	Kronecher symbol
δx	Deviation of a current value of variable from its mean
$\delta\nu_{sel}$	Longitudinal mode selector half bandpass
$\delta\omega_c$	Cavity bandpass
$\delta\omega_D$	Doppler line halfwidth
$\delta\omega_0$ $(\delta\nu_0)$	Homogeneous line halfwidth
$\delta\omega_{inh}$ $(\delta\nu_{inh})$	Inhomogeneous line halfwidth
$\Delta\omega$	Intermode frequency spacing
Δ	Dimensionless intermode frequency spacing
Δ_{ck}	Relative detuning of the k-th cavity mode from the spectral line centre
Δ_{cr}	Critical detuning at the instability threshold
Δ_k	Relative detuning of the k-th component of the field from the gain line centre
Δ_{sel}	Laser spectrum halfwidth

Δ_{las}	Detuning from the selector band centre
Δ_{NR}	Phase nonreciprosity of a ring cavity
Δ_0	Detuning of the field frequency from the line centre
Δ'	Relative interval between frequencies of two components of gain line
Δ'_0	Relative detuning of oscillating frequency from frequency of a weaker component of the gain line
Δ_c	Detuning of field frequency from the cavity mode
$\Delta_c^{\pm}$	Detuning of field frequency from the ring cavity modes
$\Delta\vartheta$	Angular divergence of laser radiation
$\Delta\tau_k$	Delay of the k-th mode with respect to the reference mode
$\Delta\omega_D$	Doppler frequency shift
ε	Permittivity of material
ε_{sp}	Average rate of spontaneous emission into cavity mode
$\varepsilon_{1,2}$	Parameters of ellipticity of eigenfunctions
ζ	Dimensionless Cartesian coordinate
η	Index of refraction
η_0, η_2	Coefficients of field power expansion of refraction index
ϑ_A, ϑ_B	Decrements of phase-sensitive relaxation oscillations
ϑ_0	Rate of aperiodic transitional process in a gas laser
ϑ_1	Decrement of the fundamental relaxation oscillation
$\vartheta_k, k > 1$	Decrements of low-frequency relaxation oscillations
ϑ	Angle
$\vartheta_{\pm}$	Phases of the scattering coefficients of counterrunning waves in a ring cavity
κ	Photon decay rate in the resonator
$\hat{\kappa}$	Dimensionless photon decay rate
$\tilde{\kappa}$	Ratio of field relaxation constant to polarization constant
$\kappa_{\pm}$	Photon decay rates for the modes of a ring cavity

λ	Wavelength
Λ_j	Parameter of the j-th level excitation by a pumping source
$\Lambda(u)$	Differential parameter of pumping of a inhomoheneously (Doppler) broadened medium
Λ_{thr}	Threshold value of the differential pumping parameter of an inhomogeneously broadened laser
μ	Permeability of material
ν	Frequency
ν_{sel}	Frequency of a selector
ξ	Filling factor
$\xi_{\pm}$	Coupling (scattering) coefficients of counterrunning waves in a ring cavity
Π_{loss}	Relative single-pass cavity losses
Π_{diffr}	Relative single-pass diffraction losses
ρ	Density matrix
$\rho_{\pm}$	Counterrunning wave complex coupling coefficients
σ	Conductivity
σ, σ_{mn}	Amplitude of the density matrix element
σ_0^{mn}	Density matrix in the state of thermal equilibrium
σ_{tr}	Transition cross-section
$\tilde{\tau}$	Normalized time
τ_0	Time spacing between pulses
τ_d	Delay time
τ_p	Pulse duration
φ, ϕ, Φ	Phases
φ_e	Phase of the electric field
φ_p	Phase of the atomic polarization
χ	Susceptibility
Y	Opto-electronic feedback coefficient

ψ_k	Cavity eigenfunctions
ω	Frequency in radian frequency units
$\tilde{\omega}, \Omega$	Dimensionless frequency
ω_c, ω_{ck}	Cavity eigenfrequency
$\omega_c^{\pm}$	Eigenfrequencies of a ring cavity
ω_0	Line centre frequency
ω_R	Rabi frequency
$\Omega_{A,B}$	Frequencies of phase-sensitive relaxation oscillations in a ring laser
Ω_1	Frequency of the fundamental relaxation oscillation
$\Omega_k, k > 1$	Frequencies of antiphase relaxation oscillations

GLOSSARY

Abbreviated equations are equations for slowly changing amplitudes and phases of a oscillatory system which are obtained by averaging an initial set of equations with respect to the period of high reference frequency having the same scale as the frequency of eigen oscillations of the system.

Adiabatic rapid passage is a process of achieving population inversion by sweeping the mismatch between a pump radiation frequency and a quantum transition frequency with passing through resonance. The lower limit of the detuning rate is the relaxation constants of the atomic system, whereas its upper limit is the adiabatic condition, which retains quasi-equilibrium between the state of the atomic system and radiation field during the whole process.

Andronov-Hopf bifurcation is crossing the boundary of instability of a system at which the sign of the real part of the complex characteristic root reverses and above which the undamped self-modulation regime sets in. The bifurcation can be either *subcritical* when the finite amplitude of oscillations jumps to set in immediately above a bifurcation point or *supercritical* when the amplitude of intensity oscillations is infinitely small at the instability threshold but grows as the control parameter moves away from the bifurcation point.

Attractor is a set in the phase space where the trajectories remain. This zone of the phase space attracts all the trajectories originating inside a certain region called its *basin of attraction*.

Attractor crisis is the phenomenon of a sudden expansion, contraction or disappearance of the attractor when a control parameter is varied. It may result from the collision (or coalescence) of two attractors or a collision between an attractor and the unstable manifold.

Attractor dimension is a notion which is best illustrated by specific examples. An infinite set of generalized dimensions d_k ($k = 0,1,2...$) can be defines as

$$d_k = \lim_{l\to 0}\frac{\log\sum_i p_i^k}{(k-1)\log l},$$

Here l is the edge length of each hyper-cubic cell on which the phase space under consideration is divided, p_i is the probability of finding a phase space trajectory in the cell number i.

The d_0 dimension is so-called *Hausdorff dimension*: $d_0 = \lim_{l\to 0}\dfrac{\log N(l)}{\log l}$

Where $N(l)$ is the minimum number of cells required to cover the object. For regular objects, such as a point, a line segment, a surface or a body in ordinary space, the value of d_0 is given by an integer number 0, 1, 2, 3, i.e., the Hausdorff dimension coincides with the geometric or topological dimension. When an attractor is fractial, it is called a *strange attractor*.

The first-order dimension $d_1 = \lim_{l \to 0} \frac{\sum_i p_i \log p_i}{\log l}$

involves an entropy-like quantity $\sum_i p_i \log p_i$. For this reason, d_1 is called the information dimension.

The second-order dimension is given by $d_2 = \lim_{l \to 0} \frac{\log \sum_i p_i^2}{\log l}$

For large *N*, the sum $\sum_i p_i^2$ represents the probability that any pair of points belonging to a given set lie in the same cell. This approximately coincides with the correlation integral which gives the probability that any pair of points is separated by a distance smaller than *l*. for this reason, d_2 is known as correlation dimension. The practical algorithm for computing d_2 was developed by Grassberger and Procaccia [375,376].

Bad cavity condition consists in the fact that the bandwidth of a laser cavity is of the same order or higher than the gain linewidth of an atomic system.

Bifurcation is a transition from one structurally stable phase portrait to another through a structurally unstable state as the control parameter is changed.

Bistability is the situation when the phase space of a dynamical system contains two attractors.

Bloch's vector is the generalization of a magnetization vector of a paramagnet for a two-level system of arbitrary nature. Components of the Bloch's vector for two-level medium with electrodipole transition include the difference in population of the levels and the polarization.

Coherent interaction is an interaction of a field with an atomic system that is in a coherent state.

Coherent state is a thermodynamic non-equilibrium state of an atomic system characterized by the presence of a non-zero nondiagonal element of density matrix (of coherence), which is equivalent to the presence of a transverse component of the Bloch's vector or, in other words, polarization of the medium oscillating with the quantum transition frequency.

Glossary

Combination mode locking is the phenomenon which is conditioned by the combination tone mode-mode coupling and consists in a nonlinear frequency shift due to which the equidistant spectrum of generation is achieved in spite of an initially non-equidistant spectrum of eigenfrequencies of the laser cavity.

Combination tone mode-mode coupling is the phenomenon in which each pair of laser modes initiates population oscillations of laser levels at a beat frequency and due to scattering of the field of the third mode at these oscillations, there occurs a frequency-shifted component – the combination tone. When in the spectral vicinity of a fourth mode, the combination tone acts as a driving force; as a result, all the four modes become coupled.

Dynamical or *deterministic* chaos is a nonperiodic process in noise-free systems with a finite number of degrees of freedom which is distinguished by the high sensitivity of the individual realization to the initial conditions.

Feigenbaum scenario (or route to chaos) consists of a sequence of period-doubling bifurcations as a control parameter is changed. This is characterized by certain scaling of the control parameter range for each type of periodic orbit and of the relative strength of different spectral components.

Fixed point is a point in the phase space which corresponds to a steady state (time independent solution) of a dynamical system.

Four-wave mixing, see Combination tone mode-mode coupling.

Fractals are geometrical objects having a value for their dimension which is not an integer.

Free running mode of operation is a mode of lasing in the absence of external stresses on a laser.

Generalized multistability is the situation when the phase space of a dynamical system contains more than two attractors and at least one of them is not a fixed point.

Giant pulse is a powerful short pulse emitted by a laser after an abrupt switching off the excess losses which were purposefully introduced into a laser cavity to prevent generation until a required population of the upper laser level is achieved.

Ground state is a state of a molecule (atom) having the lowest energy.

Hard excitation of oscillations occurs in case of optical bistability when the absence of lasing corresponds to one of stable steady-state. In this case laser action can be initiated only by injecting a quite powerful seed radiation into a laser cavity.

Hard mode locking occurs when a threshold of multimode generation is higher than the mode locking threshold. For this laser to be in the regime of stable generation of ultrashort pulses, it is necessary to inject an initiating pulse in its cavity.

Intermittency route to chaos from a periodic solution is characterized by the fact that when the control parameter exceeds a critical value, the regular oscillations in the dynamical system appear to be interrupted at random times by bursts of irregular behavior. The duration of the chaotic phases is fairly regular and weakly dependent on the control parameter, but the mean duration of the regular phases decreases as this parameter increases beyond its critical value, and eventually they disappear.

Inverse problems of laser dynamics are ways of determining laser parameters by peculiarities in the dynamical behavior of a laser.

Intracavity laser spectroscopy is a highly sensitive technique of absorption spectroscopy in which a medium with narrow spectral lines is placed inside the cavity of a multimode broadband laser. The absorption spectrum being studied is displayed in the spectral profile of laser generation.

Kinematic modulation is a modulation of laser field at a Doppler shift frequency, resulting from the appearance of an intracavity component that is reflected from a moving surface.

Linear stability analysis is a method used to investigate the stability of regular (steady-state or periodic) solutions of dynamical systems. This method consists in linearization of the equations around the solution in question and in a subsequent exploration whether an initial perturbation is damped or increases with time.

Limit cycle is a closed orbit in the phase space which corresponds to periodic oscillations in a dynamical system.

Low-frequency coherence is a non-diagonal element of density matrix which characterizes the superposition of closely situated levels of a quantum system.

Lyapunov exponents give a measure of the rates at which the components of the distance between two close phase space trajectories change with time. The magnitudes and signs of the Lyapunov exponents allow one to distinguish among periodic, quasi-periodic and chaotic attractors. In the first cases these is no exponential separation of initially close trajectories and all Lyapunov exponents are negative or zero. Existence of at least one positive Lyapunov exponent is a clear sign of a strange attractor (chaos).

Multistability is the situation when the phase space of a dynamical system contains more than two attractors.

Natural fluctuations are fluctuations in a system which are caused by sources that cannot be avoided in principle, such as spontaneous transitions in an active medium and thermal radiation in a laser cavity.

Nonisochronity is the dependence of oscillation frequency in a nonlinear system on amplitude of oscillations.

Nonlinear lens is a nonuniform profile of the refractive index which is induced in a transparent nonlinear medium.

Nonreciprocity is the situation when conditions of counterrunning wave propagation in a ring cavity are different. At amplitude nonreciprocity, losses of counterrunning waves are unequal, whereas at phase nonreciprocity their phase rates are unequal. The phase nonreciprocity results from, for example, the Sagnac effect in a rotating ring laser cavity or an interferometer. The amplitude nonreciprocity can be obtained by introducing an isolator with a Faraday cell into a laser cavity.

Passive Q-modulation is a change of Q-factor due to saturation of absorption in a medium inside a laser cavity by generating radiation.

Phase diagram is a separation of the space of parameters of a dynamical system into areas with qualitatively different behavior.

Phase space trajectory is a path connecting the variables in the evolution of the system on which the direction of motion is defined.

Phase portrait of a dynamical system is a family of integral curves in the phase space.

Quasiperiodisity is an oscillatory process whose spectrum includes components with incommensurable frequencies.

Rabi oscillations are nutational oscillations of the Bloch's vector which are excited when atomic system is placed in a fairly strong resonance field of radiation. Frequency of Rabi oscillations is proportional to an amplitude of the field and a dipole moment of the quantum transition. The oscillations are observed if the Rabi frequency exceeds the rate of all relaxation processes in the medium.

Relaxation oscillations are damped oscillations of radiation intensity and population inversion when the steady-state lasing mode in class-B lasers sets in.

Resonance modulation of laser parameters is a modulation of parameters with frequency coinciding with some of eigenfrequencies of a system, for example, with a relaxation oscillation frequency, intermode beat frequency, etc.

Roell-Takens scenario is a transition to chaos characterized by the presence of a stage of quasiperiodic oscillations on the route from regular to chaotic behavior.

Secondary beats are the difference of frequencies of beats (beats of beats) of spectral components of radiation. They occur in multimode lasers with a non-equidistant lasing spectrum.

Second laser threshold is a bifurcation value of a control parameter which corresponds to an instability threshold of steady-state lasing.

Strange attractor is an attracting set in the phase space with no stable trajectories on it. This means that any trajectories started from closely located points of the phase space diverge so that the distance between them increases exponentially with time. Since the usual trajectories cannot intersect, a strange attractor exists only in a phase space with three or dimensions although the strange attractor itself may have a smaller dimension (though greater than two). A strange attractor corresponds to a nonperiodic process in a nonlinear dynamical system.

Superradiance is radiation of an electromagnetic field of a quantum system being in a coherent state. Intensity of the superradiance is proportional to the square of the number of active molecules.

Technical fluctuations are stochastic changes in radiation characteristics caused by fluctuations of laser parameters whose mechanism may be mechanical vibrations of units, occasional changes of temperature, etc.

References

1. Yariv A. Quantum Electronics, New York: Wiley, 1988.
2. Sargent M., Scully M.O., Lamb W.E. Laser Physics, London: Addison-Wesley, 1974.
3. Klyshko D.N. Physical Principles of Quantum Electronics, Moscow: Nauka, 1986, In Russian).
4. Svelto O. Principles of Lasers, New York: Plenum, 1984.
5. Pantell R.H., Puthoff H.E. Principles of Quantum Electronics, New York: Wiley, 1968.
6. Fain V.M., Khanin Ya.I. Kvantovaja Radiofizika, Moscow: Sov.Radio, 1965 (in Russian); Quantum Electronics, Oxford: Pergamon Press, 1969.
7. Shimoda K. Introduction to Laser Physics, Berlin: Springer, 1983.
8. Siegman A. Lasers, Mill Valley: Univ. Sci. Books, 1986.
9. Milloni P., Eberly J. Lasers, N.Y.: John Valey & Sons, 1988.
10. Basov N.G., Prokhorov A.M. Zh., Eksp. Teor. Fiz. 1954. V. 27. P. 431.
11. Gordon J.P., Zeiger H.J., Townes C.H. , Phys. Rev. 1954. V. 95. P. 282.
12. Schawlow A.L., Townes C.H., Phys. Rev. 1958. V. 112. P. 1940.
13. Maiman T.H., Nature. 1960. V. 187. P. 493.
14. Javan A, Bennett W.R., Herriott D.R., Phys. Rev. Lett. 1961. V. 6. P. 106.
15. Kocharovskaja O.A., Khanin Ya.I., Pisma Zh. Eksp. Teor. Fiz. 1988. V. 48. P. 581; [Sov. Phys. JETP Lett. V.48 P.630].
16. Harris S., Phys. Rev. Lett. 1989. V. 62. P. 1033.
17. Scully M.O., Zhu S.Y., Gavrielides A., Phys. Rev. Lett. 1989. V.62. P.2813.
18. Popov A.K., Introduction to Nonlinear Spectroscopy, Novosibirsk: Nauka, 1983.
19. Arkhipkin V.G., Heller Yu.I., Phys. Lett. 1983. V. A-98. P. 12.
20. NottelmannA., Peters C., Lange W., Phys. Rev. Lett. 1993. V. 70. P. 1783.
21. Zibrov A.S., Lukin M.D., Nikonov D.E. et al., Phys. Rev. Lett. 1995. V. 72. P. 1499.
22. Baldwin G.C., Solem J., Rev. Mod. Phys. 1998. V. 69. P. 1085.
23. Arecchi F.T., Harrison R.G., Instabilities and Chaos in Quantum Optics. /Eds F.T.Arecchi and R.G. Harrison, N.Y. etc.: Springer Verlag: 1987. P.6.
24. Khanin Ya.I., Dynamics of Time-Dependent Operation of Solid-State Lasers: Doctoral Thesis, Gorky, Institute of Applied Physics AN SSSR, 1981.
25. Lichtenberg F., Liberman M., Regular and Stochastic Motion, Berlin: Springer, 1984.
26. Schuster H.G., Deterministic Chaos, Weinheim: Physik-Verlag, 1988.
27. Oraevsky A.N., Molecular Oscillators, Moscow: Nauka, 1964 (In Russian).
28. Belyanin A.A., Kocharovsky V.V., Kocharovsky Vl.V., Izv. RAN. Ser. Fiz. 1998. V. 62. P. 372.
29. Belyanin A.A., Kalugin N.G., Kocharovsky V.V., Kocharovsky Vl.V., Izv. RAN. Ser. Fiz. 1999. V.63. P.3769.
30. Willett C.S., Handbook of Lasers with Selected Data On Optical Technology /Ed. Pressley, Cleveland: Chemical Rubber Co. 1971.

31. Casperson L.W., J. Opt. Soc. Am. 1985. V. B-2. P. 62.
32. Hoffer L.M., Chyba T.H., Abraham N.B., J. Opt. Soc. Am. 1985. V. B-2. P. 102.
33. Gioggia R.S., Abraham N.B,Opt. Commun. 1983. V.47. P.278.
34. Lefebvre V., Dangoisse D., Glorieux P., Phys. Rev. 1984. V. A-29. P. 2758.
35. Lawandy N.M., Koepf G.A., IEEE J. Quant. Electron. 1980. V. QE-16. P. 701.
36. Vass A., Wood R.A., Davis B.M., Pigeon C.R., Appl. Phys. 1982. V. B-27. P. 187.
37. Weiss C.O., Klische W., Opt. Commun. 1984. V.50. P.413.
38. Weiss C.O., Klische W., Opt. Commun. 1984. V.51. P.47.
39. Weiss C.O., Klische W., Eging P.S., Cooper M., Opt. Commun. 1985. V.52. P.405.
40. Klische W., Weiss C.O., Phys. Rev. 1985. V. A-31. P. 4049.
41. Weiss C.O., Abraham N.B., Hubner U., Phys. Rev. Lett. 1988. V. 61. P. 1587.
42. Weiss C.O., Instabilities and Chaos in Quantum Optics II /Eds N.B. Abraham, F.T. Arecchi, L.A. Lugiato—N.Y.: Plenum Press, 1988. P.41.
43. Watts R.K., J. Opt. Soc. Am. 1971. V. 61. P. 123.
44. Lau K.Y., Opt. & Photon. News. 1997. V.8. #4. P.27.
45. Broude V.L., Mashkevich V.S., Prikhot'ko A.F., Prokopyuk N.F., Soskin M.S., Fiz. Tverd. Tela. 1962. V. 4. P.2976.
46. Powel R.C., Xi Lin, Gang Xu, Quarles G.J., Phys. Rev. 1985. V. B-32. P. 2788.
47. Moulton P.F., J. Opt. Soc. Am. 1986. V. B-3. P. 125.
48. Gubin M.A., Popov A.I., Protsenko E.D., Opt. Spektr. 1968. V. 25. P. 736.
49. Ahn T.D., Dietel W., Opto-Electronics. 1973. V. 5. P. 243.
50. Krivoshchekov G.V., Telegin G.G., Folin K.G., Izv. Vyss. Ucheb. Zaved. Radiophys. 1967. V. 10. P. 1656.
51. Epifanov V.P., Mazan'ko I.P., Yaroshenko N.G., Radiotechn. Elektron. 1967. V. 12. P. 1517.
52. Bambini A., Burlamacchi P., Appl. Phys. Lett. 1963. V. 10. P. 189.
53. McFarlane R.A., Bennett W.R.,Jr, Appl. Opt. 1963. V. 2. P. 189.
54. Bennett W.R.,Jr. Phys. Rev. 1962. V. 126. P. 580.
55. Chetverikov V.I., Pisma Zh. Techn. Fiz. 1985. V. 11. P. 460.
56. Halas N.J., Liu S.-N., Abraham N.B., Phys. Rev. 1983. V. A-28. P. 2915.
57. Weiss C.O., King H., Opt. Commun. 1982. V. 44, P. 59.
58. Weiss C.O., Godone A., Olafsson A., Phys. Rev. 1983. V. B-28. P. 892.
59. Hillman L.W., Krasinski J., Boyd R.W., Stroud C.R., Jr, Phys. Rev. Lett. 1984. V. 52. P.1605.
60. Stroud C.R.,Jr., Koch K., Chakmakjian S., Opt. Instabilities /Eds R.G. Boyd, M.G. Raymer, L.M. Narducci, Cambridge etc: Cambridge Univ. Press, 1986. P. 274.
61. Hillman L.W., Krasinski J., Koch K., Stroud C.R., Jr, J. Opt. Soc. Am. 1985. V. B-2. P. 211.
62. Aivasjan Yu.M., Baev V.M., A.A. Kachanov and S.A. Kovalenko, Kvant. Elektron. 1986. V. 13. P. 1723 [Sov. J. Quant. Electron. V. 16. P. 1133].
63. Vinogradov C.E., Kachanov A.A., Kovalenko S.A., Sviridenkov E.A., Pisma Zh. Eksp. Teor. Fiz. 1988. V. 55. P. 581.
64. Baev V.M., Belikova T.P., Kovalenko S.A., Sviridenkov E.A., Suchkov A.F., Kvant. Elektron. 1980. V. 7. P. 903.

65. Aivasjan Yu.M., Ivanov V.V., Kovalenko S.A., Baev V.M., Sviridenkov E.A., Atmanspacher H., Sceingraber H., Appl. Phys. 1988. V. B-2. P. 175.
66. Kuhlke D., Acta Phys. Polon. 1982. V. B-61. P. 547.
67. Lisitsyn V.N., Troshin B.I., 1967. Optika Spektr. V. 22. P.363 [Opt. Spectrosc. (USSR) 22. 363].
68. Danileiko M.V., Kravchuk A.A., Nechiporenko V.N., Tselinko A.M., Yatsenko L.P., Kvant. Elektron. 1986. V. 13. P. 2147 [Sov. J. Quant. Electron. V. 16. P. 1420].
69. Kikuchi C., Lambe J., Makhov G., Terhune R.W., J. Appl. Phys. 1959. V. 30. P. 1061.
70. Makhov G., Cross L.G., Terhune R.W., Lambe J., J. Appl. Phys. 1960. V. 31. P. 936.
71. Li T., Sims S.D, Proc. IRE. 1962. V. 50. P. 464.
72. Evtuhov E., Neeland J.K., Appl. Optics. 1962. V. 1. P. 517.
73. Korobkin V.V., Leontovich A.M., Zh. Eksp. Teor. Fiz. 1963. V. 44. P. 1847 [Sov Phys. JETP. V. 17. P.1242].
74. McMurtry B.J., Siegman A.E., Appl. Optics. 1962. V. 1. P. 51.
75. Birnbaum M., Stocker T.L., Appl. Phys. Lett. 1963. V. 3. P. 164.
76. Korobkin V.V., Leontovich A.M., Zh. Eksp. Teor. Fiz. 1965. V. 49. P. 10 [Sov Phys. JETP. V. 22. P.6].
77. Huges T.P., Young K.M., Nature. 1962. V. 196. P. 332.
78. Leontovich A.M., Veduta A.P., Zh. Eksp. Teor. Fiz. 1964. V. 46. P. 71 [Sov Phys. JETP. V. 19. P.51].
79. Stickley C.M, IEEE J. Quant. Electron. 1966. V. QE-2. P.511.
80. Evtuhov E., Appl. Phys. Lett. 1965. V. 6. P. 141.
81. Ridgway S.L., Clark G.L., York C.M., J. Opt. Soc. Am. 1963. V. 53. P. 700.
82. Adamson M.C., Huges T.P., Young K.M., Electron. Quant. 3 Conf. Internat. Paris 2, N.Y.: Columbia Univ. Press, 1964. P. 1459.
83. Kubarev A.M., Piskarev V.I., Zh. Eksp. Teor. Fiz. 1964. V. 46. P. 508 [Sov Phys. JETP. V. 19. P.345].
84. Kubarev A.M., Piskarev V.I., Zh. Eksp. Teor. Fiz. 1965. V. 48. P. 1233 [Sov Phys. JETP. V. 21. P.823].
85. Gurs K, Z., Angew. Math. Phys. 1965. V.16. P. 49.
86. Kornienko L.S., Krawtsov N.V., Lariontsev E.G., Naumkin N.I., Pisma Zh. Eksp. Neor. Fiz. 1970. V. 11. P. 585; [Sov. Phys. JETP Lett. V. 11. P. 404].
87. Kravchenko V.I., Soskin M.S., In : Kvant. Elektron, Ed. M.P. Lisitsa. –Kiev. Naukova Dumka. 1969. #4. P. 42.
88. Ready J.P., Hardwick D.L., Proc. IRE. 1962. V. 50. P. 2483.
89. Broude V.L., Zaika V.V., Kravchenko V.I., Soskin M.S., Zh. Prikl. Spetr. 1968. V. 12. P. 388.
90. Freund I., Appl. Phys. Lett. 1968. V. 12. P. 388.
91. Collins R.W., Braun L.O., Dean D.R., Appl. Phys. Lett. 1963. V. 10. P. 392.
92. Sokolov A.K., Zubarev T.N., Fiz. Tverd. Tela. 1964. V. 6. P.2590 [Sov. Phys. Solid State. V. 6. P. 2065].
93. Sacchi C.A., Svelto O., IEEE J. Quant. Electron. 1965. V. QE-1. P. 398.

94. Daneu V., Polloni R., Sacchi C.A., Svelto O., Nuovo Chimento. 1965. V. B-40. P. 446.
95. Globus M.E., Naboikin Yu.V., Ratner A.M., Rom-Krichevskaya I.A., Tiunov Yu.A., Zh. Eksp. Teor. Fiz. 1967. V. 52. P. 857 [Sov Phys. JETP. V. 25. P. 562].
96. Vinokurov G.N., Galactionova N.M., Egorova V.F., Mak A.A., Sedov B.M., Khanin Ya.I., Zh. Eksp. Teor. Fiz. 1971. V. 60. P. 489 [Sov Phys. JETP. V. 33. P. 262].
97. Strozyk J.W., IEEE J. Quant. Electron. 1967. V. QE-3. P.343.
98. Ratner A.M., Quantum Oscillators of Light with High Bean Divergence, Kiev: Naukova Dumka, 1970 (in Russian).
99. Pole P.V., Wieder H., Appl. Optics. 1964. V. 3. P. 1086.
100. Wieder H, J., Appl. Phys. 1966. V. 31. P. 93.
101. Evtuhov V., Neeland J.K., IEEE J. Quant. Electron. 1965. V. QE-1. P.7.
102. Laig-Horsterbrock H., Wieder H., Z. Angew. Phys. 1967. V. B-23. P.1.
103. Greenhow R.C. Schmidt A.T., Appl. Phys. Lett. 1968. V. 12. P. 390.
104. Davis B.I., Keller D.V., Appl. Phys. Lett. 1964. V. 5. P. 80.
105. Korobkin V.V., Leontovich A.M., Smirnova M.N., Zh. Eksp. Teor. Fiz. 1965. V. 48. P. 78 [Sov Phys. JETP. V. 21. P. 53].
106. Katzman M., Strozyk J.W., J. Appl. Phys. 1964. V. 35. P. 80.
107. Leontovich A.M., Popova M.N., Zh. Prikl. Spetr. 1967. V. 6. P. 735.
108. Roess D., Appl. Phys. Lett. 1966. V. 8. P. 109.
109. Tang C.L., Statz H., DeMars G., Appl. Phys. Lett. 1963. V. 2. P. 222.
110. Mak A.A., Sedov B.M., Zh. Tekhn. Fiz. 1968. V. 38. P. 2119.
111. Antsiferov V.V., Ghiner A.V., Komarov K.P., Pivtsov V.S., Folin K.G., Zh. Prikl. Spetr. 1976. V. 24. P. 18.
112. Antsiferov V.V., Pivtsov V.S., Ugozhaev V.D., Folin K.G., Kvant. Electron, Ed. N.G.Basov, M.: Sov. Radio, 1973, #3, P. 57.
113. Malyshev V.I., Markin A.S., Sychev A.A., Zh. Tekhn. Fiz. 1969. V. 39. P. 334.
114. Antsiferov V.V., Kravchenko V.I., Control of Ruby Laser Output in Free Running Regime. Preprint. Institute of Physics of the Ukrainian Academy of Sciences, Kiev, 1988, #4.
115. Folin K.G., Ghiner A.V., Dynamics of Free Running Solid-State Lasers, Novosibirsk: Nauka 1979.
116. Free J., Korpel A., Proc. IEEE. 1964. V. 52. P. 90.
117. Livshits B.L., Tsykunov V.N., Dokl. Akad. Nauk SSSR. 1965. V. 163. P. 870.
118. Livshits B.L., Nazarov V.P., Sidorenko L.K., Tsykunov V.N., Pisma Zh. Eksp. Neor. Fiz. 1965. V. 1. P. 23; [Sov. Phys. JETP Lett. V. 1. P. 136].
119. Livshits B.L., Nazarov V.P., Sidorenko L.K., Tursunov A.G., Tsykunov V.N., Pisma Zh. Eksp. Neor. Fiz. 1966. V. 3. P. 279; [Sov. Phys. JETP Lett. V. 3. P. 179].
120. Livshits B.L., Tursunov A.G., Zh. Eksp. Teor. Fiz. 1970. V. 58. P. 1518 [Sov Phys. JETP. V. 31. P. 812].
121. Livshits B.L., Usp. Fiz. Nauk. 1969. V. 98. P. 393.
122. Danielmeyer H.G., Nilsen W.G., Appl. Phys. Lett. 1970. V. 16. P. 124.
123. Malota F., Opto-Electronics. 1970. V. 2. P. 99.
124. Gerber E.A., Ahlstrom, J. Appl. Phys. 1964. V. 35. P. 2546.
125. Gerber E.A., Ahlstrom, IEEE J. Quant. Electron. 1969. V. QE-5. P.403.

126.Folin K.G., Antsiferov V.V., Zh. Eksp. Teor. Fiz. 1968. V. 88. P. 122 [Sov Phys. JETP. 1969 V. 28. P. 65].

127.Antsiferov V.V., Krivoshchekov G.V., Folin K.G., Zh. Eksp. Teor. Fiz. 1969. V. 56. P. 526 [Sov Phys. JETP. V. 29. P. 289].

128.Folin K.G., Antsiferov V.V., Anikeev B.V., Ugozhaev V.D., Zh. Eksp. Teor. Fiz. 1970. V. 58. P. 1156 [Sov Phys. JETP. V. 31. P. 613].

129.Danielmeyer H.G., Turner E.H., Appl. Phys. Lett. 1970. V. 17. P. 519.

130.Antsiferov V.V., Pivtsov V.S., Ugozhaev V.D., Folin K.G., Optika Spektr. 1972. V. 32. P. 1159 [Opt. Spectrosc. (USSR) 32. 628].

131.Antsiferov V.V., Pivtsov V.S., Ugozhaev V.D., Folin K.G., Avtometrija. 1972. #5. P. 98.

132.Antsiferov V.V., Kuchjanov A.S., Pivtsov V.S., Ugozhaev V.D., Folin K.G., Optika Spektr. 1975. V. 38. P. 599 [Opt. Spectrosc. (USSR) 38. 339].

133.Antsiferov V.V., Ghiner A.V., Komarov K.P., Folin K.G., Kvant. Elektron. 1975. V. 2. P. 591 [Sov. J. Quant. Electron. 1976. V. 5. P. 332].

134.Antsiferov V.V., Ruby Laser Dynamics and Methods of Its Control in Various Regimes: Ph. D. Thesis, Novosibirsk, 1973.

135.Gulev V.S., Pivtsov V.S., Folin K.G., Radiotechn. Elektron. 1980. V. 25. P. 573.

136.Antsiferov V.V., Golyaev Yu.D., Optika Spektr. 1982. V. 25. P. 706.

137.Potapov C.E., Neodymium Glass Laser Dynamics.: Ph. D. Thesis, Leningrad, USSR, 1972.

138.Keen W.H., Wess J.A., Appl. Optics. 1963. V. 3. P. 545.

139.Mashkevich V.S., Soskin M.S., Pisma Zh. Eksp. Teor. Fiz. 1967. V. 5. P. 456.

140.Snitzer E., Woodcock R., IEEE J. Quant. Electron. 1966. V. QE-2. P.627.

141.Shiner W., Snitzer E., Woodcock R., Phys. Lett. 1966. V. 21. P. 412.

142.Kotsubanov V.D., Naboikin Yu.V., Ratner A.M., Rom-Krichevskaya I.A., Optika Spektr. 1967. V. 23. P. 986 [Opt. Spectrosc. (USSR) 23. 537].

143.Bonch-Bruevich A.M., Potapov S.E., Khanin Ya.I., Optika Spektr. 1970 V. 28. P. 203 [Opt. Spectrosc. (USSR) 28. 110].

144.Landry R.J., Snitzer E., Bartram R.H., J. Appl. Phys. 1971. V. 42. P. 3827.

145.Kurnosov V.D., Magalyas V.I., Pleshkov A.A., Rivlin L.A., Tukhan V.G., Tsvetkov V.V., Pisma Zh. Eksp. Teor. Fiz. 1966. V. 4. P. 449; [Sov. Phys. JETP Lett. V. 4. P. 303].

146.D'Asaro L.A, Cherlow J.M., Paoli T.L., IEEE J. Quant. Electron. 1968. V. QE-4. P. 164.

147.Nikitin V.V., Semenov A.S., Strakhov, Pisma Zh. Eksp. Teor. Fiz. 1969. V. 9. P. 516; [Sov. Phys. JETP Lett. V. 9. P. 313].

148.Zakharov Yu.P., Kompanets I.N., Nikitin V.V., Semenov A.S., Fiz. Tekhn. Poluprovodn. 1969. V. 3. P. 864.

149.Paoli T.L., Ripper J.E., Appl. Phys. Lett. 1969. V. 15. P. 105.

150.Paoli T.L., Ripper J.E., Phys. Rev. Lett. 1969. V.22. P. 1085

151.Basov N.G., Nikitin V.V., Semenov A.S., Usp. Fiz. Nauk. 1969. V. 97. P. 561.

152.Rivlin L.A., Dynamics of Oscillation of Semiconductor Lasers, Moscow: Sovetskoe Radio, 1976.

153.Rivlin L.A., Semenov A.T., Yakubovich S.D., Dynamics and Emission Spectra of Semiconductor Lasers, Moscow: Radio Svjaz', 1983.

154.Kurbatov L.N., Molochev V.I., Nikitin V.V., Sharin A.I., Kvant. Electron. /Ed. N.G.Basov. M.: Sov. Radio, 1971, #6, P. 110.
155.Lippi G.L., Abraham N.B., Puccioni G.P., Arecchi F.T., Tredicce J.R., Phys. Rev. 1987. V. A-35. P. 3978.
156.Lippi G.L., Abraham N.B., Puccioni G.P., Tredicce J.R., Narducci L.N., Arecchi F.T., Proc. SPIE Int. Soc. Opt. Eng. 1985. V. 667. P.41.
157.Kornienko L.S., Kravtsov N.V., Shelaev A.N., Opt. Spektr. 1973 V. 35. P.775 [Opt. Spectrosc. (USSR) 35. 449].
158.Klochan E.L., Kornienko L.S., Kravtsov N.V., Lariontsev E.G., Shelaev A.N., Zh. Eksp. Teor. Fiz. 1973. V. 65. P. 1344 [Sov Phys. JETP. 1974. V. 38. P. 669].
159.Klochan E.L., Kornienko L.S., Kravtsov N.V., Lariontsev E.G., Shelaev A.N., Radiotechn. Elektron. 1974. V. 19. P. 2096 [Sov. Phys. Rad. Electron. V. 19. P. 58].
160.Klochan E.L., Kornienko L.S., Kravtsov N.V., Lariontsev E.G., Shelaev A.N., Pisma Zh. Eksp. Teor. Fiz. 1973. V. 17. P. 405; [Sov. Phys. JETP Lett. V. 17. P. 289].
161.Mak A.A., Ustyugov V.I., Pisma Zh. Eksp. Teor. Fiz. 1973. V. 18. P. 253; [Sov. Phys. JETP Lett. V. 18. P. 151].
162.Lippi G.L., Tredicce J.R., Abraham N.B., Arecchi F.T., Opt. Commun. 1985. V. 53, P. 129.
163.Tredicce J.R., Lippi G.L., Arecchi F.T., Abraham N.B., Phil. Trans. R. Soc. Lond. 1984. V. A-313. P. 411.
164.Malcolm G.P.A., Ferguson A.I., Contemp. Phys. 1991. V. 32. P. 305.
165.Otsuka K., Georgiou M., Mandel P., Jap. J. Appl. Phys. 1992. V. 31. P. L1250.
166.Mandel P., Georgiou M., Otsuka K., //Opt. Commun. 1993. V. 106, P. 341.
167.Khandokhin P.A., Khanin Ya.I., Dangoisse D., Celet J.-C., Glorieux P, Opt. Commun. 1998. V. 123. P. 372.
168.Bielawski S., Derozier D., Glorieux P, Phys. Rev. 1992. V. A-46. P. 2811.
169.Lehers R., Francois P.L., Stephan G, Opt. Lett. 1993. V. 19. P. 275.
170.Khandokhin P., Khanin Ya., Mamaev Yu., Milovsky N., Shirokov E., Bielawski S., Derosier D., Glorieux P., Opt. Commun. 1998. V. 10, P. 97.
171.Martin-Regalado J., Balle S., San Miguel M., Valle A., Pesquera L., Quantum Semiclass. Opt, 1997. V. 9. P. 713.
172.Martin-Regalado J., San Miguel M., Abraham N.B., Opt. Lett. 1996. V. 21. P. 351.
173.Khandokhin P., Khanin Ya., Pisma Zh. Techn. Fiz. 1979. V. 5. P. 35 [Tech. Phys. Lett. V. 5. P. 14].
174.Parfenov V.A., Khandokhin P., Khanin Ya., Kvant. Elektron. 1988. V. 15. P. 1985 [Sov. J. Quant. Electron. V. 18. P. 1243].
175.Abraham N.B., Weiss C.O., Opt. Commun. 1988. V. 68, P. 437.
176.Haken H., Phys. Lett. 1975. V. A-53. P. 77.
177.Lorenz E., J. Atm. Sci. 1963. V. 20. P. 130.
178.Klische W., Weiss C.O., Al-Soufi W., Huttmann G., In: Optical Instabilities /Eds. R.W. Boyd, M.G.Raimer, L.M. Narducci, Cambridge etc: Cambridge Univ. Press, 1986) p. 274.
179.Weiss C.O., Brock J., Phys. Rev. Lett. 1986. V.57. P. 2804.

180.Hogenboom E.H.M., Klische W., Weiss C.O., Godone A., Phys. Rev. Lett. 1985. V. 55. P. 2571.
181.Biswas D.J., Harrison R.G., Weiss C.O., Dangoisse D., Glorieux P., and Lawandy N.M., In: Instabilities and Chaos in Quantum Optics, Eds. F.T.Arecchi, R.G.Harrison (Springer Verlag, N.Y., 1987) p. 109.
182.Weiss C.O., Klische W., Abraham N.B., Hubner U., Appl. Phys. 1989. V. B-49. P. 241.
183.Tang D.Y., Li M.Y., Weiss C.O., Phys. Rev. 1991. V. A-44. P. 7597.
184.Tang D.Y., Pujiol J., Weiss C.O., Phys. Rev. 1991. V. A-44. P. R35.
185.Tang D.Y., Li M.Y., Weiss C.O., Phys. Rev. 1992. V. A-46. P. 676.
186.Weiss C O., Vilaseca R., Abraham N.B., Corbalan R., Roldan E., de Valcarcel G.J., Pujol J., Hubner U., and Tang D.Y., Appl. Phys. 1995. V. B-61. P. 223.
187.Letokhov V.S., Chebotaev V.P., Principles of Nonlinear Laser Spectroscopy, Moscow.: Nauka, 1975.
188.Lawandy N.M., J. Opt. Soc. Am. 1985. V. B-2. P. 108.
189.Abraham. N.B., Dangoisse D., Glorieux P., Mandel P., J. Opt. Soc. Am. 1985. V. B-2. P. 23.
190.Harrison R.G., Al-Saidi I.A., Opt. Commun. 1985. V. 54, P. 107.
191.Biswas D.J., Harrison R.G., Opt. Commun. 1985. V.54, P. 112.
192.Biswas D.J., Harrison R.G., In: Optical Instabilities /Eds. R.W. Boyd, M.G.Raimer, L.M. Narducci, Cambridge etc: Cambridge Univ. Press, 1986) p. 227.
193.Biswas.D.J., Harrison R.G., Al-Saidi I.A., In: Optical Instabilities, Eds. R.W. Boyd, M.G.Raimer, L.M. Narducci, Cambridge etc: Cambridge Univ. Press, 1986 p. 231.
194.Casperson L.W., Phys. Rev. 1980. V. A-21. P. 911.
195.Casperson L.W., Yariv A., IEEE J. Quant. Electron. 1972. V. QE-8. P. 69.
196.Casperson L.W., IEEE J. Quant. Electron. 1978. V. QE-14. P. 756.
197.Bentley J., Abraham N.B., Opt. Commun. 1982. V. 41, P. 52.
198.Maeda M., Abraham N.B., Phys. Rev. 1982. V. A-26. P. 3395.
199.Tarroja M.F.H., Abraham N.B., Bandy D.K., Narducci L.M., Phys. Rev. 1986. V. A-34. P. 3148.
200.Casperson L.W., Laser Physics, Eds J.D. Harvey D.F. Walls: Springer Lecture Notes in Physics. V. 182, Berlin etc: Springer Verlag, 1983. P. 88.
201.Abraham N.B., ChybaT., Colemann T. et al, Laser Physics, Eds J.D. Harvey D.F. Walls: Springer Lecture Notes in Physics. V. 182, Berlin etc: Springer Verlag, 1983. P. 107.
202.Gioggia R.S., Abraham N.B., Phys. Rev. Lett. 1983. V.51. P. 650.
203.Fain V.M., Zh. Eksp. Teor. Fiz. 1957. V. 33. P. 945 [Sov Phys. JETP. 1958. V. 6. P. 726].
204.Oraevsky A.N., Radiotechn. Elektron. 1959. V. 4. P. 718.
205.Talanov V.I., Pisma Zh. Eksp. Teor. Fiz. 1965. V. 2. P. 218; [Sov. Phys. JETP Lett. 1966. V. 2. P. 303].
206.Sukhorukov A.P., Khokhlov R.V., Vestnik Moskov. Univers., Ser. Fiz. Astron. 1966, V. 7. !1. P. 95.
207.Slater J., Ultrahigh Frequency Electronics, Moscow: Sovetskoe Radio (Translation from English), 1948.

208. Vainstein L.A., Electromagnetic Waves. Moscow: Radio Svjaz', 1988.
209. Landau L.D., Lifshits M., Electrodynamics of Condensed Media, M.: Nauka (in Russian), 1982 [Electrodynamics of Condensed Media. 1984, 2nd ed., New York: Pergamon].
210. Vainstein L.A., Open Resonators and Open Waveguides, Moscow: Sovetskoe Radio, 1966.
211. Andronov A.A., Vitt A.A., Khaikin. Theory of Oscillations (in Russian), M.: Nauka, 1981 [Theory of Oscillations, Oxford: Pergamon Press, 1966].
212. Bogolyubov N.N., Mitropol'sky, Asymptotic Methods in the Theory of Nonlinear Oscillations, Moscow: Nauka, 1974.
213. Efanova I.P., Lariontsev E.G., Zh. Eksp. Teor. Fiz. 1967. V. 58. P. 1532 [Sov Phys. JETP. 1969. V. 28. P. 802].
214. Klochan E.L., Kornienko L.C., Kravtsov N.V., Lariontsev E.G., Shelaev A.N., Dokl. Akad Nauk SSSR. 1974. V. 215. P. 313.
215. Wave and Fluctuational Processes in Lasers, Ed. Yu.L. Klimontovich, Moscow: Nauka, 1974.
216. Khaldre Kh. Yu., Khokhlov R.V., Izv. Vyss. Ucheb. Zaved. Radiophys. 1958. V. 1. #5-6. P. 80.
217. Akhmanov S.A., Usp. Fiz. Nauk. 1986. V. 149. P.361; [Sov. Phys. Usp. 29(7). 589].
218. Haken H., Z. Phys. 1975. V. B-21. P. 506; Z. Phys. 1975. V. B-21. P. 69.
219. Haken H., Synergetics. 1978, Springer: Berlin.
220. Haken H., In: Optical Instabilities, Eds. R.W. Boyd, M.G.Raimer, L.M. Narducci, Cambridge etc: Cambridge Univ. Press, 1986) p. 227.
221. Lugiato L.A., Mandel P., Narducci L.M., Phys. Rev. 1984. V. A-29. P. 1438.
222. Van Campen N.G., Phys. Reports. 1985. V. 124. P.69.
223. Macomber J.D., The Dynamics of Spectroscopic Transitions. New York: Wiley, 1976.
224. Allen L., Eberly J.H., Optical Resonance and Two-Level Atoms.—New York: Wiley, 1975.
225. Bloch F,, Phys. Rev. 1946. V. 70. P. 460.
226. Feynman R.P., Vernon F.L., Hellwarth R.W., J. Appl. Phys. 1957. V. 28. P. 49.
227. Fain V.M., Khanin Ya.I., Yashchin, Izv. Vyss. Ucheb. Zaved. Radiophys. 1962. V. 5. P. 697.
228. Newell A.C., Moloney J.U., Nonlinear Optics, New York etc: Addison Wesley Publishing Company, 1992)
229. Mikaelyan A.L., Ter-Mikaelyan M.L., Turkov Yu.G., Solid-State Optical Oscillators, Moscow: Sovetskoe Radio, 1967.
230. Davis L.W., Proc. IRE. 1963. V. 51. P. 76.
231. Statz H., DeMars G., Quantum Electronics, N.Y.: Columbia Univ. Press, 1960, P.530.
232. Genkin V.N., Khanin Ya.I., Izv. Vyss. Ucheb. Zaved. Radiophys. 1962. V. 5. P. 423.
233. Tang C.L., J. Appl. Phys. 1963. V. 34. P. 2935.
234. Khanin Ya.I., Izv. Vyss. Ucheb. Zaved. Radiophys. 1966. V. 9. P. 697 [Sov Radiophys. Quant. Electron. V. 9. P. 416].

235.Oppo G.L., Politi A., Instabilities and Chaos in Quantum Optics II, Eds N.B. Abraham, F.T. Arecchi, L.A. Lugiato, N.Y.: Plenum Press, 1988. P. 363.
236.Korn G.A., Korn T.M., Mathematical Handbook for Scientists and Engineers, New York: McGraw-Hill, 1968.
237.Oraevsky A.N., Uspensky A.V., Trudy FIAN. 1965. V. 31. P. 96.
238.Kaiser W., Garrett C.G.B., Wood D.I., Phys. Rev. 1961. V. 123. P. 766.
239.Makhov G. J., Appl. Phys. 1962. V. 33. P. 202.
240.Sinnet D.M., J. Appl. Phys. 1960. V. 31. P. 1578.
241.Post E.J., Appl. Optics. 1962. V. 1. P. 165.
242.Bespalov V.I., Gaponov A.V., Izv. Vyss. Ucheb. Zaved. Radiophys. 1965. V. 8. P. 70 [Sov Radiophys. Quant. Electron. V. 8. P. 49].
243.Menne T.J., IEEE J. Quant. Electron. 1966. V. QE-2. P. 38.
244.Dunsmuir R., J. Electron. Control. 1961. V. 10. P. 453.
245.Clendenin W.W., J. Appl. Phys. 1964. V. 35. P. 2277.
246.Ratner A.M., Spectral, Spatial and Temporal Properties of Lasers, Kiev: Naukova Dumka, 1968 [New York: Plenum Press, 1972].
247.Abraham B.B., Mandel P., Narducci L.M., Instabilities and Pulsations in Lasers, Prrogress in Optics, Ed. E. Wolf, Amsterdam etc: North-Holland. 1988. V. 25. P. 1.
248.Gurtovnik A.S., Izv. Vyss. Ucheb. Zaved. Radiophys. 1958. V. 1. #5-6, P. 83 [translated to Applied Physics Laboratory Library Bull. (Johns Hopkins University, Balimore].
249.Collins R.J., Nelson D.F., Schawlow A.L., Bond W., Garrett C.G.B., Kaiser W., Phys. Rev. Lett. 1960. V.5. P. 303.
250.Snitzer E., Phys. Rev. Lett. 1961. V.7. P. 444.
251.Gurs K., Z. Angew. Math Phys. 1965. V. 16. P. 49.
252.Korobkin V.V., Uspensky A.V., Zh. Eksp. Teor. Fiz. 1963. V. 45. P. 1003 [Sov Phys. JETP. 1964. V. 18. P. 693].
253.Uspensky A.V., Radiotechn. Elektron. 1964. V. 9. P. 747 [Sov. Phys. Rad. Electron. V. 9. P. 605].
254.Grasyuk A.Z., Oraevsky A.N., Technical Digest: 4th Int. Congr. On Microwave Tubes. Scheveningen, Holland, 1964.
255.Grasyuk A.Z., Oraevsky A.N., Radiotechn. Elektron. 1964. V. 9. P. 524 [Sov. Phys. Rad. Electron. V. 9. P. 424].
256.Buley E.R., Cummings F.W., Phys. Rev. 1964. V. 134. P. A1455.
257.Yakubovich E.I., Zh. Eksp. Teor. Fiz. 1968. V. 58. P. 304 [Sov Phys. JETP. 1969. V. 28. P. 160].
258.Idiatulin V.S., Uspensky A.V., Radioyechn. Elektron. 1973. V. 18. P. 580 [Sov. Phys. Rad. Electron. V. 18. P. 422].
259.Fradkin E.E., In: Tunable Lasers, Kiev: Institut of Physics Ukrainian Acad Sci. 1973. P. 324.
260.Casperson L.W., Phys. Rev. 1981. V. A-23. P. 248.
261.Minden M., Casperson L.W., IEEE J. Quant. Electron. 1982. V. QE-18. P.1952.
262.Mandel P., Opt. Commun. 1983. V. 44, P. 400.
263.Mandel P., Opt. Commun. 1983. V. 45, P. 269.

264.Mandel P., Coherence and Quantum Optics V., Eds L. Mandel, E. Wolf, N.Y.: Plenum Publ. Corp. 1984. P. 579.

265.Mandel P., J. Opt. Soc. Am. 1985. V. B-2. P. 112.

266.Abraham N.B., Ligiato L.A., Mandel P., Narducci L.M., Bandy D.K., J. Opt. Soc. Am. 1985. V. B-2. P. 35.

267.Hendow S., Sargent M., Opt. Commun. 1982. V. 40, P. 385.

268.Hendow S., Sargent M., Opt. Commun. 1982. V. 43, P. 59.

269.Hendow S., Sargent M., J. Opt. Soc. Am. 1985. V. B-2. P. 84.

270.Zhang J., Haken H., Ohno H., J. Opt. Soc. Am. 1985. V. B-2. P. 141.

271.Bandy D.K., Narducci L.M., Lugiato L.A., Abraham N.B., J. Opt. Soc. Am. 1985. V. B-2. P. 56.

272.Lugiato L.A., Narducci L.M., Bandy D.K., Abraham N.B., Opt. Commun. 1983. V. 46, P. 115.

273.Narducci L.M., Bandy D.K., Lugiato L.A., Abraham N.B., Coherence and Quantum Optics V. /Eds L. Mandel, E. Wolf, N.Y.: Plenum Publ. Corp. 1984. P. 217.

274.Casperson L.W., J. Opt. Soc. Am. 1985. V. B-2. P. 73.

275.Casperson L.W., J. Opt. Soc. Am. 1988. V. B-5. P. 958.

276.Shin M.L., Milloni P.W., Ackerhalt J.R., J. Opt. Soc. Am. 1985. V. B-2. P. 130.

277.Dupertuis M.A., Salomaa R.R.E., Siegrist M.R., Opt. Commun. 1986. V. 57, P. 410.

278.Mehendale S.C., Harrison R.G., Opt. Commun. 1986. V. 60, P. 257.

279.Mehendale S.C., Harrison R.G., Phys. Rev. 1986. V. A-34. P. 1613.

280.Moloney J.V., Uppal J.S., Harrison R.G., Phys. Rev. Lett. 1987. V. 59. P. 2868.

281.Frysiak F., Harrison R.G., Moloney J.V., Phys. Rev. 1989. V. A-39. P. 421.

282.Lawandy N.M., Ryan J.C., Opt. Commun. 1987. V. 63, P. 53.

283.Ryan J.C., Lawandy N.M., Opt. Commun. 1987. V. 64, P. 54.

284.Khandokhin P.A., Khanin Ya.I., Koryukin I.V., Opt. Commun. 1988. V. 65, P. 367.

285.Koryukin I.V., Khandokhin P.A., Khanin Ya.I., Izv. Akad. Nauk SSSR. Ser. Fiz. 1988. V. 52 P. 1081.

286.Pujiol J., Laguarta F., Vilaseca R., Corbalan R., J. Opt. Soc. Am. 1985. V. B-5. P. 1004.

287.Corbalan R., Laguarta F., Pujiol J., Vilaseca R., Opt. Commun. 1989. V. 71, P. 290.

288.Oraevsky A.N., Toronov V.Yu., Kvant. Elektron. 1989. V. 19. P. 1327 [Sov. J. Quant. Electron. V. 16. P. 1420].

289.Khanin Ya.I., Dynamics of Quantum Oscillators, Moscow: Sovetskoe Radio. 1975.

290.Narducci L.M., Sadiky H., Lugiato L.A., Abraham N.B., Opt. Commun. 1985. V. 55, P. 370.

291.Koryukin I.V., Khandokhin P.A., Khanin Ya.I., Periodic and Chaotic Pulsations in a single-mode Homogeneously Broadened Laser. Prepint. Institut of Applied Physics, Gorky 1987. !161.

292.Zeglache H., Mandel P., J. Opt. Soc. Am. 1985. V. B-2. P. 18.

293.Oraevsky A.N., Trudy FIAN. 1986. V. 9. P. 3. [In//Research on Laser Theory /Ed. A.N. Oraevsky, Commack: Nova, 1988.

294.Kuznetsova T.I., Trudy FIAN. 1968. V. 43. P. 116.

295.Karlov N.V., Kuznetsova T.I., Radiotekhn. Elektron. 1964. V. 9. P. 524.
296.Harrison R.G., Biswas D.J., Progr. Quant. Electron. 1985. V. 127. P. 147.
297.Casperson L.W., Yariv A., Appl. Phys. Lett. 1970. V. 17. P. 259.
298.Mandel P., Zeglache H., Opt. Commun. 1983. V. 47, P. 146.
299.Vilaseca R., De Valcarcel G.J., Roldan E., Phys. Rev. 1990. V. A-41. P. 2569.
300.Zeglache H., Mandel P., Abraham N.B., Weiss C.O., Phys. Rev. 1988. V. A-38. P. 3128.
301.Kontorovich V.M., Prokhorov A.M., Zh. Eksp. Teor. Fiz. 1957. V. 107. P. 1428.
302.Javan A., Phys. Rev. 1957. V. 107. P. 1579.
303.Moloney J.V., Forysiak W., Uppal J.S., Harrison R.G., Phys. Rev. 1989. V. A-39. P. 1277.
304.Harrison R.G., Lu W., Gupta P.K., Phys. Rev. Lett. 1989. V. 63. P. 1372.
305.Lu W., Harrison R.G., J. Mod. Opt. 1990. V. 37. P. 585.
306.Lu W., Harrison R.G., Phys. Rev. 1990. V. A-41. P. 6563.
307.Heppner J., Weiss C.O., Hubner U., Shinn G., IEEE J. Quant. Electron. 1980. V. QE-16. P.392.
308.Hubner U., Abraham N.B., Weiss C.O., Phys. Rev. 1989. V. A-40. P. 6355.
309.Arijona M., Corbalan R., Laguarta F., Pujol J., Vilaseca R., Phys. Rev. 1990. V. A-41. P. 4559.
310.Corbalan R., Laguarta F., Pujol J., Vilaseca R., Opt. Commun. 1989. V. 71, P. 290.
311.Casperson L.W., J. Opt. Soc. Am. 1985. V. B-2. P. 993.
312.Casperson L.W., Opt. Quant. Electron. 1987. V. 19. P. 29.
313.Casperson L.W., In: Optical Instabilities, Eds. R.W. Boyd, M.G.Raimer, L.M. Narducci, Cambridge etc: Cambridge Univ. Press, 1986. P. 58.
314.Gioggia R.S., Abraham N.B., Lange W., Tarroja M.F.H., Wesson J.C., J. Opt. Soc. Am. 1988. V. B-5. P. 992.
315.Lamb W.E., Jr, Phys. Rev. 1964. V. 134. P. 1429.
316.Tang C.L., Statz H., DeMars G., J. Appl. Phys. 1963. V. 34. P. 2289.
317.Khanin Ya.I., Zh. Eksp. Teor. Fiz. 1971. V. 60. P. 1282 [Sov Phys. JETP. V. 33. P. 694].
318.Haken H., Laser Light Dynamics (Light, V. 2), Amsterdam: North-Holland. 1985.
319.Akhmanov S.A., Khokhlov R.V., Problems of Nonlinear Optics, Moscow: All-Union Information Center (VINITI), Acad. Sci. USSR. 1964 [Problems of Nonlinear Optics, Electromagnetic Waves in Nonlinear Dispersive Media, New York: Gordon and Breach. 1972.
320.Ratner A.M., Fisher A.M., Izv. Vyss. Ucheb. Zaved. Radiophys. 1973. V. 16. P. 150. 239.
321.Anan'ev Yu.A., Sedov B.M., Zh. Eksp. Teor. Fiz. 1965. V. 48. P. 782 [Sov Phys. JETP. V. 21. P. 517].
322.Statz H., Tang C.L., J. Appl. Phys. 1964. V. 35. P. 1377.
323.Lifshits B.L., Tsykunov V.N., Zh. Eksp. Teor. Fiz. 1965. V. 49. P. 1843 [Sov Phys. JETP. V. 22. P. 1260].
324.Akhmanov S.A., Golyaev Yu.D., Dmitriev V.G., Zh. Eksp. Teor. Fiz. 1972. V. 62. P. 133 [Sov Phys. JETP. V. 35. P. 70].
325.Ronchi L., Nuovo Chimento. 1966. V. B-44. P. 372.

326.Tang C.L., Appl. Optics. 1962. V. 1. P. 768.

327.Anan'ev Yu.A., Zh. Tekhn. Fiz. 1967. V. 37. P. 139.//"$. 1967. ". 37. !. 139.

328.Anan'ev Yu.A., Mak A.A., Sedov B.M., Zh. Eksp. Teor. Fiz. 1967. V. 52. P. 12 [Sov Phys. JETP. V. 25. P. 6].

329.Ghiner A.V., Komarov K.P., Folin K.G., Opt. Commun. 1975. V. 19, P. 350.

330.Meller A.S., Khandokhin P.A., Khanin Ya.I., Kvant. Elektron. 1986. V. 13. P. 2278 [Sov. J. Quant. Electron. V. 16. P. 1502].

331.Kovalenko S.A., Pugovkin A.V., Izv. Vyss. Ucheb. Zaved. Radiophys. 1968. V. 11. P. 232.

332.Golyaev Yu. D., Lantratov S.V., Kvant. Elektron. 1979. V. 6. P. 2361 [Sov. J. Quant. Electron. V. 19. P. 1390].

333.Azarova V.V., Galaktionova N.M., Mak A.A., Orlov O.A., Ustyugov V.I., Kvant. Elektron. 1979. V. 6. P. 2339 [Sov. J. Quant. Electron. V. 9. P. 1378].

334.Evdokimova O.N., Kaptsov L.N., Kvant. Elektron. 1989. V. 16. P. 1557 [Sov. J. Quant. Electron. V. 19. P. 1001].

335.Otsuka K., Mandel P., Bielawski S., Derozier D., and Glorieux P., Phys. Rev. 1992. V. A-46. P. 1692.

336.Pieroux D., Mandel P., Opt. Commun. 1994. V. 107, P. 245.

337.Erneux T., Mandel P., Phys. Rev. 1995. V. A-52. P. 4137.

338.Mandel P., Wang Jing-Yi, Phys. Rev. Lett. 1995. V. 75. P. 1923.

339.Mandel P., Otsuka K., Wang Jing-Yi, Pieroux D., Phys. Rev. Lett. 1996. V. 76. P. 2694.

340.Nguyuen B.A., Mandel P., Phys. Rev. 1996. V. A-54. P. 1638.

341.Khandokhin P.A., Mandel P., Koryukin I.V., Nguyen B.A., Khanin Ya.I, Phys. Lett. 1997. V. A-235. P. 248.

342.Lacot E., Stoeckel F., J. Opt. Soc. Am. 1996. V. B-13. P. 2034.

343.Mandel P., Nguyuen B.A., Otsuka K., Quantum Semiclass. Opt, 1997. V. 9. P. 1.

344.Nguyuen B.A., Mandel P., Wang Jing-Yi, Pieroux D., Phys. Rev. Lett. 1995. V. 75. P. 1923.

345.Khanin Ya.I., Chaos. 1996. V. 6. P. 373.

346.Zhislina V.G., Khandokhin P.A., Khanin Ya.I., Radiophys. Quantum Electron. 1996. V.39. P511. [Izv. Vyss. Ucheb. Zaved. Radiofiz. 1996. V. 39. P. 771].

347.Abraham N.B., Khanin Ya.I., Proc. SPIE Int. Soc. Opt. Eng. 1996. V. 2792. P. 2.

348.Khandokhin P.A., Mandel P., Koryukin I.V., Nguyen B.A., Khanin Ya.I., Radiophys. Quantum Electron. 1997. V.40. P103. [Izv. Vyss. Ucheb. Zaved. Radiofiz. 1997. V. 40. P. 161].

349.Pieroux D., Mandel P., Quantum Semiclass. Opt, 1997. V. 9. P. L17.

350.Abraham N.B., Sekaric L., Carson L.L., Seccareccia V., Khandokhin P.A., Khanin Ya.I. Koryukin I.V., Zhislina V.G., Phys. Rev. A. 2000. V. 62. 013810.

351.Abraham N.B., Everett L.L., Iwata C., Janicki M.B., Proc. SPIE Int. Soc. Opt. Eng. 1993. V. 2095. P. 16.

352.McCumber D.B., Phys. Rev. 1966. V. 141. P. 306.

353.Sushchik M.M., Freidman G.I., Izv. Vyss. Ucheb. Zaved. Radiofiz. 1966. V. 9. P. 919.

354.Sooy W.R., Appl. Phys. Lett. 1965. V. 7. P. 36.

355.Leontovich A.M., Churkin V.I., Zh. Eksp. Teor. Fiz. 1970. V. 59. P. 7 [Sov Phys. JETP. 1971. V. 32. P. 4].
356.Bondarenko N.G., Eremina I.V., Zh. Prikladn. Spectr. 1968. V. 8. P. 599.
357.Suchkov A.F., Zh. Eksp. Teor. Fiz. 1965. V. 49. P. 1495 [Sov Phys. JETP. 1966. V. 22. P. 717]; //Trudy FIAN. 1968. V. 43. P. 161.
358.Baev V.M., Belikova T.P., Sviridenkov E.A., Suchkov A.F., Zh. Eksp. Teor. Fiz. 1978. V. 74. P. 43 [Sov Phys. JETP. V. 47. P. 21].
359.Gitlin M.S., Khanin Ya.I., Izv. Vyss. Ucheb. Zaved. Radiofiz. 1985. V. 28. P. 978.
360.Mironenko V.R., Yudson V.I., Zh. Eksp. Teor. Fiz. 1980. V. 79. P. 1174 [Sov Phys. JETP. V. 52. P. 504].
361.Kagan A.G., Khanin Ya.I., Kvant. Elektron. 1983. V. 10. P. 149 [Sov. J. Quant. Electron. V. 13. P. 88].
362.Rivlin L.A., Kvant. Electron. /Ed. N.G.Basov.—M.: Sov. Radio, 1972, #5, P. 94.
363.Hollinger F., Jung Chr, J. Opt. Soc. Am. 1985. V. B-2. P. 218.
364.Hollinger F., Jung Chr., Weber H, In: Optical Instabilities /Eds. R.W. Boyd, M.G.Raimer, L.M. Narducci, Cambridge etc: Cambridge Univ. Press, 1986. P. 20.
365.Hollinger F., Jung Chr., Weber H., Opt. Commun. 1990. V. 75, P. 84.
366.Victorov E.A., Sokolov V.A., Tkachenko E.V., Ustyugov V.I., Opt. Spektr. 1990. V. 68. P.920 [Opt. Spectrosc. (USSR) 68. 537].
367.McMackin I., Radzewicz C., Beck M., Raymer M.G., Phys. Rev. 1987. V. A-38. P. 820.
368.Beck M., Raymer M.G., McMackin I., Phys. Rev. 1989. V. A-40. P. 2410.
369.Mironenko V.R., Yudson V.I., Opt. Commun. 1980. V. 34, P. 397.
370.Kovalenko S.A., Kvant. Elektron. 1981. V. 8. P. 1271 [Sov. J. Quant. Electron. V. 11. P. 759].
371.Baev V.M., Gaida G., Schroder H., Toschek P.E., Opt. Commun. 1981. V. 38, P. 309.
372.Atmanspatcher H., Scheingraber H., Baev V.M., Phys. Rev. 1987. V. A-35. P. 142.
373.Atmanspatcher H., Scheingraber H., Vidal C.R., Phys. Rev. 1986. V. A-32. P. 1052.
374.Atmanspatcher H., Scheingraber H., Phys. Rev. 1986. V. A-34. P. 253.
375.Grassberger P., Procaccia I., Phys. Rev. Lett. 1983. V. 50. P. 346.
376.Procaccia I., Phys. Scr. 1985. V. T-9. P. 40.
377.Ben-Mizrachi A. Procaccia I., Grassberger P., Phys. Rev. 1984. V. A-29. P. 975.
378.Mashkevich V.S., Ukr. Fiz. Zh. 1967. V. 12. P. 1731.
379.Mashkevich V.S., Zh. Eksp. Teor. Fiz. 1967. V. 53. P. 1003 [Sov Phys. JETP. 1968. V. 26. P. 601].
380.Mashkevich V.S., Kinetic Theory of Lasers, Moscow: Nauka. 1971.
381.Tolstoy N., Shapovalov V.N., Opt. Spektr. 1967. V. 23. P. 648 [Opt. Spectrosc. (USSR) 23. V. 352].
382.Matorin I.I., Khanin Ya.I., Izv. Vyss. Ucheb. Zaved. Radiofiz. 1973. V. 16. P. 386.
383.Pachomycheva I.A., Sviridenkov E.A., Suchkov A.F., Titova L.V., Churilov S.S., Pisma Zh. Eksp. Teor. Fiz. 1970. V. 12. P. 60; [Sov. Phys. JETP Lett. 1972. V. 12. P. 43].

384.Belikova T.P., Sviridenkov E.A., Suchkov A.F., Titova L.V., Churilov S.S., Zh. Eksp. Teor. Fiz. 1972. V. 62. P. 2060 [Sov Phys. JETP. V. 35. P. 1076].

385.Hansch T.W., Schawlov A.L., Toschek P.E., IEEE J. Quant. Electron. 1972. V. QE-8. P. 802.

386.Harris S.J., Appl. Optics. 1984. V. 23. P. 1311.

387.Ratner A.M., Zh. Tekhn. Fiz. 1968. V. 38. P. 1377.

388.Risken H., Nummedal K., Phys. Lett. 1968. V. A-26. P. 275.

389.Risken H., Nummedal K., J. Appl. Phys. 1968. V. 39. P. 4662.

390.Graham R., Haken H., Z. Phys. 1968. V. 213. P. 420.

391.Risken H., In: Optical Instabilities /Eds. R.W. Boyd, M.G.Raimer, L.M. Narducci, Cambridge etc: Cambridge Univ. Press, 1986. P. 20.

392.Lugiato L.A., Narducci L.M., Phys. Rev. 1985. V. A-32. P. 1563.

393.Mandel P., Opt. Commun. 1985. V. 53, P. 249.

394.Gerber P.R., Butticker M., Z. Phys. 1979. V. 53. P. 219.

395.Mayr M., Risken H., Vollmer H.D., Opt. Commun. 1981. V. 36, P. 480.

396.Milovsky N.D., Phys. Lett. 1970. V. A-33. P. 492.

397.Milovsky N.D., Izv. Vyss. Ucheb. Zaved. Radiofiz. 1971. V. 14. P. 93.

398.Pessina E.M., Bonfrate G., Fontane F., Lugiato L.A., Phys. Rev. 1997. V. A-56. P. 4086.

399.Roldan E., Opt. Commun. 1997. V. 143, P. 235.

400.Ostrovsky L.A., Zh. Eksp. Teor. Fiz. 1965. V. 48. P. 1087; [Sov. Phys. JETP. 21(4)].

401.Ostrovsky L.A., Zh. Eksp. Teor. Fiz. 1965. V. 48. P. 1087; [Sov. Phys. JETP. .21(6)].

402.Fleck J.A., Jr., Kidder R.E, J. Appl. Phys. 1964. V. 35. P. 2825.

403.Fleck J.A., Jr., Kidder R.E, J. Appl. Phys. 1965. V. 36. P. 2327.

404.Roldan E., Opt. Commun. 1997. V. 143, P. 235.

405.Basov N.G., Morosov V.N., Oraevsky A.N., Zh. Eksp. Teor. Fiz. 1965. V. 49. P. 895; [Sov. Phys. JETP. 1966. 22(3)].

406.Mandel P., Etrich C., Otsuka K., IEEE J. Quant. Electron. 1993. V. QE-29. P. 836.

407.Khandokhin P.A., Khanin Ya.I., Koryukin I.V., Mandel P., IEEE J. Quant. Electron. 1995. V. QE-31. P. 647.

408.Riedel E.P., Baldwin G.D., J. Appl. Phys. 1967. V. 38. P. 2720.

409.Butylkin V.S., Kaplan A.E., Khronopulo Yu.G., Yakubovich E.I., Resonance Interaction of Light with Matter, Moscow: Nauka. 1977.

410.Antipov O.L., Kuzhelev A.S., Luk'yanov A.Ya., Zinov'ev A.P., Kvant. Elektron. 1998. V. 25. P. 891

411.Henry C.H., Semiconductor Injection Lasers V. 1., Ed. W.T. Tsang (Semiconductors and Semimaterials, V. 22), Orlando: Academic Press. 1985.

412.Henry C.H., IEEE J. Quant. Electron. 1982. V. QE-18. P. 259.

413.Kravtsov N.V., Lariontsev E.G., Kvant. Elektron. 1994. V. 21. P. 903.

414.Klimontovich Yu.L., Landa P.S., Lariontsev E.G., Zh. Eksp. Teor. Fiz. 1967. V. 52. P. 1616.

415.Murina T.A., Theoretical analysis of the Methods for Control of Single-Mode Laser Rsdiation: Ph. D. Thesis, Leningrad. 1982.

416.Mandel P., Abraham N.B., Opt. Commun. 1984. V. 51, P. 87.

417. Zeglache H., Mandel P., Abraham N.B., Hoffer L.M., Lippi G.L., Mello T., Phys. Rev. 1988. V. A-37. P. 470.
418. Vladimirov A., Opt. Commun. 1998. V. 149, P. 67.
419. Mandel P., Agraval G.P., Opt. Commun. 1982. V. 42, P. 269.
420. Zhelnov B.A., Smirnov V.S., Fadeev F.P., Opt. Spektr. 1970. V. 28. P. 744 [Opt. Spectrosc. (USSR) 28. 400].
421. Krivoshchekov G.V., Makukha V.K., Semibalamut V.M., Smirnov V.S., Kvant. Elektron. 1976. V. 3. P. 1782 [Sov. J. Quant. Electron. V. 6. P. 965].
422. Khandokhin P.A., Khanin Ya.I., J. Opt. Soc. Am. 1985. V. B-2. P. 226.
423. Hoffer L.M., Lippi G.L., Abraham N.B., Mandel P., Opt. Commun. 1988. V. 66, P. 219.
424. Anokhov S.P., Marusii T.Ya., Soskin M.S., Tunable Lasers, Moscow: Radio Svyaz'. 1982.
425. Golyaev Yu.D., Evtuhov K.N., Kaptsov L.N., Smyshlyaev S.P., Kvant. Elektron. 1981. V. 8. P. 230 [Sov. J. Quant. Electron. V. 11. P. 1427].
426. Sekita M., Kimura S., J. Appl. Phys. 1983. V. 54. P. 3415.
427. Polushkin I.N., Khandokhin P.A., Khanin Ya.I., Kvant. Elektron. 1983. V. 10. P. 1461 [Sov. J. Quant. Electron. V. 13. P. 950].
428. Koryukin I.V., Khandokhin P.A., Khanin Ya.I., Kvant. Elektron. 1990. V. 17. P. 978. [Sov. J. Quant. Electron. V. 20. P. 895].
429. Khandokhin P.A., Radiophys. Quantum Electron. 1979. V.22. P.564. [Izv. Vyss. Ucheb. Zaved. Radiofiz. 1979. V. 22. P. 813].
430. Kushima T., Marcos H.M., Geusic J.E., Phys. Rev. 1968. V. A-167. P. 289.
431. Koryukin I.V., Khandokhin P.A., Khanin Ya.I., Mandel P., Kvant. Elektron. 1995. V. 22. P. 1081.
432. Perevedentsetva G.V., Khandokhin P.A., Khanin Ya.I., Kvant. Elektron. 1980. V. 7. P. 128 [Sov. J. Quant. Electron. V. 10. P. 71].
433. Murina T.A., Rosanov N.N., Abstracts of Reports, First All-Union Conference Optics of Laser, Leningrad; GOI. 1977. P. 154.
434. Khandokhin P.A., Khanin Ya.I., Kvant. Elektron. 1989. V. 16. P. 1993. [Sov. J. Quant. Electron. V. 18. P. 1248].
435. Zolotoverkh I.I., Kravtsov N.V., Lariontsev E.G., MakarovA.A., Firsov V.V., Opt. Commun. 1994. V. 113, P. 249.
436. Khanin Ya.I., J. Opt. Soc. Am. 1988. V. B-5. P. 889.
437. Azarova V.V., Intensity Noise in the CW Garnet Laser Radiation: Ph. D, Leningrad, 1980.
438. Khandokhin P.A., Low-Frequency Processes in a Solid-State Ring Laser.: Ph. D thesis, Gorky, USSR, 1985.
439. Khandokhin P.A., Khanin Ya.I., Kvant. Elektron. 1996. V. 23. P. 29.
440. Khandokhin P.A., Khanin Ya.I., Koryukin I.V., Opt. Commun. 1991. V. 81, P. 297.
441. Lacot E., Stoeckel F., J. Opt. Soc. Am. 1996. V. B-13. P. 2034.
442. Leners R., Stephan G., Quantum Semiclass. Opt, 1995. V. 7. P. 757.
443. Bielawski S., Derozier D., Glorieux P., Phys. Rev. 1992. V. A-46. P. 284.
444. Paddon P., Sjerve E., May D., Baurouis M., Stephan G., J. Opt. Soc. Am. 1992. V. B-9. P. 574.

445.Zeglache H., Boulnois A., Phys. Rev. 1995. V. A-52. P. 4229.
446.Khandokhin P., Khanin Ya., Mamaev Yu., Milovsky N., Shirokov E., Bielawski S., Derosier D., Glorieux P., Quantum Semiclass. Opt, 1998. V. 10. P. 97.
447.Khandokhin P., Khanin Ya., Mamaev Yu., Milovsky N., Shirokov E., Bielawski S., Derosier D., Glorieux P., Kvant. Elektron. 1995. V. 25. P. 517.
448.San Miguel M., Feng O., Moloney J., Phys. Rev. 1995. V. A-52. P. 1728.
449.Martin-Regalado J., Prati F., San Miguel M., Abraham N.B., IEEE J. Quant. Electron. 1997. V. QE-33. P. 765.
450.Belenov E.M., Markin E.P., Morozov V.N., Oraevsky A.N., Pisma Zh. Eksp. Teor. Fiz. 1966. V. 3. P. 54.
451.Belenov E.M., Morozov V.N., Oraevsky A.N., Trudy FIAN. 1970. V. 52. P. 237.[In: Quantum Electronics in Masers and Lasers /Ed. D.V. Scobel'tsyn, New York: Consaltants Bureau, 1972.
452.Bayer E., Hellwege K.H., Schaack //Zs. Naturforsch. 1965. V. 20A. P. 1181.
453.Morozov V.N., Oraevskii A.N., Izv. VUZ, Radiofiz. 1966. V.9. P. 710.
454.Tsykunov V.N., Zh. Eksp. Teor. Fiz. 1958. V. 70. P. 1646 [Sov Phys. JETP. V. 31. P. 883].
455.Livshits B.L., Zh. Eksp. Teor. Fiz. 1963. V. 59. P. 516 [Sov Phys. JETP. V. 32. P. 283].
456.Vinokurov G.N., Opt. Spektr. 1971. V. 22. P. 5399 [Opt. Spectrosc. (USSR) 31. 251].
457.Tredicce J.P., Abraham N.B., Puccioni G.P., Arecchi F.T, Opt. Commun. 1985. V. 55, P. 131.
458.Stamatesku L., Hamilton M.W, Phys. Rev. 1997. V. A-55. P. 2115.
459.Pieroux D., Mandel P., Otsuka K., Opt. Commun. 1993. V. 108, P. 173.
460.Matorin I.I., Pikowsky A.S., Khanin Ya.I., Kvant. Elektron. 1984. V. 11. P. 2096 [Sov. J. Quant. Electron. V. 14. P. 1401].
461.Toda M., Phys. Rep. 1975. V. 18. P. 1.
462.Oppo G.L., Politi A., Z. Phys. 1985. V. B-2. P. 111.
463.Migulin V.V., Medvedev V.I., Mustel' E.R., Parygin V.N., Principles of the thory of Oscillations, Moscow: Nauka. 1978.
464.Ivanov D.V., Khanin Ya.I., Matorin I.I., Pikovsky A.S., Phys. Lett. 1982. V. 89. P. 229.
465.Arecchi F.T., Meucci R., Puccioni G.P., Tredicce J.R., Phys. Rev. Lett. 1982. V. 49. P. 1271.
466.Puccioni G.P., Poggi A., Gadomski W., Tredicce J.R., Arecchi F.T., Phys. Rev. Lett. 1985. V. 55. P. 339.
467.Tredicce J.R., Arecchi F.T. Puccioni G.P., Poggi A., Gadomski W., Phys. Rev. 1986. V. A-34. P. 2073.
468.Dangoisse D., Glorieux P., Hennequin D., Phys. Rev. 1987. V. A-36. P. 4775.
469.Midavaine T., Dangoisse D., Glorieux P., Phys. Rev. Lett. 1985. V. 55. P. 1989.
470.Dangoisse D., Glorieux P., Hennequin D., Phys. Rev. Lett. 1986. V. 57. P. 2657.
471.Khandokhin P., Khanin Ya. I., Kvant. Elektron. 1984. V. 11. P. 1483 [Sov. J. Quant. Electron. V. 14. P. 1004].
472.Mandel P., Nardone P., Erneux T., J. Opt. Soc. Am. 1988. V. B-5. P. 1113.

473.Bezaeva L.G., Kartsov L.N., Sharikov I.Z., Kvant. Elektron. 1985. V. 12. P. 1743 [Sov. J. Quant. Electron. V. 15. P. 1152].
474.Evdokimova O.N., Kartsov L.N., Kvant. Elektron. 1990. V. 17. P. 901 [Sov. J. Quant. Electron. V. 20. P. 824].
475.Samson A.M., Turovets S.I., Zh. Prikladn. Spectr. 1988. V. 48. P. 384.
476.Erneux T., Mandel P., Phys. Rev. 1989. V. A-40. P. 286.
477.Arecchi F.T., Gadomski W., Meucci R., Roversi J.A., Opt. Commun. 1988. V. 65, P. 47.
478.Landau L.D., Lifshits E.M., Mechanics, Moscow: Nauka, 1988.
479.Kimura T., Otsuka K, J. Appl. Phys. 1969. V. 40. P. 5399.
480.Kimura T., Otsuka K, IEEE J. Quant. Electron. 1970. V. QE-6. P.764.
481.De Maria A.J., Gagosz R., Proc. IRE. 1962. V. 50. P. 1522.
482.Kozlova V.K., Mustel' E.R., Parygin V.N., Radiotekhn. Elektron. 1969. V. 14. P. 875.
483.Livshits B.L., Dokl. Akad. Nauk SSSR. 1970. V. 191. P. 1298.
484.Feher G., Gordon J.P., Buechler E., Gere E. A., Thurmond C.D., Phys. Rev. 1958. V. 109. P. 221.
485.Chester P.F., Wagner P.E., Castle J.P., Phys. Rev. 1958. V. 110. P. 281.
486.Thorp J.S., Advances in Quantum Electronics, N.Y.: Columbia Univ. Press, 1960. P. 602.
487.Firth I.M., Bijl D., Nature. 1961. V. 192. P. 860.
488.Shamfarov Ya.L., Smirnova T.A., Radiotekhn. Elektron. 1963. V. 8. P. 1567.
489.Kemp J.C., J. Appl. Phys. 1959. V. 30. P. 1451.
490.Yariv A., J. Appl. Phys. 1960. V. 31. P. 740.
491.Wang S., Singer J.R., J. Appl. Phys. 1961. V. 32. P. 1371.
492.Firth I.M., Physica. 1963. V. 29. P. 857.
493.Singer J.R., Masers, New York: Chapman and Hall, 1959.
494.Kemp J.C., Phys. Rev. Lett. 1961. V. 7. P. 21.
495.Mukhtarov Ch.K., Dokl. Akad. Nauk SSSR. 1966. V. 171. P. 1301 [Sov Phys. Doklady. V. 11. P. 1083].
496.Golyaev Yu.D., Dynamics of Oscillations of Quasi-CW Neodymium YAG Lasers: Ph. D. Thesis, Moscow, 1973.
497.Jeppesen M.A., J. Opt. Soc. Am. 1958. V. 48. P. 629.
498.Kirsanov B.P., Leontovich A.M., Mozharovsky A.M., Kvant. Elektron. 1974. V. 1. P. 2211 [Sov. J. Quant. Electron. 1975. V. 4. P. 1232].
499.Golyaev Yu.D., Lantratov S.V., Kvant. Elektron. 1974. V. 1. P. 2197 [Sov. J. Quant. Electron. 1975. V. 4. P. 1224].
500.Marshall F.R., Roberts O.L., Proc. IRE. 1962. V. 50. P. 2108.
501.Collins R.J., Kisliuk P., J. Appl. Phys. 1962. V. 33. P. 2009.
502.Statz H., DeMars G.A., Wilson D.T., Tang C.L., J. Appl. Phys. 1965. V. 36. P. 1510.
503.Keller D.V., Davis B.I., IEEE J. Quant. Electron. 1966. V. QE-2. P.179.
504.Paoli T., Ripper J.E., Appl. Phys. Lett. 1970. V. 16. P. 96.
505.Berzing E.G., Naboikin Yu.V., Rom-Krichevskaya I.A., Tiunov Yu.A., Opt. Spektr. 1967. V. 22. P. 503.

506.Brzhazovsky Yu.V., Vasilenko L.S., Rautian S.G., Popova G.S., Chebotaev V.P., Zh. Eksp. Teor. Fiz. 1971. V. 61. P. 500 [Sov Phys. JETP. V. 34 1972. P. 265].
507.Wood D.R. Schwarz S.E., Appl. Phys. Lett. 1967. V. 11. P. 88.
508.Karlov N.V., Kuzmin G.P., Petrov Yu.M., Prokhorov A.M., Pisma Zh. Eksp. Teor. Fiz. 1968. V. 7. P. 174; [Sov. Phys. JETP Lett. V. 7. P. 134].
509.Powell H.T., Wolga G.J., IEEE J. Quant. Electron. 1971. V. QE-7. P.213.
510.Jacques A., Glorieux P., Opt. Commun. 1982. V. 40, P. 455.
511.Arimondo E., Casagrande F., Glorieux P., Lugiato L.A., Appl. Phys. 1983. V. B-30. P. 57.
512.Arimondo E., Dinelli B.M., Opt. Commun. 1983. V. 44, P. 277.
513.Arimondo E., Menchi E., Appl. Phys. 1985. V. B-37. P. 55.
514.Arimondo E., Bootz P., Glorieux P., Menchi E., J. Opt. Soc. Am. 1985. V. B-2. P. 193.
515.Glorieux P., Dangoisse D., IEEE J. Quant. Electron. 1985. V. QE-21. P.1486.
516.Arimondo E., Gabannini C., Menchi E., Zambon B., Proc. SPIE 1986. V. 667. P. 234.
517.Arimondo E., Dangoisse D., Gabannini C., Menchi E., Papoff F., J. Opt. Soc. Am. 1987. V. B-4. P. 892.
518.Arimondo E., Instabilities and Chaos in Quantum Optics II /Eds N.B. Abraham, F.T. Arecchi, L.A. Lugiato, N.Y.: Plenum Press, 1988. P. 69.
519.Dangoisse D., Bekkali A., Papoff F., Glorieux P., Europhys. Lett. 1988. V. 6. P. 335.
520.Bekkali A., Papoff F., Dangoisse D., Glorieux P., J. de Phys. 1988. V. 49. P.C2-349.
521.Hennequin D., Lefranc M., Bekkali A., et al, Measures and Complexity and Chaos /Ed. N.B.Abraham, N.Y.; L: Plenum Press, 1989. P. 299.
522.Gryaznov Yu.M., Chastov A.A., Zh. Prikladn. Spectr. 1972. V. 16. P. 658.
523.Berzing E.G., Naboikin Yu.V., Tiunov Yu.A., Chernov V.S., Zh. Prikladn. Spectr. 1972. V. 13. P. 404.
524.Bespalov V.I., Yakubovich E.I., Izv. Vyss. Ucheb. Zaved. Radiofiz. 1965. V. 8. P. 909.
525.Roshchin N.V., Izv. Vyss. Ucheb. Zaved. Radiofiz. 1973. V. 16. P. 1006.
526.Soffer B.H., J. Appl. Phys. 1964. V. 35. 2551.
527.She C.V., Ang-Tick Tan, Appl. Phys. Lett. 1966. V. 9. P. 198.
528.Dovger L.S., Ermakov B.A., Lukin A.V., Shklover L.P., Opt. Spektr. 1966. V. 20. P.903.
529.Erickson L.E., Szabo A., J. Appl. Phys. 1966. V. 37. 4953.
530.Erneux T., J. Opt. Soc. Am. 1988. V. B-5. P. 1063.
531.Shil'nikov L.P., Dokl. Akad. Nauk SSSR. 1965. V. 160. P. 558.
532.Tachikava M., Tanii K., Kajita M., Shimizu T., Appl. Phys. 1986. V. B-39. P. 83.
533.Tachikava M., Tanii K., Shimizu T., J. Opt. Soc. Am. 1988. V. B-5. P. 1077.
534.Tatarkova S.N., Tuchin V.V., Laser Physics. 1993. V.2. P. 482.
535.Carmichael C.H., Simpson G.N., Nature. 1964. V.202. P.4934.
536.Birnbaum M., Stocker T.L., IEEE J. Quant. Electron. 1966. V. QE-2. P.184.
537.Elissev P.G., Man'ko M.A., Zh. Prikladn. Spectr. 1968. V. 36. P. 2215.
538.Shen Y.R., Phys. Rev. Lett. 1966. V. 20. P. 378.

539.Duguay M.A., Hansen J.W., Shapiro S.L., IEEE J. Quant. Electron. 1970. V. QE-6. P. 725.
540.Giuliano C.R., Marburger J.H., Phys. Rev. Lett. 1971. V. 27. P. 905.
541.Katzenstein J., Magyar G., Selden A.C., Opto-Electronics. 1969. V. 1. P. 13.
542.Gaponov S.V., Goncharov A.G., Kraftmakher G.A., Khanin Ya.I., Pisma Zh. Eksp. Teor. Fiz. 1970. V. 11. P. 370; [Sov. Phys. JETP Lett. V. 11. P. 248].
543.Drobnik A., Wolf L., Kvant. Elektron. 1978. V. 5. P. 482 [Sov. J. Quant. Electron. V. 8. P. 274].
544.Gaponov S.V., Paramonov L.V., Salashchenko N.N., Khanin Ya.I., Kvant. Elektron. 1980. V. 7. P. 2432 [Sov. J. Quant. Electron. V. 10. P. 1415].
545.Mikaelyan A.L., Kuprishov V.F., Turkov Yu.G., Anlreev Yu.V., Shcherbakova A A., Pisma Zh. Eksp. Teor. Fiz. 1970. V. 11. P. 244; [Sov. Phys. JETP Lett. V. 11. P. 155], Kvant. Electron. /Ed. N.G.Basov.—M.: Sov. Radio, 1971, #1, P. 102. [Sov. J. Quant. Electron. V. 1. P. 74].
546.Birnbaum M., Fincher C.L., Proc. IEEE. 1969. V. 57. P. 804.
547.Allakhverdyan R.G., Morozov V.N., Oraevsky A.N., Suchkov A.F., Kvant. Electron. /Ed. N.G.Basov, M.: Sov. Radio, 1971, #6, P. 53. [Sov. J. Quant. Electron. V. 1. P. 102].
548.Bogdankevich O.V., Kovalenko A.N., Mestverishvili A.N., Nasibov A.S., Pechenov A.N., Ryabov E.A., Suchkov A.F., Zh. Eksp. Teor. Fiz. 1971. V. 60. P. 132 [Sov Phys. JETP. V. 33. 1972. P. 7455].
549.Gurevich G.L., Ingel' L.Kh., Khanin Ya.I., Kvant. Electron., Ed. N.G.Basov, M.: Sov. Radio, 1972, #3, P. 45. [Sov. J. Quant. Electron. V. 2. P. 230].
550.Talanov V.I., Izv. Vyss. Ucheb. Zaved. Radiofiz. 1965. V. 8. P. 260.
551.Kelley P.L., Phys. Rev. Lett. 1965. V. 15. P. 1005.
552.Kaminsky A.A., Mak A.A., Pashinin P.P., Popov Yu.M., In: Laser Handbook. Ed. A.M. Prokhorov, Moscow: Sovetskoe Radio, 1978. V. 1. P. 237.
553.Mocker H.W., Collins R.J., Appl. Phys. Lett. 1965. V. 7. P. 210.
554.Malyshev V.I., Markin A.S., Sychev A.A., Pisma Zh. Eksp. Teor. Fiz. 1967. V. 6. P. 190; [Sov. Phys. JETP Lett. V. 6. P. 34].
555.Schmackhfeffer A., Weber Y., Phys. Lett. 1967. V. 24. P. 190.
556.De Maria A.J., Stetser D.A., Heunau H., Appl. Phys. Lett. 1966. V. 8. P. 174.
557.De Maria A.J., Glenn W.H., Brienza M.J., Mack M.E., Proc. IEEE. 1969. V. 57. P. 2.
558.New G.H.C., Rep. Progr. Phys. 1983. V. 46. P. 877.
559.Statz H., Tang C.L., J. Appl. Phys. 1965. V. 36. P. 3923.
560.Statz H., De Mars G.A., Tang C.L., J. Appl. Phys. 1967. V. 38. P. 2212.
561.Statz H., J. Appl. Phys. 1967 V. 38 P. 4648.
562.Tang C.L., Statz H., J. Appl. Phys. 1967. V. 38. P. 2963.
563.Statz H., Bass M., J. Appl. Phys. 1969. V. 40. P. 377.
564.Schwarz S.E., Gordon P.L., J. Appl. Phys. 1969. V. 40. P. 377.
565.Kravtsov N.V., Yatsenko Yu.P., Zh. Tekhn. Fiz. 1978. V. 47. P. 2433.
566.Haus H.A., J. Appl. Phys. 1975. V. 46. P. 3049.
567.Bondarenko A.N., Krivoshchekov G.V., Semibalamut V.M., Smirnov V.A., Stupak M.F., Izv. Vyss. Ucheb. Zaved. Radiofiz. 1971. V. 14. P. 1615.

568.Gusev A.A., Kruzhalov S.V., Pakhomov L.N., Petrún'kin V.Yu., Pisma Zh. Techn. Fiz. 1978. V. 4. P. 1250.
569.Spence D.E., Kean P.N., Sibbert W., Opt. Lett. 1991. V. 16. P. 43.
570.Liu L.Y., Huxley J.M., Ippen E.P., Haus H.A., Opt. Lett. 1990. V. 15. P. 553.
571.Malcolm G.P.A., Fergusson A.J., Opt. Lett. 1992. V. 17. P. 877.
572.Piche M., Salin F., Opt. Lett. 1993. V. 18. P. 1041.
573.Ru P., Jakobsen P.K., Moloney J.V., Technical Digest "Nonlinear Dynamics in Optical Systems", Alpbach, Austria (Optical Society of America, Washington D.C., 1992) P. 48.
574.Gurevich G.L., Izv. Vyss. Ucheb. Zaved. Radiofiz. 1970. V. 13. P. 1019.
575.New G.H.C., Rea T.H., J. Appl. Phys. 1976. V. 47. P. 3107.
576.Garside B.K., Lim T.K., J. Appl. Phys. 1973. V. 44. P. 2335.
577.Khanin Ya.I., Matorin I.I., Opt. and Quant. Electron. 1981. V. 13. P. 439.
578.New G.H.C., IEEE J. Quant. Electron. 1974. V. QE-10. P.115.
579.Arthurs E.G., Bradley D.J., Puntambeckar P.N., Ruddock I.S., Glenn T.J., Opt. Commun. 1974. V. 12, P. 360.
580.McClung F.J., Hellwarth R.W., J. Appl. Phys. 1962. V. 33. P. 828.
581.Basov N.G., Zuev V.S., Kryukov P.G., Zh. Eksp. Teor. Fiz. 1962. V. 43. P. 353 [Sov Phys. JETP. 1963 V. 16. P. 254].
582.Macker P.D., Terhune R.W., Savage C.M., In 2nd Int. Conf. Electron. Quant.. Paris 1964. N.Y.: Columbia Univ. Press. P. 1559.
583.Bespalov V.I., Bogatyrev Yu.K., Izv. Vyss. Ucheb. Zaved. Radiofiz. 1966. V. 9. P. 525.
584.Pilipovich V.A., Morgun Yu.F., Zh. Prikladn. Spectr. 1967. V. 7. P. 225.
585.Landsberg G.V., Optics, Moscow: Nauka,1976 (in Russian).
586.Balashov I.F., Berzing B.G., Ermakov B.A., Konyshev V.V., Zh. Prikladn. Spectr. 1967. V. 7. P. 354.
587.Geusic J.E., Hensel V.I., Smith R.G., Appl. Phys. Lett. 1966. V. . P. 175.
588.Smith R.G., Galvin M.F., IEEE J. Quant. Electron. 1967. V. QE-3. P. 406.
589.Covacs M.A., Flynn G.W., Javan A., Appl. Phys. Lett. 1966. V. 8. P. 62.
590.Ohtsuka Y., Tsukamoto M., Masutani K., Yoshinaga H., Jap. J. Appl. Phys. 1968. V. 39. P. 2157.
591.McClung F.J., Weiner D., IEEE J. Quant. Electron. 1967. V. QE-1. P. 94.
592.Ambartsumyan R.V., Basov N.G., Zuev V.S., Kryukov P.G., Letokhov V.S., Shataberashvili O.B., Zh. Eksp. Teor. Fiz. 1963. V. 51. P. 406 [Sov Phys. JETP. 1964 V. 24. P. 272].
593.Korobkin V.V., Leontovich A.M., Popova M.N., Shchelev M.Ya., Pisma Zh. Eksp. Teor. Fiz. 1967. V. 5. P. 314 [Sov Phys. JETP. Lett. 1968 V. 8. P. 257].
594.Gibbs W.E.K., Whitcher, Appl. Optics. 1967. V. 6. P. 1957.
595.Bradley D.J., Magyar G., Richardson M.C., Nature. 1966. V. 212. P. 63.
596.Flamholz A., Wolga G.J., J. Appl. Phys. 1968. V. 39. P. 2723.
597.Prokhorov A.M., Radiotechn. Elektron. 1963. V. 8. P. 1073.
598.Mikaelyan A.L., Turkov Yu.G., Radiotechn. Elektron. 1964. V. 9. P. 743.
599.Wagner W.G., Lenguel B.A., J. Appl. Phys. 1968. V. 39. P. 2040.
600.Wang C.C., Proc. IEEE. 1963. V. 51. P. 1767.
601.Wuylsteke, J. Appl. Phys. 1963. V. 34. P. 1615.

602.Menat M, J., Appl. Phys. 1965. V. 36. P. 73.
603.Markov V.B., Marusy T.Ya., Soskin M.S., Khizhnyak A.I,, Zh. Eksp. Teor. Fiz. 1973. V. 64. P. 1538 [Sov Phys. JETP. V. 37. P. 778].
604.Letokhov V.S., Suchkov A.F,, Zh. Eksp. Teor. Fiz. 1966. V. 50. P. 1148 [Sov Phys. JETP. V. 23. P. 764].
605.Bespalov V.I., Freidman G.I,, Izv. Vyss. Ucheb. Zaved. Radiofiz. 1966. V. 9. P. 513.
606.Letokhov V.S., Suchkov A.F., Zh. Eksp. Teor. Fiz. 1967. V. 52. P. 282 [Sov Phys. JETP. V. 25. P. 182].
607.Letokhov V.S., Suchkov A.F., Trudy FIAN. 1968. V. 43. P. 169.
608.Grant D.G., Proc. IRE. 1963. V. 51. P. 604.
609.Masters J.I., Ward J., Hartount E., Rev. Sci. Instr. 1963. V. 34. P. 365.
610.Masters J.I., Murray E.M.E., Proc. IRE. 1965. V. 53. P. 76.
611.Rivlin L.A., Zh. Eksp. Teor. Fiz. 1964. V. 47. P. 624 [Sov Phys. JETP. 1965. V. 20. P. 416].
612.Bret G., Gires F., Appl. Phys. Lett. 1964. V. 4. P. 173.
613.Sorokin P.P., Luzzi J.J., Lankard J.R., Petti G.D., IBM J. Res. Dev. 1964. V. 8. P. 182.
614.Kafalas P., Masters J.I., Murray E.M.E., J. Appl. Phys. 1964. V. 35. P. 2349.
615.Soffer B.H., Hoskins R., Nature. 1964. V. 204. P. 276.
616.Daneu V., Sacchi c.a., Svelto O., IEEE J. Quant. Electron. 1966. V. QE-2. P. 290.
617.Leontovich A.M., Popova M.N., Shchelev M.Ya., Izv. Vyss. Ucheb. Zaved. Radiofiz. 1966. V. 9. P. 513.
618.Korobkin V.V., Leontovich A.M., Popova M.N., Shchelev M.Ya., Pisma Zh. Eksp. Teor. Fiz. 1966. V. 3. P. 301 [Sov Phys. JETP lett. V. 3. P. 194].
619.Forrest M.J., Magyar G., Opto-Electronics. 1970. V. 2. P. 48.
620.McFarland B.B., Hoskins R., Soffer B.H., Nature. 1965. V. 207. P. 1180.
621.Theoretical Approaches for Optical Quantum Oscillators, Ed. B.I. Stepanov, Minsk: Nauka i Tekhnika V. 1. 1966.
622.Erickson L.E., Szabo A., J. Appl. Phys. 1966. V. 37. P. 4953.
623.Szabo A., Stein R.A., J. Appl. Phys. 1965. V. 36. P. 1562.
624.Borovich B.L., Zuev V.S., Shcheglov V.A., Zh. Eksp. Teor. Fiz. 1965. V. 49. P. 624 [Sov Phys. JETP. 1966. V. 22. P. 717].
625.Burak I., Houston P.L., Sutton D.G., Steinfeld J.I., IEEE J. Quant. Electron. 1971. V. QE-7. P. 73.
626.Khanin Ya.I., Koryukin I.V., Proc. SPIE 1992. V.1840 P.148.

Index